Keys

ELEMENTARY LINEAR ALGEBRA

ELEMENTARY LINEAR ALGEBRA

Robert S. Johnson
Washington and Lee University

Thomas O. Vinson, Jr.
Washington and Lee University

HBJ

Harcourt Brace Jovanovich, Publishers
and its subsidiary, Academic Press

San Diego New York Chicago Austin Washington, D.C.
London Sydney Tokyo Toronto

ISBN: 0-15-521082-3
Library of Congress Catalog Card Number: 86-80759
Printed in the United States of America

PREFACE

This book provides an introduction to linear algebra at the freshman or sophomore level. A prerequisite of one year of single-variable calculus is helpful but not necessary, for our development of linear algebra in no way presupposes calculus. Until the concluding sections, we confine our attention to real scalars and focus mostly on $\mathbf{R}^n$.

We have made every effort to treat the subject of linear algebra in a mathematical way: the theory receives as much attention as the techniques, and we have included a substantial number of theorems and proofs. Moreover, we have worked to make the propositions and proofs detailed and thorough, so students can read them and profit from them as much as from the examples.

The problem sets provide a great number of computational exercises, but we also have included many problems that require elaboration, completion, or even extension of proofs given in the text. (Many such exercises have accompanying hints, and answers to selected computational exercises are supplied at the end of the book.) This solid mathematical treatment of linear algebra creates an effective bridge to more rigorous and abstract courses. It is also a source of useful techniques and illustrations.

The first section, "$\mathbf{R}$ and $\mathbf{R}^n$: Preliminary Ideas," catalogs for future reference the essential properties of the real number system. To build on knowledge students bring to an introductory linear algebra course, the development begins with systems of linear equations and introduces matrices as an efficient way to organize solutions of systems. Later, elementary geometry increasingly motivates new ideas. In general, the topics covered are in an order typical for a first course in linear algebra. There is surely more material, however, than can be dealt with in a one-term course, and an instructor will encounter no difficulty if some later topics must be omitted or treated very lightly.

Aside from the thorough treatment of theory, *Elementary Linear Algebra* offers other distinguishing features. We have deviated from the customary matrix notation by denoting the i-j entry of a matrix A by the symbol A_{ij}. Not only is this more satisfactory from a logical point of view, but it also greatly facilitates the confirmation—often omitted in other textbooks—of the algebraic properties of matrix operations, as is readily apparent in Chapter 1. Partly as a consequence of this, we have written cof A_{ij} for the i-j cofactor of A; this notation has the advantage of being mnemonic, and its meaning is thus easy to recall. We have attached similarly conceptual labels to vector-space properties, inner product properties, and the like.

Determinants (written as det A, etc.) are presented by way of cofactor expansion because we find permutations to be an annoying digression. Nevertheless, the permutation definition is in an appendix, along with a proof that the two versions do yield the same determinant value. The application of diagonalizing symmetric matrices to the rotation of conic sections brings the development of change of basis to a satisfying and accessible conclusion.

This book also provides three topics not always found in introductory linear algebra books: a brief introduction to complex numbers and complex vector spaces; solutions of systems of linear differential equations; and an introduction to mathematical induction, found in an appendix. (The appendices can easily be made an integral part of a course because each comes with its own set of problems.)

We are grateful for comments and suggestions from colleagues at Washington and Lee University who taught from preliminary versions of this text. They are Michael Collins (visiting from Oxford), Eung Cho, and Wayne Dymacek. Professor Collins was particularly generous in providing us with a detailed, section-by-section critique. Another colleague, Bill Smith, kindly volunteered to read the page proof with fresh eyes and detected several errors which we had overlooked. Two of our advanced students, Charles Atkinson and David Harrar, worked with diligence and enthusiasm at the tasks of finding answers for computational exercises, constructing additional examples and exercises, and advising us on the overall readability of our exposition. We are in their debt. Many other students at Washington and Lee University had for their textbooks preliminary versions of this work and helped us—often with ill-disguised glee—to find and correct numerous errors.

We thank the following reviewers for worthy suggestions and criticisms that resulted in many improvements as we progressed from our original manuscript to the present text: Gerald Bradley (Claremont McKenna College), Lloyd Davis (College of San Mateo), Hiram Paley (University of Illinois at Urbana-Champaign), and Franklin E. Schroeck, Jr. (Florida Atlantic University).

Finally, we extend our thanks to our editor, Richard Wallis, our manuscript editor, Audrey Thompson, and our production editor, Jon Preimesberger, all of Harcourt Brace Jovanovich. Their encouragement and expertise proved invaluable. Our local HBJ representative, Nancy Peck, expressed unreserved enthusiasm for our project from the start and we thank her for initiating our collaboration with HBJ; it has been a congenial and gratifying association throughout.

CONTENTS

ELEMENTARY LINEAR ALGEBRA

1

Systems of Equations and Matrices

1.1

R and $\mathbf{R}^n$: Preliminary Ideas

We assume familiarity with the *real number system,* which we denote by **R**. **R** is closed under the two operations of addition and multiplication, and the following rules are obeyed.

N1 *The commutative law of addition.* For any real numbers a and b,

$$a + b = b + a$$

N2 *The associative law of addition.* For any real numbers a, b, and c,

$$a + (b + c) = (a + b) + c$$

N3 *The existence of an additive identity.* There is a real number 0 such that, for any real number a,

$$a + 0 = a$$

N4 *The existence of additive inverses.* For any real number a there is a real number $-a$ such that

$$a + (-a) = 0$$

N5 *The commutative law of multiplication.* For any real numbers a and b,

$$ab = ba$$

N6 *The associative law of multiplication.* For any real numbers a, b, and c,

$$a(bc) = (ab)c$$

N7 *The existence of a multiplicative identity.* There is a real number 1 such that for any real number a,

$$a \cdot 1 = a$$

N8 *The existence of multiplicative inverses.* For any real number $a \neq 0$ there is a real number a^{-1} such that

$$a \cdot (a^{-1}) = 1$$

N9 *The distributive laws.* For any real numbers a, b, and c,

$$a(b + c) = ab + ac \quad \text{and} \quad (a + b)c = ac + bc$$

We also recall the notion of the *absolute value* $|a|$ of a real number a, defined by

$$|a| = \begin{cases} a & \text{if } a \geq 0 \\ -a & \text{if } a < 0 \end{cases}$$

Absolute value is subject to the following rules.

N10 For any real numbers a and b,

$$|a + b| \leq |a| + |b| \quad \text{and} \quad |ab| = |a||b|$$

N11 For any real number a,

$$|-a| = |a|$$

Observe that for any real number a, $|a| = \sqrt{a^2}$, since the radical sign means *nonnegative* square root.

We shall have little occasion to mention complex numbers in this text until the final sections, so, for economy, "number" will mean "real number" unless specifically indicated otherwise.

Just as we assume familiarity with the correspondence between numbers and points on a line, we assume awareness of the relationship between ordered pairs of numbers and points in a plane. An ordered pair (x, y) is characterized by specifying a *pair* of numbers x and y and their *order*. Thus $(x, y) = (u, v)$ if and only if $x = u$ and $y = v$. The set of all ordered pairs of numbers will be denoted by $\mathbf{R}^2$.

By analogy we shall use the symbol $\mathbf{R}^n$ to denote the set of all *ordered n-tuples* of numbers. Here n can be any positive integer. Thus $(1, 0, -2)$ is in the set $\mathbf{R}^3$, for instance, since $(1, 0, -2)$ is an ordered triple of numbers. If n is not specified, we will write $(x_1, x_2, \ldots, x_n)$ or similar expressions to denote arbitrary members of $\mathbf{R}^n$, but for specific small n we can write (x, y), (u, v, w), and the like if we wish to avoid subscripts. The ordered triples in $\mathbf{R}^3$ can be used to indicate points in three-dimensional space. Finally, $\mathbf{R}^1$ need not be distinguished from $\mathbf{R}$ itself.

Eventually we will introduce algebraic operations on $\mathbf{R}^n$ and discover properties analogous to N1 through N11 above.

1.2 Systems of Linear Equations

A set of equations such as

$$\begin{aligned} x + 3y - 2z &= 1 \\ 2y - z &= 0 \end{aligned}$$

is called a *system of two linear equations in three unknowns*. The "unknowns" are x, y, and z. For such a system, the aim will be to find numerical values for x, y, and z which satisfy both equations. More specifically, to *solve* such a system is to find all possible ordered triples (x, y, z) in $\mathbf{R}^3$ which meet the requirements imposed by the two equations; each such triple is a *solution* and the set of all solutions is called the *solution set*. In this example the solution set can be described, as we shall see later, by writing

$$(x, y, z) = (1 + \tfrac{1}{2}t, \tfrac{1}{2}t, t) \qquad t \text{ arbitrary}$$

meaning

$$x = 1 + \frac{1}{2}t$$

$$y = \frac{1}{2}t$$

$$z = t \qquad\qquad t \text{ arbitrary}$$

In this case there are infinitely many solutions, one for each numerical value of t; that is, there are infinitely many triples in the solution set. The following definition expands the concepts above.

Definition

A *system of* m *linear equations in* n *unknowns* is a set of equations of the form

$$\begin{aligned} A_{11}x_1 + A_{12}x_2 + \cdots + A_{1n}x_n &= b_1 \\ A_{21}x_1 + A_{22}x_2 + \cdots + A_{2n}x_n &= b_2 \\ &\cdots \\ A_{m1}x_1 + A_{m2}x_2 + \cdots + A_{mn}x_n &= b_m \end{aligned}$$

The numbers $A_{11}, A_{12}, \ldots, A_{mn}$ are called the *coefficients*, and the *unknowns* are $x_1, x_2, \ldots, x_n$. A *solution* for the system is an n-tuple in $\mathbf{R}^n$ satisfying all of the equations simultaneously.

To solve such systems, at least for small m and n, we might proceed as in the following examples.

Example 1

To solve the system

$$\begin{aligned} x + 4y &= -2 \\ 2x - 3y &= 7 \end{aligned}$$

we can subtract twice the first equation from the second to "eliminate" the unknown x; this results in

$$-11y = 11$$

Therefore $y = -1$, and by substituting into the first equation we get $x = 2$. Thus the (only) solution is the pair $(2, -1)$. □

Example 2

We solve the system

$$\begin{aligned} 2x \qquad + \ z &= 1 \\ x - 3y - 2z &= 3 \\ 3x + \ y + \ z &= 2 \end{aligned}$$

Here we shall take what will seem like an unnecessarily lengthy route to the solution, but the reason for it will eventually become apparent. We subtract twice the second equation from the first (and replace the first with the result):

$$\begin{aligned} 6y + 5z &= -5 \\ x - 3y - 2z &= 3 \\ 3x + y + z &= 2 \end{aligned}$$

Next we subtract 3 times the second equation from the third:

$$\begin{aligned} 6y + 5z &= -5 \\ x - 3y - 2z &= 3 \\ 10y + 7z &= -7 \end{aligned}$$

Subtracting the first equation from the third yields the new system

$$\begin{aligned} 6y + 5z &= -5 \\ x - 3y - 2z &= 3 \\ 4y + 2z &= -2 \end{aligned}$$

Multiplying the third equation by $\frac{1}{4}$ gives

$$\begin{aligned} 6y + 5z &= -5 \\ x - 3y - 2z &= 3 \\ y + \frac{1}{2}z &= -\frac{1}{2} \end{aligned}$$

Now we use the third equation to eliminate the y in the other two; this is achieved in two steps by adding 3 times the third equation to the second and -6 times the third to the first. The result is

$$\begin{aligned} 2z &= -2 \\ x \qquad - \frac{1}{2}z &= \frac{3}{2} \\ y + \frac{1}{2}z &= -\frac{1}{2} \end{aligned}$$

Although the solution may be apparent to the reader by now, we continue, multiplying the first equation by $\frac{1}{2}$ and then using it to eliminate z from the other two:

$$\begin{aligned} z &= -1 \\ x \qquad\quad &= 1 \\ y \quad &= 0 \end{aligned}$$

For ease of readability we re-arrange the order of the equations and get

$$\begin{aligned} x \qquad\quad &= 1 \\ y \quad &= 0 \\ z &= -1 \end{aligned}$$

and finally conclude that the (unique) solution is the triple $(1, 0, -1)$. □

We resorted to such a lengthy procedure in this example to show that, by means of certain "operations" performed on a system of equations, the system can be replaced by progressively simpler systems, each *equivalent* to the original in the sense that the solution set of each new system is identical to that of the original, until the point is reached where the solution is obvious. The operations which are allowable—that is, which yield equivalent systems—are of three types:

I. Interchanging two equations.

II. Multiplying an equation by a nonzero number.

III. Adding a multiple of one equation to another. (It should be noted that this alters *only the second equation*.)

Though some steps were combined in the examples above, it can be verified that each operation was of one of these three types.

It is important to note that the effect of each operation can be "undone" by another operation of the same type. For instance, if 5 times the first equation is added to the fourth, then we can recover the original system by adding -5 times the first equation to the (altered) fourth. This ability to recover the original system explains why extraneous solutions are not introduced, and why legitimate solutions are not lost by the process just described.

This method of *solution by elimination* therefore consists of applying a sequence of these operations with the intent of transforming the system into an equivalent one in which the values of the unknowns are explicitly displayed. Though finding an explicit solution is not always possible, the next two examples illustrate that the method nevertheless reveals all we need to know about a system's solution.

Example 3

We return to the system illustrated at the beginning of the section:

$$\begin{aligned} x + 3y - 2z &= 1 \\ 2y - z &= 0 \end{aligned}$$

Since x appears in only one equation, we need not "eliminate" it from the other. By adding $-\frac{3}{2}$ times the second equation to the first, we obtain

$$\begin{aligned} x \qquad - \frac{1}{2}z &= 1 \\ 2y - z &= 0 \end{aligned}$$

and one more step gives

$$\begin{aligned} x \qquad - \frac{1}{2}z &= 1 \\ y - \frac{1}{2}z &= 0 \end{aligned}$$

Now we can do no more. We cannot get unique numerical values for each of x, y, and z. But we can see from our work that a solution does exist since there are values of x, y, and z satisfying these two equations. It is a simple matter to describe the solutions: it is clear that z can have any value whatsoever, but that the value of z determines the values for x and y. Therefore we can write

$$(x, y, z) = (1 + \tfrac{1}{2}t, \tfrac{1}{2}t, t) \qquad t \text{ arbitrary}$$

as we asserted earlier. □

Example 4

By using the elimination method, the system

$$\begin{aligned} x - 2y + z &= 0 \\ x \qquad + 3z &= 5 \\ x + 4y + 7z &= 2 \end{aligned}$$

becomes (verify)

$$\begin{aligned} x \qquad + 3z &= 5 \\ 2y + 2z &= 5 \\ 0 &= -13 \end{aligned}$$

From this we can deduce that the system has no solution, since any solution would have to satisfy the impossible condition of the third equation. □

Example 5

Applying the elimination method to the system

$$\begin{aligned} 2x - 4y - 3z - 7w &= -13 \\ 3x - 6y + z + 6w &= 8 \end{aligned}$$

we multiply the first equation by $\frac{1}{2}$:

$$\begin{aligned} x - 2y - \frac{3}{2}z - \frac{7}{2}w &= -\frac{13}{2} \\ 3x - 6y + z + 6w &= 8 \end{aligned}$$

We add -3 times the first equation to the second:

$$\begin{aligned} x - 2y - \frac{3}{2}z - \frac{7}{2}w &= -\frac{13}{2} \\ \frac{11}{2}z + \frac{33}{2}w &= \frac{55}{2} \end{aligned}$$

We multiply the second equation by $\frac{2}{11}$:

$$\begin{aligned} x - 2y - \frac{3}{2}z - \frac{7}{2}w &= -\frac{13}{2} \\ z + 3w &= 5 \end{aligned}$$

We add $\frac{3}{2}$ times the second equation to the first:

$$\begin{aligned} x - 2y \qquad + w &= 1 \\ z + 3w &= 5 \end{aligned}$$

At this point, no more unknowns can be eliminated; the simplest way to describe the solutions is to let those unknowns not coming first in any equation be arbitrary and to express the others in terms of those:

$$(x, y, z, w) = (1 + 2s - t, s, 5 - 3t, t) \qquad s, t \text{ arbitrary}$$

We have, therefore, infinitely many solutions. □

From these examples, it might be reasonably inferred that, for a given system, there are three possible outcomes:

a. The system has a unique solution.

b. The system has infinitely many solutions.

c. The system has no solution.

Before we verify this inference, we recall that $Ax + By = C$ (with at least one of A and B nonzero) is an equation for a straight line in the xy plane, and that we can interpret the solution(s) of a system of, say, two equations in two unknowns as the point(s) common to two straight lines. The three outcomes listed above correspond exactly to the three ways we know from elementary geometry in which two lines can intersect (or not intersect) each other (Figure 1.1).

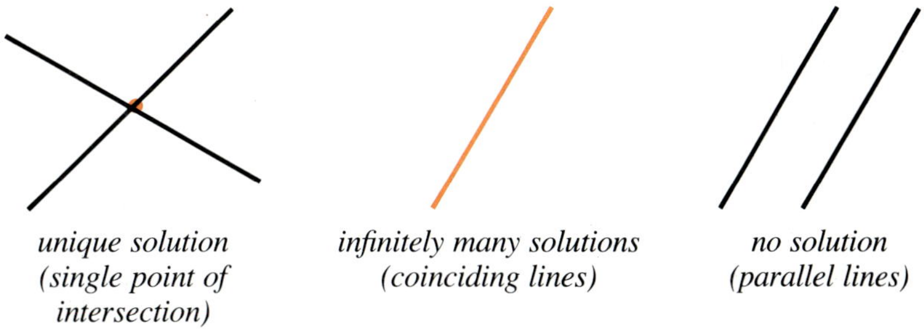

Figure 1.1

We close this section by singling out a special type of system.

Definition If, in a system as defined at the beginning of this section, all the numbers $b_1, b_2, \ldots, b_m$ are equal to 0, then the system is said to be *homogeneous*.

Example 6

We alter Example 2 to make it a homogeneous system:

$$\begin{aligned} 2x \qquad\quad + z &= 0 \\ x - 3y - 2z &= 0 \\ 3x + y + z &= 0 \end{aligned}$$

If we follow the same procedure we used before, we will wind up with

$$\begin{aligned} x &= 0 \\ y &= 0 \\ z &= 0 \end{aligned}$$

which is to say that the only solution is (0, 0, 0). □

Example 7

We adapt Example 3 to obtain the homogeneous system

$$\begin{aligned} x + 3y - 2z &= 0 \\ 2y - z &= 0 \end{aligned}$$

This time, we arrive at

$$\begin{aligned} x \qquad - \frac{1}{2}z &= 0 \\ y - \frac{1}{2}z &= 0 \end{aligned}$$

and we write the solutions in the form $(\frac{1}{2}t, \frac{1}{2}t, t)$, t arbitrary. □

Example 8

By making the system in Example 4 homogeneous we find the new system,

$$\begin{aligned} x - 2y + z &= 0 \\ x \qquad + 3z &= 0 \\ x + 4y + 7z &= 0 \end{aligned}$$

which yields

$$\begin{aligned} x \qquad + 3z &= 0 \\ y + z &= 0 \\ 0 &= 0 \end{aligned}$$

so we write

$$(x, y, z) = (-3t, -t, t) \qquad t \text{ arbitrary}$$ □

Note that in none of these homogeneous versions of our earlier examples did we find a case of no solution. This should not be surprising. Because of its special nature, *a homogeneous system must have at least the "trivial" solution in which all the unknowns have the value zero.*

Thus, for *homogeneous* systems, the number of possible outcomes is reduced to two:

a. The system has a unique solution (that is, only the trivial one).

b. The system has infinitely many solutions.

1.2 Problem Set

1. Solve the following systems.

a. $$\begin{aligned} 2x - 3y + 2z &= -1 \\ x + y - 4z &= 2 \end{aligned}$$

b. $$\begin{aligned} x + y + z &= 1 \\ x - y &= 0 \\ y + 2z &= -1 \\ x - z &= 2 \end{aligned}$$

c. $$\begin{aligned} x - y &= 2 \\ 2x - 3y &= 6 \end{aligned}$$

d. $$\begin{aligned} 2x + 3y + z &= 1 \\ x - y + 4z &= 0 \\ 4x + 11y - 5z &= 1 \end{aligned}$$

e. $$\begin{aligned} 2x + 2y + 3z + w &= 1 \\ 3x - y + z + 3w &= 2 \\ -2x + 3y - z - 2w &= 4 \\ x + 5y + 3z - 3w &= 2 \\ 2x + 7y + 3z - 2w &= 8 \end{aligned}$$

2. Solve these homogeneous systems.

a. $$\begin{aligned} x + y + z &= 0 \\ x - y &= 0 \\ y + z &= 0 \end{aligned}$$

b. $$\begin{aligned} x + y + z &= 0 \\ x - y &= 0 \\ 2x + z &= 0 \end{aligned}$$

c. $$\begin{aligned} x - 2y + z &= 0 \\ 2x + y &= 0 \end{aligned}$$

d. $$\begin{aligned} x_1 + x_2 + x_3 + x_4 &= 0 \\ x_1 - x_2 + x_3 - x_4 &= 0 \end{aligned}$$

e. $$\begin{aligned} 2x + y + z + w &= 0 \\ 2x + y - w &= 0 \end{aligned}$$

3. Find, by inspection of the solution of Example 5 the solution of the *homogeneous* version of that system.

4. For a system of at least four equations, describe for each of the following operations of the "inverse" operation which will "undo" the effect of the given operation.

a. Multiplying the second equation by $\frac{1}{2}$.

b. Adding 5 times the second equation to the fourth.

c. Interchanging the first and third equations.

5. Suppose we have a *homogeneous* system of two equations in three unknowns. Suppose also that each of (u_1, u_2, u_3) and (v_1, v_2, v_3) is a solution and that r is a number.

    a. Show that $(u_1 + v_1, u_2 + v_2, u_3 + v_3)$ is also a solution of the same system.

    b. Show that (ru_1, ru_2, ru_3) is also a solution of the same system.

6. Suppose that we have a *nonhomogeneous* system of two equations in three unknowns. If each of (u_1, u_2, u_3) and (v_1, v_2, v_3) is a solution of this system, show that $(u_1 - v_1, u_2 - v_2, u_3 - v_3)$ is a solution of the homogeneous version of the system.

7. If a certain homogeneous system has fewer equations than unknowns, what can be said about the number of solutions of the system? Why?

8. Solve these systems.

    a.
$$\begin{array}{rcl} 2x - y + z + 3w &=& 1 \\ 3x - 2y + 4z - 6w &=& -6 \\ -7x - 3y - 5z - 9w &=& 1 \\ 9x + 5y + 6z + 15w &=& 0 \\ 2x - 4y + 3z - 12w &=& -5 \end{array}$$

    b.
$$\begin{array}{rcl} x - y &=& -6 \\ -y + z &=& 5 \\ z - w &=& -4 \\ x - w &=& -15 \end{array}$$

    c.
$$\begin{array}{rcl} 2x + 3y - z + 4w &=& 13 \\ -4x - 15y + 5z - 8w &=& -47 \end{array}$$

    d.
$$\begin{array}{rcl} 3x - 7y &=& 14 \\ 14x + 6y &=& 12 \end{array}$$

    e.
$$\begin{array}{rcl} x + z &=& 2 \\ 2x - y + 4z &=& 6 \end{array}$$

    f.
$$\begin{array}{rcl} 7x - 14y &=& 2 \\ -5x + 10y &=& 3 \end{array}$$

1.3 MATRICES

Now that we have considered systems of linear equations, discussed a method for solving them, and made several conjectures about the results, we turn our attention to formalizing these ideas.

It should be clear that the operations we performed in Section 1.2 affected only the coefficients and the numbers on the right side of the equations in the systems. From another viewpoint, we can assert that the solution of a system of equations hinges only on what these numbers are and where they are located in the system. This naturally leads us to suppress the superfluous distractions in a system of equations and to focus only on the essentials.

Definition

A *matrix* is a rectangular array of numbers.

We shall write a matrix A in the form

$$A = \begin{bmatrix} A_{11} & A_{12} & A_{13} & \cdots & A_{1n} \\ A_{21} & A_{22} & A_{23} & \cdots & A_{2n} \\ & & \cdots & & \\ A_{m1} & A_{m2} & A_{m3} & \cdots & A_{mn} \end{bmatrix}$$

Each number A_{ij} in a matrix A is called an *entry* of the matrix. Those entries in one horizontal line constitute a *row* of the matrix; those in a vertical line constitute a *column*. Thus, our matrix A above has m rows and n columns. We say that this matrix has *dimensions* $m \times n$ ("m by n"). A_{ij} is the entry in the i-th row and the j-th column, the "i-j" entry.

Example 1

Let

$$A = \begin{bmatrix} 2 & 1 & -1 & 0 \\ 5 & \frac{1}{2} & 0 & 6 \\ 1 & 1 & 0 & 3 \end{bmatrix}$$

This matrix is 3×4; it has three rows and four columns. Some examples of its entries are $A_{11} = 2$, $A_{13} = -1$, $A_{22} = \frac{1}{2}$, $A_{23} = 0$. (Note that there is an entry A_{24} but there is no entry A_{42}.) □

We remark that $\mathbf{R}^n$ may be presented as the set of all $1 \times n$ matrices or as the set of all $n \times 1$ matrices. The state of being an ordered n-tuple is retained for elements of $\mathbf{R}^n$ when they are displayed in matrix form.

We shall now introduce an algebraic structure on matrices by defining certain operations. *It is important to note that, by definition, the essence of a matrix lies in what its entries are and where they are. Thus, a particular matrix is specified by describing its i-j entry for each possible i and j. Moreover, for matrices A and B of the same size,* $A = B$ *if and only if* $A_{ij} = B_{ij}$ *for all i and j.*

Definition

Suppose that each of A and B is an $m \times n$ matrix. Their *sum* $A + B$ is that matrix given by

$$(A + B)_{ij} = A_{ij} + B_{ij} \qquad i = 1, 2, \ldots, m; j = 1, 2, \ldots, n$$

That is, $A + B$ is defined to be that $m \times n$ matrix whose i-j entry is the sum of the i-j entry of A and the i-j entry of B.

Note that two matrices cannot be added by our formula unless their dimensions are the same. Moreover, the sum of two matrices will be a matrix of the same size as the two matrices.

Example 2

Let

$$A = \begin{bmatrix} 1 & -1 & -\frac{1}{2} \\ 2 & 0 & 1 \end{bmatrix} \quad \text{and} \quad B = \begin{bmatrix} 0 & 3 & 1 \\ -1 & 2 & 1 \end{bmatrix}$$

Then their sum is given by

$$A + B = \begin{bmatrix} 1 & 2 & \frac{1}{2} \\ 1 & 2 & 2 \end{bmatrix}$$

□

Theorem 1.1

Matrix addition obeys the following laws.

M1 *Commutative law.* If each of A and B is an $m \times n$ matrix, then

$$A + B = B + A$$

M2 *Associative law.* If each of A, B, and C is an $m \times n$ matrix, then

$$A + (B + C) = (A + B) + C$$

M3 *Existence of an additive identity.* There is an $m \times n$ matrix O such that, for any $m \times n$ matrix A,

$$A + O = A$$

M4 *Existence of additive inverses.* If A is an $m \times n$ matrix, then there is an $m \times n$ matrix $-A$ such that

$$A + (-A) = O$$

Proof. We shall prove M1 and M3 and leave proof of the other parts as exercises.

M1: To establish equality of two matrices, we must show that the corresponding entries are the same. For the *i*-*j* entry, we get

$$\begin{aligned}(A + B)_{ij} &= A_{ij} + B_{ij} && \text{(definition of matrix sum)}\\ &= B_{ij} + A_{ij} && \text{(number-property N1)}\\ &= (B + A)_{ij} && \text{(definition of matrix sum)}\end{aligned}$$

Thus, $A + B = B + A$.

M3: To produce the required additive identity matrix O, we let O be the matrix whose entries are all 0; that is, $O_{ij} = 0$, for all i and j. Then

$$\begin{aligned}(A + O)_{ij} &= A_{ij} + O_{ij} && \text{(definition of matrix sum)}\\ &= A_{ij} + 0 && \text{(definition of } O)\\ &= A_{ij} && \text{(number-property N3)}\end{aligned}$$

Thus, $A + O = A$.

Definition

Hereafter, the symbol O shall denote any matrix whose entries are all 0. (If the size of the matrix is important, it will be clear from the context.)

Definition

Suppose that A is an $m \times n$ matrix and that c is a number (called a *scalar* in this context). The matrix cA is given by

$$(cA)_{ij} = c(A_{ij}) \qquad i = 1, 2, \ldots, m; j = 1, 2, \ldots, n$$

Example 3

If

$$A = \begin{bmatrix} 4 & -1 & 2 \\ 1 & 0 & 3 \end{bmatrix}$$

then

$$7A = \begin{bmatrix} 28 & -7 & 14 \\ 7 & 0 & 21 \end{bmatrix}$$

□

Theorem 1.2 For any $m \times n$ matrices A and B and any scalars c and d

M5 $c(A + B) = cA + cB$

M6 $(c + d)A = cA + dA$

M7 $c(dA) = (cd)A$

Proof. For M5, we prove that matrices $c(A + B)$ and $cA + cB$ have equal corresponding entries and hence are equal matrices:

$$
\begin{aligned}
[c(A + B)]_{ij} &= c(A + B)_{ij} && \text{(definition of scalar multiple of a matrix)} \\
&= c(A_{ij} + B_{ij}) && \text{(definition of matrix sum)} \\
&= cA_{ij} + cB_{ij} && \text{(number-property N9)} \\
&= (cA)_{ij} + (cB)_{ij} && \text{(definition of scalar multiple of a matrix)} \\
&= (cA + cB)_{ij} && \text{(definition of matrix sum)}
\end{aligned}
$$

We leave the proofs of M6 and M7 as exercises.

The last of our matrix operations gives the product of two matrices.

Definition Let A be an $m \times n$ matrix and B be an $n \times p$ matrix. The *product AB* is that matrix given by

$$
\begin{aligned}
(AB)_{ij} &= A_{i1}B_{1j} + A_{i2}B_{2j} + \cdots + A_{in}B_{nj} \\
&= \sum_{k=1}^{n} A_{ik}B_{kj} \qquad i = 1, 2, \ldots, m; j = 1, 2, \ldots, p
\end{aligned}
$$

The dimensions of the product AB are $m \times p$.

In computing each i-j entry of AB, we must match the i-th row of A with the j-th column of B, as Figure 1.2 indicates.

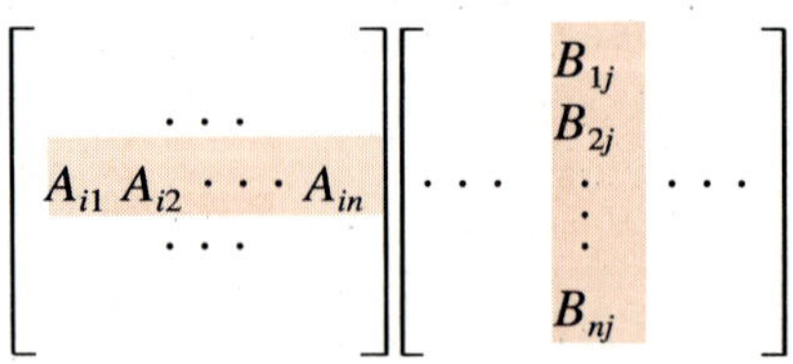

Figure 1.2

Example 4

To find the entry in row 3 and column 1 of AB, we identify row 3 from A and column 1 from B and follow the definition of product AB, multiplying corresponding entries together and adding:

$$\begin{bmatrix} 1 & -1 & 0 \\ 3 & 2 & -2 \\ 1 & \frac{1}{2} & 4 \\ 4 & 0 & 1 \end{bmatrix} \begin{bmatrix} 2 & 0 \\ 2 & -6 \\ 3 & 1 \end{bmatrix} = \begin{bmatrix} 0 & 6 \\ 4 & -14 \\ 15 & 1 \\ 11 & 1 \end{bmatrix}$$

□

Example 5

Let

$$A = \begin{bmatrix} 4 & -1 & 2 \\ 1 & 0 & 3 \end{bmatrix} \text{ and } B = \begin{bmatrix} 1 & 3 & 4 \\ -3 & 1 & 3 \\ 2 & 0 & 2 \end{bmatrix}$$

Then, by definition,

$$AB = \begin{bmatrix} 11 & 11 & 17 \\ 7 & 3 & 10 \end{bmatrix}$$

□

Note that for A and B to be compatible for multiplication (in that order), the number of *columns* of A must equal the number of *rows* of B. Note also that matrix multiplication is *not* commutative. Even when AB and BA both exist and are the same size, they are rarely equal. For example,

$$\begin{bmatrix} 2 & 3 \\ 1 & -2 \end{bmatrix} \begin{bmatrix} 1 & -1 \\ 2 & 0 \end{bmatrix} = \begin{bmatrix} 8 & -2 \\ -3 & -1 \end{bmatrix}$$

but

$$\begin{bmatrix} 1 & -1 \\ 2 & 0 \end{bmatrix} \begin{bmatrix} 2 & 3 \\ 1 & -2 \end{bmatrix} = \begin{bmatrix} 1 & 5 \\ 4 & 6 \end{bmatrix}$$

Theorem 1.3 If c is a scalar and A, B, and C are matrices compatible for the operations as required, then

M8 *Left distributive law.*

$$A(B + C) = AB + AC$$

M9 *Right distributive law.*

$$(A + B)C = AC + BC$$

M10 $c(AB) = A(cB) = (cA)B$

M11 *Associative law.*

$$A(BC) = (AB)C$$

Proof. For M8, we let A be $m \times n$ and B and C be $n \times p$. Then

$$\begin{aligned}
[A(B + C)]_{ij} &= \sum_{k=1}^{n} A_{ik}(B + C)_{kj} && \text{(definition of matrix product)} \\
&= \sum_{k=1}^{n} A_{ik}(B_{kj} + C_{kj}) && \text{(definition of matrix sum)} \\
&= \sum_{k=1}^{n} (A_{ik}B_{kj} + A_{ik}C_{kj}) && \text{(number-property N9)} \\
&= \sum_{k=1}^{n} A_{ik}B_{kj} + \sum_{k=1}^{n} A_{ik}C_{kj} && \text{(number-properties N1 and N2)} \\
&= (AB)_{ij} + (AC)_{ij} && \text{(definition of matrix product)} \\
&= (AB + AC)_{ij} && \text{(definition of matrix sum)}
\end{aligned}$$

For M11, we let A be $m \times n$, B be $n \times p$, and C be $p \times q$.

$$[A(BC)]_{ij} = \sum_{k=1}^{n} A_{ik}(BC)_{kj} = \sum_{k=1}^{n} A_{ik}\left(\sum_{h=1}^{p} B_{kh}C_{hj}\right) = \sum_{k=1}^{n}\sum_{h=1}^{p} A_{ik}B_{kh}C_{hj}$$

On the other hand,

$$[(AB)C]_{ij} = \sum_{h=1}^{p} (AB)_{ih}C_{hj} = \sum_{h=1}^{p}\left(\sum_{k=1}^{n} A_{ik}B_{kh}\right)C_{hj} = \sum_{h=1}^{p}\sum_{k=1}^{n} A_{ik}B_{kh}C_{hj}$$

With close examination, it can be seen that the two double sums contain exactly the same terms, although in different orders. The reasons for all the steps will be asked for in the exercises.

The proofs of the other two parts of the theorem are left as exercises.

Of particular interest concerning matrix multiplication is the existence of a multiplicative identity.

Definition

For any positive integer n, the $n \times n$ *identity matrix* I is given by

$$I_{ij} = \begin{cases} 1 & \text{if} \quad i = j \\ 0 & \text{if} \quad i \neq j \end{cases}$$

Clearly, for different n, we have different identity matrices; to be precise, we should write I_n for the $n \times n$ identity matrix. Thus, for instance, the 3×3 identity matrix would be

$$I_3 = \begin{bmatrix} 1 & 0 & 0 \\ 0 & 1 & 0 \\ 0 & 0 & 1 \end{bmatrix}$$

However, since the dimensions will always be clear from the context, we shall write simply I, regardless of the size. No matter what the dimensions are, an identity matrix is a square matrix with 1's on the diagonal from upper left to lower right and 0's off that diagonal.

Theorem 1.4 If A is any $m \times n$ matrix, then $AI = IA = A$. (Here, the first identity matrix is $n \times n$ and the second is $m \times m$.)

Proof.

$$\begin{aligned} (AI)_{ij} &= A_{i1}I_{1j} + A_{i2}I_{2j} + \cdots + A_{ij}I_{jj} + \cdots + A_{in}I_{nj} && \text{(definition of matrix product)} \\ &= A_{i1}0 + A_{i2}0 + \cdots + A_{ij}1 + \cdots + A_{in}0 && \text{(definition of } I) \\ &= A_{ij} \end{aligned}$$

Hence, $AI = A$, since corresponding entries are equal. Similarly, $IA = A$.

1.3 Problem Set

1. Let

$$A = \begin{bmatrix} 1 & 0 & -1 \\ 2 & 1 & 2 \end{bmatrix}, \qquad B = \begin{bmatrix} 3 & 1 & 1 \\ 0 & 1 & 1 \\ 1 & 2 & -1 \end{bmatrix},$$

$$C = \begin{bmatrix} 1 & -2 & 1 \\ 4 & 2 & 3 \end{bmatrix}, \qquad D = \begin{bmatrix} 1 \\ 0 \\ 3 \end{bmatrix}, \quad \text{and} \quad E = [4 \quad -2 \quad 1]$$

Compute, if possible, AB, BA, $A + B$, $A + C$, $(2A + C)B$, AD, $D + E$, BE, DE, ED.

2. Solve for x and y:

a. $\begin{bmatrix} 2 & -3 \\ -1 & 2 \end{bmatrix}\begin{bmatrix} x \\ y \end{bmatrix} = \begin{bmatrix} 1 \\ 2 \end{bmatrix}$ b. $\begin{bmatrix} x & 1 \\ x & 2 \end{bmatrix}\begin{bmatrix} 1 & -1 \\ 2 & 2 \end{bmatrix} = \begin{bmatrix} 1 & y \\ y & 5 \end{bmatrix}$

c. $x\begin{bmatrix} 2 & -1 \\ 0 & 1 \end{bmatrix} + y\begin{bmatrix} 5 & -3 \\ 1 & 2 \end{bmatrix} = \begin{bmatrix} 1 & -1 \\ 1 & 0 \end{bmatrix}$

3. a. For

$$A = \begin{bmatrix} 1 & -2 \\ -2 & 4 \end{bmatrix} \quad \text{and} \quad B = \begin{bmatrix} 2 & 6 \\ 1 & 3 \end{bmatrix}$$

compute AB. What is the point of this exercise?

b. For

$$A = \begin{bmatrix} 1 & -2 \\ -2 & 4 \end{bmatrix}, \quad B = \begin{bmatrix} 4 & 0 \\ 1 & 4 \end{bmatrix}, \quad \text{and} \quad C = \begin{bmatrix} 2 & -6 \\ 0 & 1 \end{bmatrix}$$

compute AB and AC. What is the point of this exercise?

4. For a square matrix M, let $M^2 = MM$, as with numbers. Let A and B be square matrices of the same size.

a. Is the familiar number-formula $(A + B)^2 = A^2 + 2AB + B^2$ valid for matrices? Justify your conclusion.

b. Is the familiar number-formula $(A + B)(A - B) = A^2 - B^2$ valid for matrices? Justify your conclusion.

5. Let A and B be matrices such that the product AB exists. Show that

a. if A has a row of zeros, then AB has a row of zeros.

b. if B has a column of zeros, then AB has a column of zeros.

6. Prove that if AB and BA both exist and are of the same size, then A and B must be square and of the same size.

7. Prove M2 and M4 of Theorem 1.1.

8. Prove M6 and M7 of Theorem 1.2.

9. Prove M9 and M10 of Theorem 1.3.

10. Complete the proof of Theorem 1.4 by proving that $IA = A$.

11. Supply reasons for the proof of M11, Theorem 1.3.

12. a. Show that there cannot be two $m \times n$ additive identity matrices (that is, matrices satisfying conditions M3) by showing that if each of O_1 and O_2 satisfies M3, then $O_1 = O_2$. (*Hint:* Consider $O_1 + O_2$.)

b. Show that there cannot be two $n \times n$ multiplicative identity matrices (that is, matrices satisfying $AI = IA = A$ for all $n \times n$ matrices A).

13. a. Show that if C is a 2×2 matrix such that $AC = CA$ for all 2×2 matrices A, then C must be of the form

$$\begin{bmatrix} c & 0 \\ 0 & c \end{bmatrix}$$

for some scalar c. (*Hint:* Let

$$C = \begin{bmatrix} x & y \\ u & v \end{bmatrix}$$

Draw conclusions about x, y, u, and v by using specific matrices such as

$$A = \begin{bmatrix} 1 & 0 \\ 0 & 0 \end{bmatrix}$$

in the condition $AC = CA$.) Generalize to the $n \times n$ case.

b. Find all 2×2 matrices which commute with

$$\begin{bmatrix} a & 0 \\ 0 & b \end{bmatrix}$$

14. An $n \times n$ matrix D such that $D_{ij} = 0$ if $i \neq j$ is called a *diagonal* matrix. Thus in a diagonal matrix, all entries off the main diagonal from upper left to lower right are 0's (as in the matrix in Problem 13b).

a. Let D be an $n \times n$ diagonal matrix and let A be any $n \times n$ matrix. Show that the entries in the i-th row of DA are obtained by multiplying the entries in the i-th row of A by D_{ii}. Show that the entries in the j-th column of AD are obtained by multiplying the entries in the j-th column of A by D_{jj}.

b. Generalize Problem 13b to the $n \times n$ case.

1.4 Systems of Equations in Matrix Form

By virtue of our operation of multiplication of matrices, a system of linear equations can be expressed as a single matrix equation of the form $AX = B$. For instance, the system

$$\begin{aligned} 3x + 5y - 2z + w &= 2 \\ 2x - y + 6z + 8w &= 5 \\ x + 4y + 4z - 3w &= 0 \end{aligned}$$

can be written in matrix form as

$$\begin{bmatrix} 3 & 5 & -2 & 1 \\ 2 & -1 & 6 & 8 \\ 1 & 4 & 4 & -3 \end{bmatrix} \begin{bmatrix} x \\ y \\ z \\ w \end{bmatrix} = \begin{bmatrix} 2 \\ 5 \\ 0 \end{bmatrix}$$

In $\mathbf{R}^4$, a solution of this system of equations is a point (x, y, z, w); in matrix format, a solution to the matrix equation is a 4×1 matrix

$$\begin{bmatrix} x \\ y \\ z \\ w \end{bmatrix}$$

(This is in line with our earlier comment that members of $\mathbf{R}^n$ can be viewed as $1 \times n$ or as $n \times 1$ matrices.)

Definition In a matrix equation $AX = B$, the matrix A is called the *coefficient matrix* (because it displays the coefficients of the corresponding system).

In the example above, the matrix

$$A = \begin{bmatrix} 3 & 5 & -2 & 1 \\ 2 & -1 & 6 & 8 \\ 1 & 4 & 4 & -3 \end{bmatrix}$$

is the coefficient matrix.

A matrix equation $AX = B$ need not correspond to a system of linear equations; that is, the matrices X and B may consist of more than one column. All that is required is that the dimensions of the matrices be such that the product and equality involved are possible. To illustrate this point, we mention here that we shall soon be concerned with the possibility of solving a matrix equation such as

$$\begin{bmatrix} a & b \\ c & d \end{bmatrix} \begin{bmatrix} x & y \\ z & w \end{bmatrix} = \begin{bmatrix} 1 & 0 \\ 0 & 1 \end{bmatrix}$$

Definition The *augmented matrix* for a matrix equation $AX = B$ is the matrix $[A \mid B]$ obtained by following the columns of A with the columns of B.

Returning once again to our example, we find that the augmented matrix is the 3×5 matrix

$$[A \mid B] = \left[\begin{array}{cccc|c} 3 & 5 & -2 & 1 & 2 \\ 2 & -1 & 6 & 8 & 5 \\ 1 & 4 & 4 & -3 & 0 \end{array}\right]$$

For a system of equations $AX = B$, it is the augmented matrix which contains all the essential information about the system, namely, the coefficients and the numbers on the right side of the equations. Recall from our introduction to matrices that we commented on this essential information and on the fact that the elimination method of solving systems of equations in reality consists of performing certain operations on this augmented matrix (although we did not use those terms then). Thus a certain economy can be effected by translating our earlier efforts to the setting of matrices.

Definition

Let M be a matrix of any dimensions. An *elementary row operation* on M is one of the following:

I. Interchanging any two rows.

II. Multiplying any one row by a nonzero scalar.

III. Adding a multiple of one row to another.

While these elementary row operations are directly analogous to the operations we performed on systems of equations, they are defined for matrices in general, whether or not the matrices arise as augmented matrices for certain systems of equations.

Definition

Let M and N be matrices of the same dimensions. M is said to be *row-equivalent* to N provided that N can be obtained from M by means of a sequence of elementary row operations.

Now, we can see that solving a system of equations amounts essentially to applying a suitable sequence of elementary row operations to the augmented matrix of that system. The goal is to obtain an augmented matrix that is row-equivalent to the original, but is of such simplicity that the solution(s), if any, will be obvious.

Example 1

We solve the system

$$\begin{aligned} 2x \qquad + \ z &= 1 \\ x - 3y - 2z &= 3 \\ 3x + \ y + \ z &= 2 \end{aligned}$$

by operating on the augmented matrix, which, in this case, is

$$\left[\begin{array}{ccc|c} 2 & 0 & 1 & 1 \\ 1 & -3 & -2 & 3 \\ 3 & 1 & 1 & 2 \end{array}\right]$$

(The sequence of row operations used is up to the individual; however, the goal is fixed by the discussion above.) We add -2 times the second row to the first:

$$\left[\begin{array}{ccc|c} 0 & 6 & 5 & -5 \\ 1 & -3 & -2 & 3 \\ 3 & 1 & 1 & 2 \end{array}\right]$$

We add -3 times the second row to the third:

$$\left[\begin{array}{ccc|c} 0 & 6 & 5 & -5 \\ 1 & -3 & -2 & 3 \\ 0 & 10 & 7 & -7 \end{array}\right]$$

We add -1 times the first row to the third:

$$\left[\begin{array}{ccc|c} 0 & 6 & 5 & -5 \\ 1 & -3 & -2 & 3 \\ 0 & 4 & 2 & -2 \end{array}\right]$$

We multiply the third row by $\frac{1}{4}$:

$$\left[\begin{array}{ccc|c} 0 & 6 & 5 & -5 \\ 1 & -3 & -2 & 3 \\ 0 & 1 & \frac{1}{2} & -\frac{1}{2} \end{array}\right]$$

We add 3 times the third row to the second:

$$\left[\begin{array}{ccc|c} 0 & 6 & 5 & -5 \\ 1 & 0 & -\frac{1}{2} & \frac{3}{2} \\ 0 & 1 & \frac{1}{2} & -\frac{1}{2} \end{array}\right]$$

We add -6 times the third row to the first:

$$\left[\begin{array}{ccc|c} 0 & 0 & 2 & -2 \\ 1 & 0 & -\frac{1}{2} & \frac{3}{2} \\ 0 & 1 & \frac{1}{2} & -\frac{1}{2} \end{array}\right]$$

We multiply the first row by $\frac{1}{2}$:

$$\left[\begin{array}{ccc|c} 0 & 0 & 1 & -1 \\ 1 & 0 & -\frac{1}{2} & \frac{3}{2} \\ 0 & 1 & \frac{1}{2} & -\frac{1}{2} \end{array}\right]$$

We add $\frac{1}{2}$ times the first row to the second:

$$\left[\begin{array}{ccc|c} 0 & 0 & 1 & -1 \\ 1 & 0 & 0 & 1 \\ 0 & 1 & \frac{1}{2} & -\frac{1}{2} \end{array}\right]$$

We add $-\frac{1}{2}$ times the first row to the third:

$$\left[\begin{array}{ccc|c} 0 & 0 & 1 & -1 \\ 1 & 0 & 0 & 1 \\ 0 & 1 & 0 & 0 \end{array}\right]$$

We interchange the first and second rows:

$$\left[\begin{array}{ccc|c} 1 & 0 & 0 & 1 \\ 0 & 0 & 1 & -1 \\ 0 & 1 & 0 & 0 \end{array}\right]$$

We interchange the second and third rows:

$$\left[\begin{array}{ccc|c} 1 & 0 & 0 & 1 \\ 0 & 1 & 0 & 0 \\ 0 & 0 & 1 & -1 \end{array}\right]$$

We have now reached our goal: we have a matrix that is row-equivalent to the original and that is, in a sense to be made explicit very shortly, as "simple" as possible. As an augmented matrix, it represents the system

$$\begin{aligned} x &= 1 \\ y &= 0 \\ z &= -1 \end{aligned}$$

so our solution is $(x, y, z) = (1, 0, -1)$. □

Generally speaking, the aim of the procedure used in Example 1 is to "simplify" a matrix by transforming the original augmented matrix into one that is as nearly like the identity matrix (as introduced in Section 1.3) as possible. This desirable state is more precisely described in the next definition.

Definition A matrix is said to be an *echelon matrix* (or to be *in echelon form*) provided that

a. Any row consisting entirely of zeros follows (that is, is below) any row not consisting entirely of zeros.

b. The first (that is, the left-most) nonzero entry in any row, if there is one, is a 1. Such an entry is called a *leading one*.

c. The leading one in any row is to the left of the leading one in any lower row.

d. All other entries in a column containing a leading one are zero.*

An identity matrix is an echelon matrix; here are a few others:

$$\begin{bmatrix} 1 & 0 & -1 & 2 \\ 0 & 1 & 0 & 5 \end{bmatrix} \qquad \begin{bmatrix} 0 & 1 & 2 & 0 & -\frac{1}{2} \\ 0 & 0 & 0 & 1 & 1 \\ 0 & 0 & 0 & 0 & 0 \end{bmatrix}$$

$$\begin{bmatrix} 0 & 1 & 1 \\ 0 & 0 & 0 \\ 0 & 0 & 0 \end{bmatrix} \qquad \begin{bmatrix} 1 & 0 & 0 \\ 0 & 1 & 0 \\ 0 & 0 & 0 \end{bmatrix}$$

The following theorem summarizes the situation.

Theorem 1.5 Any matrix is row-equivalent to one and only one echelon matrix.

(See Appendix A for a proof.)

Example 2

We present here some matrices and their echelon forms. In each case, a sequence of elementary row operations may be found that will make the transformation from one matrix to the other.

*Some mathematicians say "echelon" if the first three conditions are met and "reduced echelon" if all four are met.

(echelon form)

a. $\begin{bmatrix} 1 & 4 & -2 \\ 2 & -3 & 7 \end{bmatrix}$ $\begin{bmatrix} 1 & 0 & 2 \\ 0 & 1 & -1 \end{bmatrix}$

b. $\begin{bmatrix} 1 & 3 & -2 & 1 \\ 0 & 2 & -1 & 0 \end{bmatrix}$ $\begin{bmatrix} 1 & 0 & -\frac{1}{2} & 1 \\ 0 & 1 & -\frac{1}{2} & 0 \end{bmatrix}$

c. $\begin{bmatrix} 1 & -2 & 1 & 0 \\ 1 & 0 & 3 & 5 \\ 1 & 4 & 7 & 2 \end{bmatrix}$ $\begin{bmatrix} 1 & 0 & 3 & 0 \\ 0 & 1 & 1 & 0 \\ 0 & 0 & 0 & 1 \end{bmatrix}$ □

We have been tacitly assuming all along that if two augmented matrices are row-equivalent, then they represent systems having the same solution(s). While this assumption still awaits formal confirmation, if we apply it to the matrices in Example 2, we can see that the echelon matrices immediately give us, for the corresponding systems, the following solutions:

a. $(x, y) = (2, -1)$

b. $(x, y, z) = (1 + \frac{1}{2}t, \frac{1}{2}t, t)$, t arbitrary

c. no solution

We point out that if the coefficient portion of the echelon augmented matrix has a column *without* a leading one, the corresponding unknown should be declared arbitrary and the others expressed in terms of it; if there are more than one such unknowns, declare them all arbitrary and express the others in terms of them. (Such a case follows shortly.) Taking precedence over this principle, however, is a situation such as in Example 2c. If in the echelon matrix there is a row in which all entries are zero *except* for the last entry, then as an augmented matrix it represents a system with *no* solution.

In the case of a *homogeneous* system (see Section 1.2), turning to matrix techniques can be even more economical, for the augmented matrix of such a system will have in its final column nothing but zeros, and no amount of row operations will alter this situation. Thus the echelon form will likewise have a final column of zeros and we might as well have operated on only the coefficient matrix. Needless to say, we must keep in mind that only the coefficient matrix is on view.

Example 3

The homogeneous system

$$\begin{aligned} 2x - 4y - 3z - 7w &= 0 \\ 3x - 6y + z + 6w &= 0 \end{aligned}$$

has as its *coefficient* matrix

$$\begin{bmatrix} 2 & -4 & -3 & -7 \\ 3 & -6 & 1 & 6 \end{bmatrix}$$

It is possible by four elementary row operations to reduce this to echelon form (see Example 5 of Section 1.2):

$$\begin{bmatrix} 1 & -2 & 0 & 1 \\ 0 & 0 & 1 & 3 \end{bmatrix}$$

Bearing in mind that this is only the *coefficient* matrix for a *homogeneous* system (and not the augmented matrix), we identify the unknowns corresponding to columns not containing leading ones (see Example 2). They are y and w. Letting these be arbitrary and expressing x and z in terms of them, we display our solution as

$$(x, y, z, w) = (2s - t, s, -3t, t) \qquad s, t \text{ arbitrary}$$ □

The preceding example might well suggest an answer to a question we posed at the end of Section 1.2 and are now in a position to resolve.

Theorem 1.6 If a homogeneous system of linear equations has fewer equations than unknowns, then it must have infinitely many solutions.

Proof. The *coefficient* echelon matrix for a homogeneous system having fewer equations than unknowns will have fewer rows than columns. Since no row can contain more than one leading one, there are fewer leading ones than there are columns. Hence at least one unknown corresponds to a column not having a leading one and can be declared arbitrary, giving rise to infinitely many solutions.

1.4 Problem Set

1. Write the following systems as matrix equations (in the form $AX = B$).

a. $$\begin{aligned} 2x - 3y + 2z &= -1 \\ x + y - 4z &= 2 \end{aligned}$$

b. $$\begin{aligned} x + y + z &= 1 \\ x - y &= 0 \\ y + 2z &= -1 \\ x - z &= 2 \end{aligned}$$

c. $\begin{aligned} x - y &= 2 \\ 2x - 3y &= 6 \end{aligned}$

d. $\begin{aligned} 2x + 3y + z &= 1 \\ x - y + 4z &= 0 \\ 4x + 11y - 5z &= 1 \end{aligned}$

e. $\begin{aligned} 2x + 2y + 3z + w &= 1 \\ 3x - y + z + 3w &= 2 \\ -2x + 3y - z - 2w &= 4 \\ x + 5y + 3z - 3w &= 2 \\ 2x + 7y + 3z - 2w &= 8 \end{aligned}$

2. For each equation in Problem 1, specify the coefficient matrix and the augmented matrix.

3. Which of the following are echelon matrices and which are not? Why?

a. $\begin{bmatrix} 0 & 1 & -2 & 3 \\ 1 & 0 & 0 & 0 \\ 0 & 0 & 0 & 0 \end{bmatrix}$ b. $\begin{bmatrix} 0 & 1 & -2 & 3 \\ 0 & 0 & 2 & 0 \\ 0 & 0 & 0 & 0 \end{bmatrix}$

c. $\begin{bmatrix} 0 & 1 & -2 & 3 \\ 0 & 0 & 1 & 0 \\ 0 & 0 & 0 & 0 \end{bmatrix}$ d. $\begin{bmatrix} 0 & 1 & 0 & 3 \\ 0 & 0 & 1 & 0 \\ 0 & 0 & 0 & 0 \end{bmatrix}$

e. $\begin{bmatrix} 0 & 1 & 0 & 3 \\ 0 & 0 & 2 & 0 \\ 0 & 0 & 0 & 0 \end{bmatrix}$ f. $\begin{bmatrix} 1 & 0 & 0 & 0 \\ 0 & 0 & 1 & 0 \\ 0 & 0 & 0 & 1 \end{bmatrix}$

4. In Example 2, verify that the given matrices convert to the given echelon forms by finding suitable row operations which give the conversions.

5. a. Suppose that the matrices in Example 2 are *augmented* matrices for certain systems of equations. Write the solutions for the systems.

   b. Suppose that the matrices in Example 2 are *coefficient* matrices for certain *homogeneous* systems. Write the solutions for the systems.

6. Follow the instructions of Problem 5 for the matrices displayed immediately after the definition of echelon matrix in this section (p. 26).

7. Show that if a system of equations has more equations than unknowns, then the echelon form of its coefficient matrix will have at least one row of zeros.

8. Solve, by the methods of this section, the following systems.

a. $$\begin{aligned} 2x - 3y + 2z &= -1 \\ x + y - 4z &= 2 \end{aligned}$$

b. $$\begin{aligned} x + y + z &= 1 \\ x - y &= 0 \\ y + 2z &= -1 \\ x - z &= 2 \end{aligned}$$

c. $$\begin{aligned} x - y &= 2 \\ 2x - 3y &= 6 \end{aligned}$$

d. $$\begin{aligned} 2x + 3y + z &= 1 \\ x - y + 4z &= 0 \\ 4x + 11y - 5z &= 1 \end{aligned}$$

e. $$\begin{aligned} 2x + 2y + 3z + w &= 1 \\ 3x - y + z + 3w &= 2 \\ -2x + 3y - z - 2w &= 4 \\ x + 5y + 3z - 3w &= 2 \\ 2x + 7y + 3z - 2w &= 8 \end{aligned}$$

f. $$\begin{aligned} x + y + z &= 0 \\ x - y &= 0 \\ y + z &= 0 \end{aligned}$$

g. $$\begin{aligned} x + y + z &= 0 \\ x - y &= 0 \\ 2x + z &= 0 \end{aligned}$$

h. $$\begin{aligned} x - 2y + z &= 0 \\ 2x + y &= 0 \end{aligned}$$

i. $$\begin{aligned} x_1 + x_2 + x_3 + x_4 &= 0 \\ x_1 - x_2 + x_3 - x_4 &= 0 \end{aligned}$$

j. $$\begin{aligned} 2x + y + z + w &= 0 \\ 2x + y - w &= 0 \end{aligned}$$

k. $$\begin{aligned} 2x - y + z + 3w &= 1 \\ 3x - 2y + 4z - 6w &= -6 \\ -7x - 3y - 5z - 9w &= 1 \\ 9x + 5y + 6z + 15w &= 0 \\ 2x - 4y + 3z - 12w &= -5 \end{aligned}$$

l. $$\begin{aligned} x - y &= -6 \\ -y + z &= 5 \\ z - w &= -4 \\ x - w &= -15 \end{aligned}$$

m. $$\begin{aligned} 2x + 3y - z + 4w &= 13 \\ -4x - 15y + 5z - 8w &= -47 \end{aligned}$$

n. $$\begin{aligned} 3x - 7y &= 14 \\ 14x + 6y &= 12 \end{aligned}$$

o. $$\begin{aligned} x + z &= 2 \\ 2x - y + 4z &= 6 \end{aligned}$$

p. $$\begin{aligned} 7x - 14y &= 2 \\ -5x + 10y &= 3 \end{aligned}$$

9. Describe the elementary row operations which produce the following sequence of row-equivalent matrices.

$$\begin{bmatrix} 2 & -1 & 0 & 3 \\ 1 & 3 & 0 & 1 \\ 0 & 5 & 5 & 0 \\ -1 & -10 & 0 & 0 \\ 5 & 4 & 1 & 7 \end{bmatrix} \rightarrow \begin{bmatrix} 0 & -7 & 0 & 1 \\ 1 & 3 & 0 & 1 \\ 0 & 5 & 5 & 0 \\ -1 & -10 & 0 & 0 \\ 5 & 4 & 1 & 7 \end{bmatrix} \rightarrow$$

$$\begin{bmatrix} 0 & -7 & 0 & 1 \\ 1 & 3 & 0 & 1 \\ 0 & 5 & 5 & 0 \\ 0 & -7 & 0 & 1 \\ 5 & 4 & 1 & 7 \end{bmatrix} \to \begin{bmatrix} 1 & 3 & 0 & 1 \\ 0 & -7 & 0 & 1 \\ 0 & 5 & 5 & 0 \\ 0 & -7 & 0 & 1 \\ 5 & 4 & 1 & 7 \end{bmatrix} \to$$

$$\begin{bmatrix} 1 & 3 & 0 & 1 \\ 0 & -7 & 0 & 1 \\ 0 & 5 & 5 & 0 \\ 0 & -7 & 0 & 1 \\ 0 & -11 & 1 & 2 \end{bmatrix} \to \begin{bmatrix} 1 & 3 & 0 & 1 \\ 0 & -7 & 0 & 1 \\ 0 & 1 & 1 & 0 \\ 0 & -7 & 0 & 1 \\ 0 & -11 & 1 & 2 \end{bmatrix} \to$$

$$\begin{bmatrix} 1 & 3 & 0 & 1 \\ 0 & 1 & 1 & 0 \\ 0 & -7 & 0 & 1 \\ 0 & -7 & 0 & 1 \\ 0 & -11 & 1 & 2 \end{bmatrix} \to \begin{bmatrix} 1 & 3 & 0 & 1 \\ 0 & 1 & 1 & 0 \\ 0 & -7 & 0 & 1 \\ 0 & 0 & 0 & 0 \\ 0 & -11 & 1 & 2 \end{bmatrix}$$

Devise further elementary row operations to continue this sequence, ending with an echelon matrix. Repeat Problem 5 for this matrix.

1.5 Inverses and Transposes

A "system" of one linear equation in one unknown, such as

$$3x = 5$$

can be easily solved by multiplying both sides of the equation by $\frac{1}{3}$, which is the multiplicative inverse of the coefficient 3. In a matrix equation

$$AX = B$$

if the coefficient *matrix* A had a multiplicative inverse, we could solve the system in the same way. We now investigate this possibility.

Definition Suppose that A is a square matrix. If there is a matrix B of the same size such that $AB = BA = I$, then B is said to be an *inverse* of A. If A has an inverse, then A is called *nonsingular*; if not, then A is *singular*. (Some say "invertible" for "nonsingular.")

It is a fact, which we shall *assume* in this section, that if $AB = I$ then $BA = I$.

Clearly the identity matrix I is its own inverse. However, there are less trivial examples of matrices which have inverses. One is the matrix given by

$$A = \begin{bmatrix} 1 & 2 \\ 3 & 5 \end{bmatrix}$$

To determine whether or not A has an inverse, given our state of knowledge, we must try to solve the matrix equation

$$\begin{bmatrix} 1 & 2 \\ 3 & 5 \end{bmatrix} \begin{bmatrix} x & y \\ u & v \end{bmatrix} = \begin{bmatrix} 1 & 0 \\ 0 & 1 \end{bmatrix}$$

Because of the way in which matrices are multiplied, this corresponds to the system

$$\begin{array}{rrrrl} x & & + 2u & & = 1 \\ & y & & + 2v & = 0 \\ 3x & & + 5u & & = 0 \\ & 3y & & + 5v & = 1 \end{array}$$

We can solve this system by reducing the augmented matrix or simply by elimination; in either case, we obtain the solution $x = -5$, $y = 2$, $u = 3$, $v = -1$. That is, we have found that

$$\begin{bmatrix} 1 & 2 \\ 3 & 5 \end{bmatrix} \begin{bmatrix} -5 & 2 \\ 3 & -1 \end{bmatrix} = \begin{bmatrix} 1 & 0 \\ 0 & 1 \end{bmatrix}$$

Hence the matrix

$$\begin{bmatrix} 1 & 2 \\ 3 & 5 \end{bmatrix}$$

has for an inverse the matrix

$$\begin{bmatrix} -5 & 2 \\ 3 & -1 \end{bmatrix}$$

and thus is nonsingular.

On the other hand, it can be determined that the matrix

$$\begin{bmatrix} 1 & 2 \\ 3 & 6 \end{bmatrix}$$

is singular (has no inverse) by showing that the corresponding system of four equations in four unknowns has no solution.

If a matrix has an inverse, can it have more than one? The answer to that question is provided by our next theorem.

Theorem 1.7 If a matrix has an inverse then it has only one.

Proof. Let each of B and C be an inverse for the matrix A. We must show that $B = C$. By definition, we have

$$AB = BA = I \quad \text{and} \quad AC = CA = I$$

Then

$$\begin{aligned} B &= IB && \text{(Theorem 1.4)} \\ &= (CA)B && \text{(given)} \\ &= C(AB) && \text{(M11)} \\ &= CI && \text{(given)} \\ &= C && \text{(Theorem 1.4)} \end{aligned}$$

Now that we know that matrix inverses are unique, we can speak of *the* inverse of a matrix (if it exists).

Definition Let A be a nonsingular matrix. The (unique) inverse of A is denoted by the symbol A^{-1}. Thus we have $AA^{-1} = A^{-1}A = I$.

Theorem 1.8 For nonsingular matrices of suitable dimensions

a. $(A^{-1})^{-1} = A$

b. $(AB)^{-1} = B^{-1}A^{-1}$

Proof. a. We must show that the inverse of A^{-1} is A; that is, that $A^{-1}A = AA^{-1} = I$. But this is simply a restatement of the definition of A^{-1}.

b. To establish that $B^{-1}A^{-1}$ is the inverse of AB, we must show that $(AB)(B^{-1}A^{-1}) = I$. (It can be similarly verified that $(B^{-1}A^{-1})(AB) = I$.)

$$\begin{aligned}(AB)(B^{-1}A^{-1}) &= A(B(B^{-1}A^{-1})) && \text{(M11)}\\ &= A((BB^{-1})A^{-1}) && \text{(M11)}\\ &= A(IA^{-1}) && \text{(definition of inverse)}\\ &= AA^{-1} && \text{(Theorem 1.4)}\\ &= I && \text{(definition of inverse)}\end{aligned}$$

Corollary 1.8 A product $A_1A_2 \cdots A_r$ of nonsingular matrices is nonsingular and its inverse is given by

$$(A_1A_2 \cdots A_r)^{-1} = A_r^{-1} \cdots A_2^{-1}A_1^{-1}$$

Proof. This is an inductive extension of part b of Theorem 1.8 (see Appendix A).

Our next theorem represents another step towards the confirmation of our earlier assertion in Section 1.4 that two row-equivalent augmented matrices correspond to systems having the same solution(s). As we have said, two equations, or systems of equations, are *equivalent* if a solution for either is also a solution for the other; that is, they have the same solution set.

Theorem 1.9 If C is nonsingular, then the matrix equations $AX = B$ and $CAX = CB$ are equivalent.

Proof. If $AX = B$ then multiplying each side on the left by C gives $CAX = CB$. Conversely, if $CAX = CB$ then multiplying each side on the left by C^{-1} gives $C^{-1}CAX = C^{-1}CB$, $IAX = IB$, and hence $AX = B$.

Corollary 1.9 An $n \times n$ matrix A is nonsingular if and only if, for any $n \times p$ matrix B, the equation $AX = B$ has a unique solution.

Proof. Assuming that A is nonsingular, the one and only solution of $AX = B$ must be $X = A^{-1}B$, as can be seen by multiplying both sides of the equation on the left by A^{-1}. Conversely, if, for *any* B (of suitable dimensions), $AX = B$ has a unique solution, then in particular the equation $AX = I$ must have a unique solution. But, by definition, that solution is therefore the inverse of A, and hence A is nonsingular.

Occasionally we must deal with the matrix derived from the given matrix by converting its rows into columns (and, automatically, its columns into rows).

Definition Let A be any matrix (not necessarily square). The *transpose* A^t of A is given by

$$(A^t)_{ij} = A_{ji}$$

(Note that if A is $m \times n$ then A^t will be $n \times m$.)

Example 1
If

$$A = \begin{bmatrix} 4 & -1 & 2 \\ 1 & 0 & 3 \end{bmatrix}$$

then

$$A^t = \begin{bmatrix} 4 & 1 \\ -1 & 0 \\ 2 & 3 \end{bmatrix}$$

□

Theorem 1.10 For matrices A and B (of appropriate sizes) and any scalar c,

a. $(A^t)^t = A$

b. $(A + B)^t = A^t + B^t$

c. $(cA)^t = cA^t$

d. $(AB)^t = B^tA^t$

e. $(A^{-1})^t = (A^t)^{-1}$, if A is nonsingular

Proof.

a. $((A^t)^t)_{ij} = (A^t)_{ji} = A_{ij}$

d. $$((AB)^t)_{ij} = (AB)_{ji} = \sum_{k=1}^{n} A_{jk}B_{ki} = \sum_{k=1}^{n} B_{ki}A_{jk}$$
$$= \sum_{k=1}^{n} (B^t)_{ik}(A^t)_{kj} = (B^tA^t)_{ij}$$

We leave the other parts to the exercises.

If a square matrix is equal to its own transpose, it has a particularly attractive form. But this special "symmetry" has important significance in certain applications as well, as we shall see in later sections.

Definition A square matrix is called *symmetric* provided that it is equal to its own transpose. In terms of its entries, a matrix A is symmetric if and only if $A_{ij} = A_{ji}$ for all i and j. Thus, $A^t = A$.

Example 2
The matrices

$$\begin{bmatrix} 1 & 0 & 0 \\ 0 & 0 & 1 \\ 0 & 1 & 0 \end{bmatrix}, \begin{bmatrix} 2 & 3 & 1 \\ 3 & 1 & 2 \\ 1 & 2 & -1 \end{bmatrix}, \begin{bmatrix} 1 & 3 & 5 & 4 \\ 3 & 0 & -1 & 2 \\ 5 & -1 & 6 & 0 \\ 4 & 2 & 0 & 6 \end{bmatrix}, \text{ and } \begin{bmatrix} 1 & 0 & 0 \\ 0 & 1 & 0 \\ 0 & 0 & 1 \end{bmatrix}$$

are all symmetric. However,

$$\begin{bmatrix} 1 & 2 & 3 \\ 4 & 0 & 2 \\ 3 & 4 & 1 \end{bmatrix}$$

is not. Note that the "symmetry" involved is across the diagonal from upper left to lower right (and not across the other diagonal). □

Theorem 1.11

a. The inverse of a nonsingular symmetric matrix is symmetric.

b. If each of A and B is symmetric and of the same size, then $A + B$ is symmetric.

c. For A and B as in part b, AB is symmetric if and only if $AB = BA$.

Proof.

a. Let A be symmetric and nonsingular. We consider $(A^{-1})^t$ and obtain $(A^{-1})^t = (A^t)^{-1} = A^{-1}$. Thus A^{-1} is symmetric.

We ask for reasons and for proofs of parts b and c in the problems.

1.5 Problem Set

1. To find the inverse of the matrix

$$\begin{bmatrix} 2 & -1 \\ -3 & 2 \end{bmatrix}$$

one must solve the matrix equation $AX = I$, where

$$A = \begin{bmatrix} 2 & -1 \\ -3 & 2 \end{bmatrix}, \quad X = \begin{bmatrix} x & y \\ u & v \end{bmatrix}, \quad \text{and} \quad I = \begin{bmatrix} 1 & 0 \\ 0 & 1 \end{bmatrix}$$

Write the augmented matrix $[A \mid I]$ for this system and apply elementary row operations to solve.

2. Use the method of Problem 1 to determine whether each of the following matrices is singular or nonsingular.

a. $\begin{bmatrix} 4 & 6 \\ 2 & 3 \end{bmatrix}$ b. $\begin{bmatrix} 5 & 3 \\ 1 & 2 \end{bmatrix}$

c. $\begin{bmatrix} 6 & 3 & 4 \\ 1 & 1 & 2 \\ 0 & 3 & 2 \end{bmatrix}$ d. $\begin{bmatrix} 0 & 1 & 1 \\ 4 & 5 & 1 \\ 2 & 3 & 1 \end{bmatrix}$

3. Use the answer to Problem 1 to solve the system

$$\begin{aligned} 2x - y &= 3 \\ -3x + 2y &= 0 \end{aligned}$$

(*Hint:* If A is nonsingular, the solution for $AX = B$ is given by $X = A^{-1}B$.)

4. Show, by computing AA^{-1}, that the inverse of

$$A = \begin{bmatrix} a & b \\ c & d \end{bmatrix}$$

is given by

$$A^{-1} = \frac{1}{ad - bc} \begin{bmatrix} d & -b \\ -c & a \end{bmatrix}$$

provided that $ad - bc \neq 0$.

5. Show that if one row or one column of a square matrix consists of all zeros then the matrix cannot have an inverse. (*Hint:* If one row of A is all-zero, what can be said about the product AB, for *any* B? For the column situation, consider BA.)

6. If $ad - bc = 0$, show that

$$\begin{bmatrix} a & b \\ c & d \end{bmatrix}$$

cannot have an inverse, by showing that

a. if one of a, b, c, or d is 0, then the matrix has an all-zero row or column.

b. if none of the entries is 0, then the augmented matrix

$$\left[\begin{array}{cc|cc} a & b & 1 & 0 \\ c & d & 0 & 1 \end{array}\right]$$

can, by row operations, be converted to a form which shows that $AX = I$ has no solution for the matrix X.

7. a. Disprove, by example: The sum of two nonsingular matrices is nonsingular.

b. Decide whether or not the sum of two singular matrices can be nonsingular.

8. For square nonsingular matrices of the same size, show that

$$(AB)^{-1} = A^{-1}B^{-1}$$

if and only if $AB = BA$.

9. Supply reasons for the steps in the parts of Theorems 1.10 and 1.11 proved in the text and prove the other parts.

10. a. Show by example that the product of two symmetric matrices of the same size need not be symmetric.

b. For any matrix A, show that AA^t and A^tA both exist and both are symmetric.

11. A square matrix A is called *skew-symmetric* provided that $A^t = -A$.

a. What is peculiar about the diagonal entries of a skew-symmetric matrix? What about the off-diagonal entries?

b. If each of A and B is skew-symmetric, show that $A + B$ is skew-symmetric and that $(AB)^t = BA$ (assuming appropriate dimensions).

c. Show that a square matrix is the sum of a symmetric and a skew-symmetric matrix. (*Hint:* Consider $A + A^t$ and $A - A^t$.)

1.6 Elementary Matrices

Definition An *elementary matrix* is a matrix obtainable from an identity matrix by the application of a single elementary row operation.

Since there are three types of elementary row operations (Section 1.4), it follows that there are three types of elementary matrices. For example, the matrix

$$E_1 = \begin{bmatrix} 1 & 0 & 0 & 0 \\ 0 & 1 & 0 & 7 \\ 0 & 0 & 1 & 0 \\ 0 & 0 & 0 & 1 \end{bmatrix}$$

is elementary and can be considered of type III because it is the result of adding 7 times the fourth row of the 4×4 identity matrix to the second row of that matrix. Similarly,

$$E_2 = \begin{bmatrix} 1 & 0 & 0 & 0 \\ 0 & -2 & 0 & 0 \\ 0 & 0 & 1 & 0 \\ 0 & 0 & 0 & 1 \end{bmatrix} \quad \text{and} \quad E_3 = \begin{bmatrix} 1 & 0 & 0 & 0 \\ 0 & 0 & 0 & 1 \\ 0 & 0 & 1 & 0 \\ 0 & 1 & 0 & 0 \end{bmatrix}$$

are elementary matrices of types II and I, respectively.

The importance of elementary matrices lies in what they "do" to other matrices. For example, let

$$A = \begin{bmatrix} a & b \\ c & d \\ e & f \\ g & h \end{bmatrix}$$

(for our immediate purposes, any other matrix with four rows would do as well). Let us calculate the matrix product E_1A, where E_1 is the elementary matrix given above. It is important to note that the elementary matrix is on the *left* in the product. A straightforward computation yields

$$E_1A = \begin{bmatrix} a & b \\ c + 7g & d + 7h \\ e & f \\ g & h \end{bmatrix}$$

When we compare E_1A with the original A, we see that E_1A could have been obtained from A by adding 7 times the fourth row of A to the second row of A. But that, of course, is the very same operation we applied to the identity matrix to obtain E_1.

Let us test the products E_2A and E_3A. We find that

$$E_2A = \begin{bmatrix} a & b \\ -2c & -2d \\ e & f \\ g & h \end{bmatrix} \quad \text{and} \quad E_3A = \begin{bmatrix} a & b \\ g & h \\ e & f \\ c & d \end{bmatrix}$$

Note that in these three instances, when A is multiplied on the left by an elementary matrix, the following effect occurs on A: A is transformed by the same row operation which was used to obtain the elementary matrix itself. We summarize this phenomenon in the next theorem.

Theorem 1.12 If E is an $n \times n$ elementary matrix and A is any matrix having n rows, then the product EA is a matrix obtainable from A by the same elementary row operation which produced E from the identity matrix.

Example 1

Let E be successively

$$\begin{bmatrix} 0 & 0 & 1 \\ 0 & 1 & 0 \\ 1 & 0 & 0 \end{bmatrix}, \quad \begin{bmatrix} 1 & 0 & 0 \\ 0 & 1 & 0 \\ 0 & 3 & 1 \end{bmatrix}, \quad \text{and} \quad \begin{bmatrix} 1 & 0 & 0 \\ 0 & 1 & 0 \\ 0 & 0 & 5 \end{bmatrix}$$

Let

$$A = \begin{bmatrix} 3 & 2 & -1 \\ 1 & 1 & 0 \\ 2 & 4 & 1 \end{bmatrix}$$

Then, successively, EA will be

$$\begin{bmatrix} 2 & 4 & 1 \\ 1 & 1 & 0 \\ 3 & 2 & -1 \end{bmatrix}, \quad \begin{bmatrix} 3 & 2 & -1 \\ 1 & 1 & 0 \\ 5 & 7 & 1 \end{bmatrix}, \quad \text{and} \quad \begin{bmatrix} 3 & 2 & -1 \\ 1 & 1 & 0 \\ 10 & 20 & 5 \end{bmatrix} \quad \square$$

As we observed in Section 1.2, the effect of an elementary row operation can be undone by applying another row operation of the same type. This property carries over to elementary matrices: *if E is an elementary matrix, then there is an elementary matrix F of the same type such that $F(EA) = A$ for any compatible matrix A.* If E interchanges two rows, then the "undoing" matrix F must interchange the same two rows, so $F = E$. If E multiplies a row by $k \neq 0$, then F must multiply that row by $1/k$. Finally, if E adds k times row i to row j, then F must add $-k$ times row i to row j.

An equally important property is this: *if F undoes E, then E undoes F; that is $E(FA) = A$ for any A.* The key to this observation is that $1/(1/k) = k$ and $-(-k) = k$. These two properties combine to yield $FE = (FE)I = F(EI) = I$ and $EF = (EF)I = E(FI) = I$ when we take $A = I$; thus by definition of inverse we have $F = E^{-1}$. We cite this fact as our next theorem.

Theorem 1.13 Each elementary matrix is nonsingular; its inverse is an elementary matrix of the same type.

Example 2
For

$$E = \begin{bmatrix} 1 & 0 & 0 \\ 0 & 1 & 0 \\ 0 & 3 & 1 \end{bmatrix}$$

E^{-1} must be

$$\begin{bmatrix} 1 & 0 & 0 \\ 0 & 1 & 0 \\ 0 & -3 & 1 \end{bmatrix}$$

This is because when we compute EE^{-1} or $E^{-1}E$, the first matrix in the product will undo the operation that created the second one from I. Thus either product is I. □

Because of the correspondence between elementary row operations and elementary matrices, we are now in a position to restate and to elaborate on some of our earlier developments. We leave the proofs of the first two theorems as exercises.

Theorem 1.14 If M and N are matrices of the same dimensions, then M is row-equivalent to N if and only if there are elementary matrices

$$E_1, E_2, \ldots, E_s$$

such that

$$N = E_s \cdots E_2 E_1 M$$

Theorem 1.15 Let A, B, and C be matrices of the same dimensions. Then

a. A is row-equivalent to itself.

b. If A is row-equivalent to B, then B is row-equivalent to A.

c. If A is row-equivalent to B and B is row-equivalent to C, then A is row-equivalent to C.

Theorem 1.16 Let A be any matrix and let C be the echelon form matrix for A. Then, for certain elementary matrices $E_1, E_2, \ldots, E_k$,

$$C = E_k \cdots E_2E_1A$$

Proof. This follows from Theorem 1.5 and Theorem 1.14.

We shall now see that singular and nonsingular matrices can be distinguished by the nature of their echelon forms.

Theorem 1.17 A square matrix in echelon form is nonsingular if and only if it is an identity matrix.

Proof. An identity matrix is certainly in echelon form and is equally certainly nonsingular (since it is its own inverse). On the other hand, if C is a square matrix in echelon form but is not an identity matrix, consider the possible locations of its leading ones. By definition of echelon form, as one descends from row to row, the leading one (if any) in a subsequent row must be at least one column to the right of the column containing the leading one in the previous row. Therefore, if the matrix is not an identity matrix, at least one column does not contain a leading one. Consequently, there are fewer leading ones than there are columns and, therefore, fewer leading ones than there are rows. Thus, at least one row (specifically, the last row) is all zeros. But, by the definition of matrix multiplication, any product of C with another matrix (C being on the left) will have a row of zeros in the same location as C does. Hence the product of C and any other matrix cannot be an identity matrix; that is, C is singular. (We note for future use that our proof has shown that a square echelon matrix which is not an identity matrix has all zeros in its last row.)

We remarked after the definition of inverse that, for square matrices of the same size, $AB = I$ implies $BA = I$, although we offered no proof at that time. Thus we have in many instances claimed that $B = A^{-1}$ after establishing only that $AB = I$. Note, however, that we did establish the nonsingularity of an *elementary* matrix in full: we produced a matrix F such that $EF = FE = I$. At this point, it is relatively easy to confirm that, in general, $AB = I$ implies $BA = I$.

Corollary 1.17a If A and B are square matrices of the same size and $AB = I$, then $BA = I$.

Proof. We use Theorem 1.16 and write

$$C = E_k \cdots E_2E_1A$$

where C is the echelon form for A and $E_1, E_2, \ldots, E_k$ are appropriate elementary matrices. If $AB = I$, we have $CB = E_k \cdots E_2E_1$; thus CB is a product of nonsingular matrices and hence must be nonsingular. Now, CB cannot have a row of zeros, since, as we argued in the proof of the theorem, that condition would make CB singular. It follows that C cannot have a row of zeros, thus $C = I$. Consequently, $B = E_k \cdots E_2E_1$; thus B is nonsingular. Multiplying both sides of the equation $AB = I$ on the right by B^{-1} yields $A = B^{-1}$, so $BA = I$, as desired.

Corollary 1.17b A square matrix A is nonsingular if and only if its echelon form is an identity matrix.

Proof. Again, as in Theorem 1.16, we can write $C = E_k \cdots E_2E_1A$, where C is the echelon form for A and $E_1, E_2, \ldots, E_k$ are certain elementary matrices. If A is nonsingular, then, by the above, C is a product of nonsingular matrices, which makes C itself nonsingular (Corollary 1.8) and hence, by Theorem 1.17, C must be an identity matrix. Conversely, let us assume that C is an identity and write

$$I = E_k \cdots E_2E_1A$$

Then $A^{-1} = E_k \cdots E_2E_1$; A is nonsingular.

Corollary 1.17c A square matrix is nonsingular if and only if it is a product of elementary matrices.

Proof. A product of elementary matrices is a product of nonsingular matrices and hence is nonsingular. On the other hand, if A is nonsingular, then we have $I = E_k \cdots E_2E_1A$, for certain elementary matrices $E_1, E_2, \ldots, E_k$, as before; we obtain $A = (A^{-1})^{-1} = (E_k \cdots E_2E_1)^{-1} = E_1^{-1}E_2^{-1} \cdots E_k^{-1}$, which is a product of elementary matrices since the inverse of an elementary matrix is elementary.

We are, at last, in a position to confirm our claim of long-standing, as demonstrated in the next theorem.

Theorem 1.18 Two row-equivalent augmented matrices represent systems having the same solution set.

Proof. Suppose that M and N are the row-equivalent augmented matrices in question. By Theorem 1.14, we can write

$$N = E_s \cdots E_2E_1M$$

for certain elementary matrices $E_1, E_2, \ldots, E_s$. (Incidentally, according to Theorem 1.15, it makes no difference whether we couple the elementary matrices with M or with N.) Now, since elementary matrices are nonsingular (by Theorem 1.13), a product of such matrices is nonsingular and we can invoke Theorem 1.9 to complete our argument.

To apply the ideas of this section, we conclude with a method for finding the inverse of a nonsingular matrix by using row operations as explained by our next theorem.

Theorem 1.19 If A is a nonsingular matrix, then the same sequence of elementary row operations which transforms A into an identity matrix (that is, into its echelon form) will, when applied to an identity matrix of the same size, produce A^{-1}.

Proof. Consider the matrix equation $AX = I$, where I is the identity matrix of the same size as A (so X must be a matrix of "unknowns" of that size as well). By the definition of inverse, the unique solution of this equation must be A^{-1}. But, if we solve this system by applying row operations to the augmented matrix $[A \mid I]$, we will be transforming A to its echelon form, namely I, while simultaneously changing I to a matrix, say, B. The result is the equivalent system $IX = B$, so $B = A^{-1}$.

Let us also view this from the perspective of elementary matrices. For a nonsingular matrix A we can write

$$I = E_k \cdots E_2 E_1 A$$

for certain elementary matrices $E_1, E_2, \ldots, E_k$, as we have seen. By the definition of inverse,

$$A^{-1} = E_k \cdots E_2 E_1$$

which can be written as

$$A^{-1} = E_k \cdots E_2 E_1 I$$

Thus the same sequence of elementary matrices which changes A to I will also change I to A^{-1}.

In the next example, we determine whether or not a given matrix is nonsingular and, if it is, find its inverse and also express the given matrix as a product of elementary matrices. For bookkeeping purposes, we follow this procedure:

Set up the augmented matrix $[A \mid I]$; apply elementary row operations to the entire augmented matrix with the goal of converting A into I. If this goal is achieved, the result will be an augmented matrix $[I \mid B]$ and B will be A^{-1}. (If, in the process, one reaches a point where the echelon form of A will clearly not turn out to be I, then the conclusion is that A is singular.) Meanwhile, if one keeps a record of the elementary row operations used by recording the corresponding elementary matrices, one can produce A expressed as a product of elementary matrices by writing the inverses of these elementary matrices in reverse order, as in the proof of Corollary 1.17b. The result will be an illustration of this corollary.

Example 3

Let

$$A = \begin{bmatrix} 2 & 3 & -2 \\ 0 & -1 & 5 \\ 0 & -2 & 4 \end{bmatrix}$$

We determine whether or not A is nonsingular and, if it is, we find its inverse and express A as a product of elementary matrices. On the right, we record the elementary matrix which corresponds to the elementary row operation used in each step.

$$[A \mid I] = \left[\begin{array}{rrr|rrr} 2 & 3 & -2 & 1 & 0 & 0 \\ 0 & -1 & 5 & 0 & 1 & 0 \\ 0 & -2 & 4 & 0 & 0 & 1 \end{array}\right]$$

$$\to \left[\begin{array}{rrr|rrr} 2 & 3 & -2 & 1 & 0 & 0 \\ 0 & -1 & 5 & 0 & 1 & 0 \\ 0 & 0 & -6 & 0 & -2 & 1 \end{array}\right] \qquad E_1 = \begin{bmatrix} 1 & 0 & 0 \\ 0 & 1 & 0 \\ 0 & -2 & 1 \end{bmatrix}$$

$$\to \left[\begin{array}{rrr|rrr} 2 & 3 & -2 & 1 & 0 & 0 \\ 0 & -1 & 5 & 0 & 1 & 0 \\ 0 & 0 & 1 & 0 & \frac{1}{3} & -\frac{1}{6} \end{array}\right] \qquad E_2 = \begin{bmatrix} 1 & 0 & 0 \\ 0 & 1 & 0 \\ 0 & 0 & -\frac{1}{6} \end{bmatrix}$$

$$\to \left[\begin{array}{rrr|rrr} 2 & 3 & -2 & 1 & 0 & 0 \\ 0 & -1 & 0 & 0 & -\frac{2}{3} & \frac{5}{6} \\ 0 & 0 & 1 & 0 & \frac{1}{3} & -\frac{1}{6} \end{array}\right] \qquad E_3 = \begin{bmatrix} 1 & 0 & 0 \\ 0 & 1 & -5 \\ 0 & 0 & 1 \end{bmatrix}$$

$$\to \left[\begin{array}{rrr|rrr} 2 & 3 & -2 & 1 & 0 & 0 \\ 0 & 1 & 0 & 0 & \frac{2}{3} & -\frac{5}{6} \\ 0 & 0 & 1 & 0 & \frac{1}{3} & -\frac{1}{6} \end{array}\right] \qquad E_4 = \begin{bmatrix} 1 & 0 & 0 \\ 0 & -1 & 0 \\ 0 & 0 & 1 \end{bmatrix}$$

$$\to \left[\begin{array}{rrr|rrr} 2 & 0 & -2 & 1 & -2 & \frac{5}{2} \\ 0 & 1 & 0 & 0 & \frac{2}{3} & -\frac{5}{6} \\ 0 & 0 & 1 & 0 & \frac{1}{3} & -\frac{1}{6} \end{array}\right] \qquad E_5 = \begin{bmatrix} 1 & -3 & 0 \\ 0 & 1 & 0 \\ 0 & 0 & 1 \end{bmatrix}$$

$$\to \left[\begin{array}{rrr|rrr} 2 & 0 & 0 & 1 & -\frac{4}{3} & \frac{13}{6} \\ 0 & 1 & 0 & 0 & \frac{2}{3} & -\frac{5}{6} \\ 0 & 0 & 1 & 0 & \frac{1}{3} & -\frac{1}{6} \end{array}\right] \qquad E_6 = \begin{bmatrix} 1 & 0 & 2 \\ 0 & 1 & 0 \\ 0 & 0 & 1 \end{bmatrix}$$

$$\to \left[\begin{array}{rrr|rrr} 1 & 0 & 0 & \frac{1}{2} & -\frac{2}{3} & \frac{13}{12} \\ 0 & 1 & 0 & 0 & \frac{2}{3} & -\frac{5}{6} \\ 0 & 0 & 1 & 0 & \frac{1}{3} & -\frac{1}{6} \end{array}\right] \qquad E_7 = \begin{bmatrix} \frac{1}{2} & 0 & 0 \\ 0 & 1 & 0 \\ 0 & 0 & 1 \end{bmatrix}$$

Since A has now been converted to an identity matrix, the latter half of this final augmented matrix must be A^{-1}. Thus

$$A^{-1} = \begin{bmatrix} \frac{1}{2} & -\frac{2}{3} & \frac{13}{12} \\ 0 & \frac{2}{3} & -\frac{5}{6} \\ 0 & \frac{1}{3} & -\frac{1}{6} \end{bmatrix} = \frac{1}{12}\begin{bmatrix} 6 & -8 & 13 \\ 0 & 8 & -10 \\ 0 & 4 & -2 \end{bmatrix}$$

as can be verified by multiplying to get I.

According to the proof of Corollary 1.17b, to express A as a product of elementary matrices, we write $A = E_1^{-1}E_2^{-1}E_3^{-1}E_4^{-1}E_5^{-1}E_6^{-1}E_7^{-1}$:

$$A = \begin{bmatrix} 1 & 0 & 0 \\ 0 & 1 & 0 \\ 0 & 2 & 1 \end{bmatrix} \begin{bmatrix} 1 & 0 & 0 \\ 0 & 1 & 0 \\ 0 & 0 & -6 \end{bmatrix} \begin{bmatrix} 1 & 0 & 0 \\ 0 & 1 & 5 \\ 0 & 0 & 1 \end{bmatrix} \begin{bmatrix} 1 & 0 & 0 \\ 0 & -1 & 0 \\ 0 & 0 & 1 \end{bmatrix}$$

$$\times \begin{bmatrix} 1 & 3 & 0 \\ 0 & 1 & 0 \\ 0 & 0 & 1 \end{bmatrix} \begin{bmatrix} 1 & 0 & -2 \\ 0 & 1 & 0 \\ 0 & 0 & 1 \end{bmatrix} \begin{bmatrix} 2 & 0 & 0 \\ 0 & 1 & 0 \\ 0 & 0 & 1 \end{bmatrix}$$ □

Example 4

Let

$$A = \begin{bmatrix} 1 & 0 & 1 \\ 1 & 1 & 0 \\ 0 & 1 & -1 \end{bmatrix}$$

Then

$$[A \mid I] = \left[\begin{array}{ccc|ccc} 1 & 0 & 1 & 1 & 0 & 0 \\ 1 & 1 & 0 & 0 & 1 & 0 \\ 0 & 1 & -1 & 0 & 0 & 1 \end{array}\right]$$

$$\left[\begin{array}{ccc|ccc} 1 & 0 & 1 & 1 & 0 & 0 \\ 0 & 1 & -1 & -1 & 1 & 0 \\ 0 & 1 & -1 & 0 & 0 & 1 \end{array}\right] \qquad E_1 = \begin{bmatrix} 1 & 0 & 0 \\ -1 & 1 & 0 \\ 0 & 0 & 1 \end{bmatrix}$$

$$\left[\begin{array}{ccc|ccc} 1 & 0 & 1 & 1 & 0 & 0 \\ 0 & 1 & -1 & -1 & 1 & 0 \\ 0 & 0 & 0 & 1 & -1 & 1 \end{array}\right] \qquad E_2 = \begin{bmatrix} 1 & 0 & 0 \\ 0 & 1 & 0 \\ 0 & -1 & 1 \end{bmatrix}$$

After only two operations, we have converted A to a matrix having an all-zero row. As we have seen in proving Theorem 1.17, such a matrix must be singular, and hence no further operations can result in the identity matrix. Thus A is singular. □

Example 5

We find the inverse (if there is one) of

$$A = \begin{bmatrix} 2 & 0 & 6 & 3 \\ 1 & 4 & 0 & 1 \\ 0 & 1 & -3 & 1 \\ 1 & -2 & 0 & 1 \end{bmatrix}$$

without adopting the additional goal of writing A (if it is nonsingular) as a product of elementary matrices. Thus we need not record an elementary matrix for each and every row operation performed; we can even combine more than one row operation in a single step. (An obvious risk in trying to shorten the work, however, is that of making more errors.)

$$\left[\begin{array}{cccc|cccc} 2 & 0 & 6 & 3 & 1 & 0 & 0 & 0 \\ 1 & 4 & 0 & 1 & 0 & 1 & 0 & 0 \\ 0 & 1 & -3 & 1 & 0 & 0 & 1 & 0 \\ 1 & -2 & 0 & 1 & 0 & 0 & 0 & 1 \end{array}\right]$$ (the augmented matrix $[A \mid I]$)

$$\left[\begin{array}{cccc|cccc} 0 & -8 & 6 & 1 & 1 & -2 & 0 & 0 \\ 1 & 4 & 0 & 1 & 0 & 1 & 0 & 0 \\ 0 & 1 & -3 & 1 & 0 & 0 & 1 & 0 \\ 0 & -6 & 0 & 0 & 0 & -1 & 0 & 1 \end{array}\right]$$ (((-2) times row 2 added to row 1; (-1) times row 2 added to row 4)

$$\left[\begin{array}{cccc|cccc} 0 & -8 & 6 & 1 & 1 & -2 & 0 & 0 \\ 1 & 4 & 0 & 1 & 0 & 1 & 0 & 0 \\ 0 & 1 & -3 & 1 & 0 & 0 & 1 & 0 \\ 0 & 1 & 0 & 0 & 0 & \frac{1}{6} & 0 & -\frac{1}{6} \end{array}\right]$$ (row 4 multiplied by $-\frac{1}{6}$)

$$\left[\begin{array}{cccc|cccc} 0 & 0 & 6 & 1 & 1 & -\frac{2}{3} & 0 & -\frac{4}{3} \\ 1 & 0 & 0 & 1 & 0 & \frac{1}{3} & 0 & \frac{2}{3} \\ 0 & 0 & -3 & 1 & 0 & -\frac{1}{6} & 1 & \frac{1}{6} \\ 0 & 1 & 0 & 0 & 0 & \frac{1}{6} & 0 & -\frac{1}{6} \end{array}\right]$$ (8 times row 4 added to row 1; (-4) times row 4 added to row 2; (-1) times row 4 added to row 3)

$$\left[\begin{array}{cccc|cccc} 0 & 0 & 0 & 3 & 1 & -1 & 2 & -1 \\ 1 & 0 & 0 & 1 & 0 & \frac{1}{3} & 0 & \frac{2}{3} \\ 0 & 0 & -3 & 1 & 0 & -\frac{1}{6} & 1 & \frac{1}{6} \\ 0 & 1 & 0 & 0 & 0 & \frac{1}{6} & 0 & -\frac{1}{6} \end{array}\right]$$ (2 times row 3 added to row 1)

$$\left[\begin{array}{cccc|cccc} 0 & 0 & 0 & 1 & \frac{1}{3} & -\frac{1}{3} & \frac{2}{3} & -\frac{1}{3} \\ 1 & 0 & 0 & 1 & 0 & \frac{1}{3} & 0 & \frac{2}{3} \\ 0 & 0 & -3 & 1 & 0 & -\frac{1}{6} & 1 & \frac{1}{6} \\ 0 & 1 & 0 & 0 & 0 & \frac{1}{6} & 0 & -\frac{1}{6} \end{array}\right]$$ (row 1 multiplied by $\frac{1}{3}$)

$$\left[\begin{array}{cccc|cccc} 0 & 0 & 0 & 1 & \frac{1}{3} & -\frac{1}{3} & \frac{2}{3} & -\frac{1}{3} \\ 1 & 0 & 0 & 0 & -\frac{1}{3} & \frac{2}{3} & -\frac{2}{3} & 1 \\ 0 & 0 & -3 & 0 & -\frac{1}{3} & \frac{1}{6} & \frac{1}{3} & \frac{1}{2} \\ 0 & 1 & 0 & 0 & 0 & \frac{1}{6} & 0 & -\frac{1}{6} \end{array}\right]$$ (((-1) times row 1 added to row 2; (-1) times row 1 added to row 3)

$$\left[\begin{array}{cccc|cccc} 0 & 0 & 0 & 1 & \frac{1}{3} & -\frac{1}{3} & \frac{2}{3} & -\frac{1}{3} \\ 1 & 0 & 0 & 0 & -\frac{1}{3} & \frac{2}{3} & -\frac{2}{3} & 1 \\ 0 & 0 & 1 & 0 & \frac{1}{9} & -\frac{1}{18} & -\frac{1}{9} & -\frac{1}{6} \\ 0 & 1 & 0 & 0 & 0 & \frac{1}{6} & 0 & -\frac{1}{6} \end{array}\right] \quad \text{(row 3 multiplied by } (-\tfrac{1}{3}))$$

$$\left[\begin{array}{cccc|cccc} 1 & 0 & 0 & 0 & -\frac{1}{3} & \frac{2}{3} & -\frac{2}{3} & 1 \\ 0 & 1 & 0 & 0 & 0 & \frac{1}{6} & 0 & -\frac{1}{6} \\ 0 & 0 & 1 & 0 & \frac{1}{9} & -\frac{1}{18} & -\frac{1}{9} & -\frac{1}{6} \\ 0 & 0 & 0 & 1 & \frac{1}{3} & -\frac{1}{3} & \frac{2}{3} & -\frac{1}{3} \end{array}\right] \quad \text{(several interchanges of rows; how many interchanges?)}$$

Since I is on the left, we have A^{-1} on the right:

$$A^{-1} = \begin{bmatrix} -\frac{1}{3} & \frac{2}{3} & -\frac{2}{3} & 1 \\ 0 & \frac{1}{6} & 0 & -\frac{1}{6} \\ \frac{1}{9} & -\frac{1}{18} & -\frac{1}{9} & -\frac{1}{6} \\ \frac{1}{3} & -\frac{1}{3} & \frac{2}{3} & -\frac{1}{3} \end{bmatrix} = \frac{1}{18}\begin{bmatrix} -6 & 12 & -12 & 18 \\ 0 & 3 & 0 & -3 \\ 2 & -1 & -2 & -3 \\ 6 & -6 & 12 & -6 \end{bmatrix} \qquad \square$$

We have seen that we can solve systems of equations and find inverses of matrices by operating on augmented matrices with row operations to reduce them to echelon form. This technique is so systematic that it is entirely suitable for computer programming. The step-by-step process, as one would instruct a machine to carry it out, would be as follows:

Step 1. Search the first column of the matrix from top to bottom to find the first nonzero entry. If there is no such entry, perform this search on the second column, and so on.

Step 2. By a row interchange, reposition the nonzero entry from step 1 to the top of the first column.

Step 3. Multiply the first row by the reciprocal of the nonzero entry from step 1.

Step 4. Add appropriate multiples of the first row to all other rows (where necessary) to create all zeros in the first column below the first entry, which is now a 1.

Step 5. Move to the second column and search for the first nonzero entry *below* the first. If there is no such entry, move to the next column, and so on.

Step 6. By a row interchange, reposition the nonzero entry from step 5 to the second-from-the-top location of the second column.

Step 7. Use row operations to change the nonzero entry of step 6 to a 1 and to create zeros elsewhere in the second column.

Continue this process until the rows or columns are exhausted, whichever happens first.

This systematic procedure, in theory, will result in the echelon form of a matrix. However, computers have certain inherent limitations which might cause the end result to be far from correct (for instance, the machine may treat numbers close in value to 0 as equal to 0). Because of these limitations and also because of efficiency demands, certain refinements of this procedure have been devised which minimize the risks of error and the number of operations involved. We shall not discuss those refinements here.

In working problems by hand, shortcuts can normally be found; indeed, most of our examples demonstrate that the algorithm we have described need not be followed in a rigid manner.

1.6 Problem Set

1. Write the inverse matrix for each of the following *elementary* matrices.

a. $\begin{bmatrix} 1 & 0 & -7 \\ 0 & 1 & 0 \\ 0 & 0 & 1 \end{bmatrix}$ b. $\begin{bmatrix} 1 & 0 & 0 \\ 0 & 1 & 0 \\ 0 & 0 & \frac{1}{2} \end{bmatrix}$

c. $\begin{bmatrix} 0 & 1 & 0 \\ 1 & 0 & 0 \\ 0 & 0 & 1 \end{bmatrix}$ d. $\begin{bmatrix} \pi & 0 \\ 0 & 1 \end{bmatrix}$

2. Find the inverses of the following matrices by using the method described in this section.

a. $\begin{bmatrix} 3 & 0 & 1 \\ 2 & 3 & 0 \\ -1 & 0 & 1 \end{bmatrix}$ b. $\begin{bmatrix} 1 & 0 & 2 & 3 \\ 2 & 1 & 3 & 4 \\ -1 & -1 & 0 & 0 \\ 0 & 2 & 1 & 0 \end{bmatrix}$

c. $\begin{bmatrix} 1 & 0 & -1 & 0 \\ 0 & 2 & 4 & -2 \\ -1 & 2 & 3 & 0 \\ 2 & 1 & 0 & 3 \end{bmatrix}$ d. $\begin{bmatrix} 1 & 0 & 0 & 2 & -3 \\ 0 & 1 & 2 & 3 & 0 \\ -1 & 2 & 3 & 3 & 2 \\ -2 & 1 & -1 & -5 & 2 \\ 0 & 3 & 0 & 1 & -9 \end{bmatrix}$

3. Suppose that an elementary matrix is placed to the *right* of a matrix of suitable dimensions, and the product is computed. Describe the result in terms of how it is related to the nonelementary matrix. (Try this with the elementary matrices of Problem 1.)

4. In Example 1 of Section 1.4, write the elementary matrices which correspond to the elementary row operations performed there. Use these to express the coefficient matrix of that example as a product of certain elementary matrices.

5. Use the final result of Problem 9, Section 1.4 (p. 30), to express the matrix given there as the product of certain elementary matrices.

6. Determine whether or not each of the following matrices is nonsingular. If it is, write its inverse and express the original matrix as a product of elementary matrices.

   a. $\begin{bmatrix} 1 & 2 & 3 \\ -1 & 1 & 2 \\ 0 & 3 & 5 \end{bmatrix}$ b. $\begin{bmatrix} 1 & 1 \\ 3 & 2 \end{bmatrix}$

   c. $\begin{bmatrix} 1 & 0 & 1 \\ 1 & 1 & 0 \\ 0 & 1 & 1 \end{bmatrix}$ d. $\begin{bmatrix} 0 & 5 & 0 \\ 1 & 0 & -4 \\ 0 & 0 & 1 \end{bmatrix}$

   e. $\begin{bmatrix} 2 & 2 & 3 \\ 3 & 0 & 2 \\ 1 & 0 & 1 \end{bmatrix}$

7. Prove Theorem 1.14.

8. Prove Theorem 1.15.

9. Use Theorem 1.15 to prove:

   a. If two matrices have the same echelon form, then they are row-equivalent.

   b. If two matrices are row-equivalent, then they have the same echelon form.

10. Use the result of Problem 9 to show that two nonsingular matrices of the same dimensions are row-equivalent.

11. How many 3×3 elementary matrices are there of type I? How many 4×4 matrices?

12. Show that an elementary matrix of type I is symmetric. Show that an elementary matrix of type II is symmetric.

13. Show that an elementary matrix of type I is a product of elementary matrices of types II and III; that is, that the operation of interchanging two rows can be achieved by a certain sequence of the other types of row operations. (*Hint:* Try the 2×2 case first.)
14. If an elementary matrix of type I is written as a product of elementary matrices of types II and III, argue that more than one matrix of type III must appear in this product. (*Hint:* Use Problem 12.)

2

DETERMINANTS

2.1 DETERMINANTS*

Associated with each square matrix is a number called its *determinant*, which is useful in many interesting and important applications of matrices. How this number is obtained is given by the following definitions.

Definition Let A be an $n \times n$ matrix. For $1 \leq i, j \leq n$, let A^{ij} (called the *i-j minor matrix* of A) be that $(n - 1) \times (n - 1)$ matrix obtained from A by deleting the i-th row and j-th column from A. (Note that A^{ij} denotes a *matrix* while A_{ij} denotes a *number*, namely the i-j entry of A.)

*The proofs of many theorems in this section hinge on the technique known as *mathematical induction*. We have included in the appendices a discussion of this method of proof and those proofs omitted from the main text.

Example 1

If

$$A = \begin{bmatrix} 3 & 1 & 2 \\ 1 & -2 & 0 \\ 2 & 1 & 1 \end{bmatrix}$$

then

$$A^{32} = \begin{bmatrix} 3 & 2 \\ 1 & 0 \end{bmatrix}$$ □

Definition

The *determinant* of an $n \times n$ matrix A, denoted by $\det A$, is given inductively by the rules:

If $n = 1$, so that A has only one entry, then $\det A$ is that entry.

If $n > 1$, then

$$\begin{aligned} \det A &= A_{11}\det(A^{11}) - A_{12}\det(A^{12}) + \cdots + (-1)^{1+n}A_{1n}\det(A^{1n}) \\ &= \sum_{k=1}^{n} (-1)^{1+k}A_{1k}\det(A^{1k}) \end{aligned}$$

The factor $(-1)^{1+k}$ is simply a device to cause the signs of the terms to alternate.

Other mathematicians use an alternate, but equivalent, definition of determinant which involves the concept of permutations. For those interested, we present this alternate definition in Appendix B.

To illustrate our definition we derive a formula for the determinant of a 2×2 matrix. Let

$$A = \begin{bmatrix} a & b \\ c & d \end{bmatrix}$$

Then $\det A = A_{11}\det(A^{11}) - A_{12}\det(A^{12}) = a \det[d] - b \det[c] = ad - bc$. The significance of this number is demonstrated in Problems 4 and 6 of Section 1.5.

To obtain a formula for a 3×3 determinant according to the definition, we must employ 2×2 determinants. Thus, if A is 3×3, we have

$$\begin{aligned}\det A &= A_{11}\det(A^{11}) - A_{12}\det(A^{12}) + A_{13}\det(A^{13})\\ &= A_{11}\det\begin{bmatrix}A_{22} & A_{23}\\ A_{32} & A_{33}\end{bmatrix} - A_{12}\det\begin{bmatrix}A_{21} & A_{23}\\ A_{31} & A_{33}\end{bmatrix} + A_{13}\det\begin{bmatrix}A_{21} & A_{22}\\ A_{31} & A_{32}\end{bmatrix}\\ &= A_{11}(A_{22}A_{33} - A_{23}A_{32}) - A_{12}(A_{21}A_{33} - A_{23}A_{31})\\ &\quad + A_{13}(A_{21}A_{32} - A_{22}A_{31})\end{aligned}$$

Example 2

We apply the above formula to obtain the determinant of the given matrix.

$$\det\begin{bmatrix}3 & 1 & 2\\ 1 & -2 & 0\\ 2 & 1 & 1\end{bmatrix} = 3(-2 - 0) - 1(1 - 0) + 2(1 + 4) = 3 \qquad \square$$

Upon investigation one sees that the determinant of a 4×4 matrix will be the sum of $4 \cdot 3 \cdot 2 = 24$ terms and that, in general, the determinant of an $n \times n$ matrix is the sum of $n! = n(n - 1)(n - 2) \cdots 3 \cdot 2 \cdot 1$ terms, half with plus and half with minus signs, when expanded fully.

Our definition of determinant involves multiplying the entries in the *first* row by the determinants of corresponding minor matrices. This is called "expansion about the first row"; it is possible to prove (see Appendix B) that the very same determinant value will be the result of expansion about *any* row.

Theorem 2.1 For any i, $1 \le i \le n$, and any $n \times n$ matrix A,

$$\begin{aligned}\det A &= (-1)^{i+1}A_{i1}\det(A^{i1}) + (-1)^{i+2}A_{i2}\det(A^{i2}) + \cdots\\ &\quad + (-1)^{i+n}A_{in}\det(A^{in})\\ &= \sum_{k=1}^{n} (-1)^{i+k}A_{ik}\det(A^{ik})\end{aligned}$$

(This gives the *expansion about the i-th row.*)

Example 3

We compute a determinant using expansion about the second row.

$$\det\begin{bmatrix} 3 & 1 & 2 \\ 1 & -2 & 0 \\ 2 & 1 & 1 \end{bmatrix} = (-1)(1)\det\begin{bmatrix} 1 & 2 \\ 1 & 1 \end{bmatrix} + (+1)(-2)\det\begin{bmatrix} 3 & 2 \\ 2 & 1 \end{bmatrix}$$

$$+ (-1)(0)\det\begin{bmatrix} 3 & 1 \\ 2 & 1 \end{bmatrix}$$

$$= (-1)(-1) + (-2)(-1) + (0)(1)$$

$$= 3$$

(This corroborates Example 2.) □

Example 4

For an identity matrix I,

$$\det I = 1$$

Formal verification of this important fact involves mathematical induction (see Appendix A); however, expanding about the first row, then about the first row of what remains, and so on, provides an insight into that verification. □

An alternate notation for determinant entails replacing the bounding brackets of the matrix with vertical lines; in this way, for instance,

$$\det\begin{bmatrix} -4 & 2 \\ 6 & 3 \end{bmatrix}$$

is the same as

$$\begin{vmatrix} -4 & 2 \\ 6 & 3 \end{vmatrix}$$

We prefer $\det[\cdot\ \cdot\ \cdot]$ to $|\cdot\ \cdot\ \cdot|$ because of the danger of misreading the latter. However, we shall use the alternate notation to save space in the next example.

Example 5

We compute another determinant.

$$\begin{vmatrix} -1 & 3 & 5 & -2 \\ 6 & 1 & -4 & 0 \\ 2 & 0 & -1 & 3 \\ 4 & 1 & 6 & 2 \end{vmatrix}$$

$$= (+1)(2)\begin{vmatrix} 3 & 5 & -2 \\ 1 & -4 & 0 \\ 1 & 6 & 2 \end{vmatrix} + (-1)(0)\begin{vmatrix} -1 & 5 & -2 \\ 6 & -4 & 0 \\ 4 & 6 & 2 \end{vmatrix}$$

$$+ (+1)(-1)\begin{vmatrix} -1 & 3 & -2 \\ 6 & 1 & 0 \\ 4 & 1 & 2 \end{vmatrix} + (-1)(3)\begin{vmatrix} -1 & 3 & 5 \\ 6 & 1 & -4 \\ 4 & 1 & 6 \end{vmatrix}$$

(expansion about the third row)

$$= (2)\left[(+1)(3)\begin{vmatrix} -4 & 0 \\ 6 & 2 \end{vmatrix} + (-1)(5)\begin{vmatrix} 1 & 0 \\ 1 & 2 \end{vmatrix} + (+1)(-2)\begin{vmatrix} 1 & -4 \\ 1 & 6 \end{vmatrix}\right]$$

(expansion about the first row)

$$+ (0)[\cdot \cdot \cdot] \qquad \text{(no expansion needed)}$$

$$+ (-1)\left[(-1)(6)\begin{vmatrix} 3 & -2 \\ 1 & 2 \end{vmatrix} + (+1)(+1)\begin{vmatrix} -1 & -2 \\ 4 & 2 \end{vmatrix} + (-1)(0)\begin{vmatrix} -1 & 3 \\ 4 & 1 \end{vmatrix}\right]$$

(expansion about the second row)

$$+ (-3)\left[(+1)(4)\begin{vmatrix} 3 & 5 \\ 1 & -4 \end{vmatrix} + (-1)(1)\begin{vmatrix} -1 & 5 \\ 6 & -4 \end{vmatrix} + (+1)(6)\begin{vmatrix} -1 & 3 \\ 6 & 1 \end{vmatrix}\right]$$

(expansion about the third row)

$$\begin{aligned} &= (2)[(3)(-8) + (-5)(2) + (-2)(10)] \\ &\quad + 0 \\ &\quad + (-1)[(-6)(8) + (1)(6) + 0] \\ &\quad + (-3)[(4)(-17) + (-1)(-26) + (6)(-19)] \\ &= (2)(-54) + (-1)(-42) + (-3)(-156) \\ &= 402 \end{aligned}$$

□

The effect of elementary row operations on the determinant of a matrix is seen in the next theorem; it is proved in Appendix B.

Theorem 2.2 Let A be a square matrix and let E be an elementary matrix of the same size.

a. If E is of type I, then $\det EA = -\det A$; that is, if two rows of A are interchanged, then the determinant changes sign.

b. If E is of type II, then $\det EA = k \det A$; that is, if a row of A is multiplied by a constant k, then the determinant is multiplied by that constant.

c. If E is of type III, then $\det EA = \det A$; that is, if a multiple of one row of A is added to another row, the determinant is unchanged.

Corollary 2.2 If E is an elementary matrix, then $\det E = -1$, k, or 1, if E is of type I, II, or III, respectively.

Proof. Let $A = I$ in Theorem 2.2 and use the fact that $\det I = 1$.

Theorem 2.3 If E is elementary and A is a square matrix of the same size, then $\det(EA) = (\det E)(\det A)$.

Proof. By Corollary 2.2, $(\det E)(\det A) = -\det A$, $k \det A$, or $\det A$, if E is of type I, II, or III, respectively. But, by Theorem 2.2, this is in each case precisely $\det(EA)$.

Example 6
If we recall from earlier examples that

$$\det\begin{bmatrix} 3 & 1 & 2 \\ 1 & -2 & 0 \\ 2 & 1 & 1 \end{bmatrix} = 3$$

we may use Theorem 2.2 to conclude that

$$\det\begin{bmatrix} 3 & 1 & 2 \\ 2 & 1 & 1 \\ 1 & -2 & 0 \end{bmatrix} = -3, \qquad \det\begin{bmatrix} 3 & 1 & 2 \\ 4 & -8 & 0 \\ 2 & 1 & 1 \end{bmatrix} = 12,$$

$$\text{and} \quad \det\begin{bmatrix} 5 & -3 & 2 \\ 1 & -2 & 0 \\ 2 & 1 & 1 \end{bmatrix} = 3$$

since these three matrices were obtained by interchanging the second and third rows of the original matrix, multiplying the second row by 4, and adding twice the second row to the first row, respectively. Of course, these determinants could have been computed directly by expansion about any row. □

Corollary 2.3 For any square matrix A,

a. If two rows of A are equal, then $\det A = 0$.

b. If all entries in a row of A are zero, then $\det A = 0$.

Proof.

a. If the two equal rows were interchanged, then the matrix would be unchanged but the determinant would change sign. Since the only number equal to its negative is 0, the value of the determinant must be 0.

b. If the all-zero row were multiplied by a nonzero constant, the matrix would be unchanged but the determinant would be multiplied by that constant. Since the only number equal to every nonzero multiple of itself is 0, the value of the determinant must be 0.

Theorem 2.4 If $E_1, E_2, \ldots, E_m$ are elementary and C is a square matrix and all are the same size, then

$$\det(E_m E_{m-1} \cdots E_1 C) = (\det E_m)(\det E_{m-1}) \cdots (\det E_1)(\det C)$$

Proof. We apply Theorem 2.3 successively m times to get

$$\begin{aligned}
\det(E_m E_{m-1} \cdots E_1 C) \\
&= (\det E_m)\det(E_{m-1}E_{m-2} \cdots E_1 C) \\
&= (\det E_m)(\det E_{m-1})\det(E_{m-2} \cdots E_1 C) \\
&\quad \cdots \\
&= (\det E_m)(\det E_{m-1})(\det E_{m-2}) \cdots (\det E_1)(\det C)
\end{aligned}$$

Definition A square matrix A is *upper-triangular* if all entries below the main diagonal are zero; that is, $A_{ij} = 0$ for $i > j$. It is *lower-triangular* if all entries above the main diagonal are zero; that is, $A_{ij} = 0$ for $i < j$.

Example 7

The matrix

$$A = \begin{bmatrix} 3 & 0 & 0 & 0 \\ 5 & 2 & 0 & 0 \\ 1 & 3 & 2 & 0 \\ 2 & 1 & 1 & 4 \end{bmatrix}$$

is lower-triangular. Let us compute its determinant, expanding about the first row:

$$\det A = (3)\det\begin{bmatrix} 2 & 0 & 0 \\ 3 & 2 & 0 \\ 1 & 1 & 4 \end{bmatrix} = (3)(2)\det\begin{bmatrix} 2 & 0 \\ 1 & 4 \end{bmatrix} = (3)(2)(2)(4) = 48$$

□

Note in Example 7 that the value of det A is the product of the entries on the main diagonal of A. Expanding an *upper*-triangular determinant about its *last* row produces the same result. This property of triangular matrices is recorded in our next theorem; a formal proof is in Appendix B.

Theorem 2.5 The determinant of a triangular matrix (upper or lower) is the product of the entries on its main diagonal.

An identity matrix is triangular; we have already observed that its determinant equals 1. A diagonal matrix, as introduced in Problem Set 1.3, is also triangular; thus the determinant of a diagonal matrix is also the product of its main-diagonal entries.

Now, it is possible to combine Theorems 2.2 and 2.5 to devise a method for finding determinants. We can apply row operations, noting at each stage the effect on the determinant, to convert the given matrix to triangular form from which the determinant value can be obtained by inspection. In some cases, this will be more efficient than expanding about a given row. (Any matrix *can* be brought to triangular form by row operations, since we know such operations will produce an echelon form, which is certainly triangular. But one normally will have a triangular form well before the echelon form is reached.)

Example 8

We will use again the $|\cdots|$ notation for determinant of a 4×4 matrix.

$$\begin{vmatrix} 1 & 2 & 3 & 4 \\ 5 & 6 & 7 & 8 \\ -4 & -3 & 2 & 1 \\ 6 & 5 & 1 & -7 \end{vmatrix} = \begin{vmatrix} 1 & 2 & 3 & 4 \\ 4 & 4 & 4 & 4 \\ 0 & 5 & 14 & 17 \\ 0 & -7 & -17 & -31 \end{vmatrix} \qquad \text{(Theorem 2.2c)}$$

$$= 4\begin{vmatrix} 1 & 2 & 3 & 4 \\ 1 & 1 & 1 & 1 \\ 0 & 5 & 14 & 17 \\ 0 & -7 & -17 & -31 \end{vmatrix} \qquad \text{(Theorem 2.2b)}$$

$$= -4\begin{vmatrix} 1 & 2 & 3 & 4 \\ 1 & 1 & 1 & 1 \\ 0 & 5 & 14 & 17 \\ 0 & 7 & 17 & 31 \end{vmatrix} \qquad \text{(Theorem 2.2b)}$$

$$= -4\begin{vmatrix} 1 & 2 & 3 & 4 \\ 0 & -1 & -2 & -3 \\ 0 & 5 & 14 & 17 \\ 0 & 7 & 17 & 31 \end{vmatrix} \qquad \text{(Theorem 2.2c)}$$

$$= +4\begin{vmatrix} 1 & 2 & 3 & 4 \\ 0 & 1 & 2 & 3 \\ 0 & 5 & 14 & 17 \\ 0 & 7 & 17 & 31 \end{vmatrix} \qquad \text{(Theorem 2.2b)}$$

$$= 4\begin{vmatrix} 1 & 2 & 3 & 4 \\ 0 & 1 & 2 & 3 \\ 0 & 0 & 4 & 2 \\ 0 & 0 & 3 & 10 \end{vmatrix} \qquad \text{(Theorem 2.2c)}$$

$$= 4\begin{vmatrix} 1 & 2 & 3 & 4 \\ 0 & 1 & 2 & 3 \\ 0 & 0 & 1 & -8 \\ 0 & 0 & 3 & 10 \end{vmatrix} \qquad \text{(Theorem 2.2c)}$$

$$= 4\begin{vmatrix} 1 & 2 & 3 & 4 \\ 0 & 1 & 2 & 3 \\ 0 & 0 & 1 & -8 \\ 0 & 0 & 0 & 34 \end{vmatrix} \qquad \text{(Theorem 2.2c)}$$

$$= (4)(34) \qquad \text{(Theorem 2.5)}$$

$$= 136$$

□

The proof of the next theorem will not be offered here. We have put one version in the exercises (with hints); another can be found in Appendix B.

Theorem 2.6 $\det(A^t) = \det A$. (Recall that A^t is the transpose of A as defined in Section 1.5.)

Because of this theorem, we immediately have new versions of Theorems 2.1 and 2.2. Determinants may now be computed by expansion about any *column* as well as about any row; specifically, for the j-th column of an $n \times n$ matrix A, we have the formula

$$\det A = \sum_{k=1}^{n} (-1)^{k+j} A_{kj} \det(A^{kj})$$

Example 9

Here is the determinant from Example 3, recomputed by expansion about the second column.

$$\begin{aligned}
&\det\begin{bmatrix} 3 & 1 & 2 \\ 1 & -2 & 0 \\ 2 & 1 & 1 \end{bmatrix} \\
&= (-1)(1)\det\begin{bmatrix} 1 & 0 \\ 2 & 1 \end{bmatrix} + (1)(-2)\det\begin{bmatrix} 3 & 2 \\ 2 & 1 \end{bmatrix} + (-1)(1)\det\begin{bmatrix} 3 & 2 \\ 1 & 0 \end{bmatrix} \\
&= (-1)(1) + (-2)(-1) + (-1)(-2) \\
&= 3
\end{aligned}$$

□

Also, interchanging two *columns* of a matrix changes the sign of the determinant; adding a multiple of one *column* to another does not change the value of the determinant; multiplying a given *column* by a scalar k multiplies the value of the determinant by k.

The remainder of this section is devoted to establishing the interaction between determinants and multiplication of matrices.

Theorem 2.7 If A is singular and B is any matrix of the same size as A, then AB is singular.

Proof. Let A be singular and suppose that AB had an inverse C. Then $I = (AB)C = A(BC)$. This says that BC is the inverse of A, contradicting the assumption that A is singular. Thus AB cannot have an inverse and hence is singular.

Theorem 2.8 If A is singular, then $\det A = 0$.

Proof. Let C be the result of converting A to echelon form:

$$C = E_m E_{m-1} \cdots E_1 A$$

where $E_1, E_2, \ldots, E_m$ are appropriate elementary matrices. By the proof of Theorem 1.17, the last row of C is all zeros. By Corollary 2.3, $\det C = 0$. On the other hand, $\det C = (\det E_m)(\det E_{m-1}) \cdots (\det A)$, because of Theorem 2.4. Now, the determinant of an elementary matrix is never 0, by Corollary 2.2. Thus, in order that $\det C$ be 0, it must be the case that $\det A = 0$, as desired.

We are now in a position to prove one of the most important facts about determinants.

Theorem 2.9 Let A and B be square matrices of the same size. Then

$$\det(AB) = (\det A)(\det B)$$

Proof. If A is singular, so is AB by Theorem 2.7. Then, by Theorem 2.8, both sides of the equation are zero and hence equal.

Now suppose that A is nonsingular. By Corollary 1.17, we can write A as a product of elementary matrices: $A = E_1 E_2 \cdots E_m$. By Theorem 2.4 (using $C = I$), $\det A = (\det E_1)(\det E_2) \cdots (\det E_m)$. But $AB = E_1 E_2 \cdots E_m B$, so Theorem 2.4, putting $C = B$, gives $\det(AB) = (\det E_1)(\det E_2) \cdots (\det E_m)(\det B)$. Thus, $\det(AB) = (\det A)(\det B)$.

Theorem 2.10 A square matrix A is nonsingular if and only if $\det A \neq 0$.

Proof. The "if" part of this theorem has already been established in Theorem 2.8. For the "only-if" part, we assume A to be nonsingular. Then A^{-1} exists, with $A\,A^{-1} = I$. Taking determinants of both sides and applying Theorem 2.9 gives $(\det A)(\det A^{-1}) = \det I = 1$. Thus $\det A$ (and also $\det A^{-1}$) cannot be 0.

Corollary 2.10 If A is nonsingular, then

$$\det A^{-1} = \frac{1}{\det A}$$

Proof. This is a corollary to the proof of Theorem 2.10, for we saw there that $(\det A)(\det A^{-1}) = 1$, for a nonsingular matrix A.

Let us digress, at the close of this section, for a brief comment on the terminology *singular* and *nonsingular*, as applied to square matrices. A dictionary defines the word "singular" to mean "rare" or "unusual." Now, in a sense which can be made mathematically precise, the probability that a real number selected at random from the set of all real numbers will turn out to be some specific real number is, in fact, zero. Thus, if real numbers a, b, c, and d are arbitrarily chosen for the entries of a 2×2 matrix

$$\begin{bmatrix} a & b \\ c & d \end{bmatrix}$$

the likelihood that $ad - bc = 0$ will be zero, since d will have to be bc/a. Therefore, the matrices with zero determinant are the *rare* ones—the *singular* ones. By Theorem 2.10, these are the ones without inverses.

2.1 Problem Set

1. Find, first by expanding about the first row, and then by expanding about the second column, the determinant of each of these matrices.

   a. $\begin{bmatrix} 2 & 0 & 1 \\ 1 & -3 & -2 \\ 3 & 1 & 1 \end{bmatrix}$ b. $\begin{bmatrix} 1 & 0 & 1 \\ 1 & 1 & 0 \\ 0 & 1 & 1 \end{bmatrix}$

c. $\begin{bmatrix} 2 & 3 & -2 \\ 0 & -1 & 5 \\ 0 & -2 & 4 \end{bmatrix}$ d. $\begin{bmatrix} 1 & 1 & 0 \\ 2 & 3 & 4 \\ 5 & 6 & 4 \end{bmatrix}$

e. $\begin{bmatrix} 1 & 2 & 3 & 4 \\ 1 & 3 & 4 & 5 \\ 2 & 5 & 7 & 10 \\ 1 & 3 & 4 & 7 \end{bmatrix}$ f. $\begin{bmatrix} 2 & -1 \\ -3 & 2 \end{bmatrix}$

g. $\begin{bmatrix} 1 & 3 & 1 & -2 \\ -2 & 6 & -3 & 1 \\ 4 & 0 & -2 & 3 \\ 3 & 2 & -1 & 1 \end{bmatrix}$

2. For each matrix in Problem 1, use row operations to convert the matrix to upper-triangular form and use Theorems 2.2 and 2.5 to find the determinant.
3. Show that the determinant of a square echelon matrix is either 0 or 1. Under what circumstances is the determinant 1?
4. Suppose that

$$\det\begin{bmatrix} a & b & c \\ d & e & f \\ g & h & i \end{bmatrix} = 3$$

What are the determinants of the following matrices?

a. $\begin{bmatrix} a + 7c & 2b & c \\ d + 7f & 2e & f \\ g + 7i & 2h & i \end{bmatrix}$ b. $\begin{bmatrix} a & d & 5g \\ b & e & 5h \\ c & f & 5i \end{bmatrix}$

c. $\begin{bmatrix} g & h & i \\ a & b & c \\ d & e & f \end{bmatrix}$ d. $\begin{bmatrix} a & b & c \\ d & e & f \\ 3a & 3b & 3c \end{bmatrix}$

5. If $\det A = 3$ and $\det B = 2$, what are $\det(AB^2)$, $\det A^3$, and $\det(A^{-1}B\,A)$?
6. Show that if A is an $n \times n$ matrix and c is a scalar, then $\det(cA) = c^n \det A$.
7. Prove that if A is nonsingular, then $\det(A^{-1}B\,A) = \det B$.
8. Prove that $\det(AB) = \det(BA)$.
9. Prove that if $A^2 = A$, then either A is singular or $\det A = 1$.
10. Prove that if $A = A^{-1}$, then $\det A = \pm 1$.

11. Let each of A, B, C, and D be a 2×2 matrix; let

$$\begin{bmatrix} A & B \\ C & D \end{bmatrix}$$

be a 4×4 matrix composed of these 2×2 *sub*matrices.

a. If either B or C is

$$\begin{bmatrix} 0 & 0 \\ 0 & 0 \end{bmatrix}$$

show that

$$\det \begin{bmatrix} A & B \\ C & D \end{bmatrix} = (\det A)(\det D)$$

(This generalizes Theorem 2.5.)

b. Show, by example, that the formula

$$\det \begin{bmatrix} A & B \\ C & D \end{bmatrix} = (\det A)(\det D) - (\det B)(\det C)$$

is not valid in case neither B nor C is the 2×2 zero matrix. (It is possible to do this using only 0's and 1's; a complicated example is not necessary.)

12. Prove that $\det A^t = \det A$ (Theorem 2.6) by establishing the following:

a. The transpose of an elementary matrix is an elementary matrix.

b. If E is an elementary matrix, then $\det E^t = \det E$.

c. If C is an echelon matrix, then $\det C^t = \det C$.

d. Let A be any square matrix, let C be the echelon form for A, and write $A = E_1E_2 \cdots E_kC$, for certain elementary matrices $E_1, E_2, \ldots, E_k$. Then, by Theorem 1.9, $A^t = C^tE_k^t \cdots E_2^tE_1^t$. Apply parts b and c, above, and Theorem 2.9 to conclude that $\det A^t = \det A$.

2.2 Some Applications of Determinants

In Theorem 2.1, we asserted that the determinant of a square matrix can be computed by multiplying the entries of a given row by their corresponding minor determinants, signing these products in a certain way, and then adding them together. If, however, the entries of a given row are "mismatched" with the minor determinants of entries in *another* row, the result is dramatically different.

Theorem 2.11 Let A be an $n \times n$ matrix. If $i \neq j$, then

$$(-1)^{j+1}A_{i1}\det(A^{j1}) + (-1)^{j+2}A_{i2}\det(A^{j2}) + \cdots + (-1)^{j+n}A_{in}\det(A^{jn}) = 0$$

(i-th row entries "mismatched" with j-th row minor determinants)

Proof. If we alter the matrix A by replacing its j-th row entries with the entries in the i-th row, we will have a matrix with two equal rows and hence a determinant of value 0, by Corollary 2.3. However, the sum in question in this theorem is simply the determinant expansion of this altered matrix about its i-th row. Thus the sum is 0.

Definition Let A be an $n \times n$ matrix. For $1 \leq i, j \leq n$, the *i-j cofactor* of A is given by $\text{cof } A_{ij} = (-1)^{i+j}\det(A^{ij})$.

In light of this definition, we can combine Theorem 2.1 and Theorem 2.11 to obtain:

$$\sum_{r=1}^{n} A_{ir} \text{ cof } A_{jr} = \begin{cases} \det A & \text{if } i = j \\ 0 & \text{if } i \neq j \end{cases}$$

Definition For an $n \times n$ matrix A, the *adjoint* of A is that $n \times n$ matrix given by

$$\text{adj } A = \begin{bmatrix} \text{cof } A_{11} & \text{cof } A_{21} & \cdots & \text{cof } A_{n1} \\ \text{cof } A_{12} & \text{cof } A_{22} & \cdots & \text{cof } A_{n2} \\ & & \cdots & \\ \text{cof } A_{1n} & \text{cof } A_{2n} & \cdots & \text{cof } A_{nn} \end{bmatrix}$$

Thus, the adjoint of A is obtained by replacing each entry of A with its cofactor and then taking the transpose. More concisely,

$$(\text{adj } A)_{ij} = \text{cof } A_{ji}$$

We shall now employ the adjoint to obtain a formula for the inverse of a nonsingular matrix.

Theorem 2.12 For an $n \times n$ matrix A,

$$A(\operatorname{adj} A) = (\det A)I$$

Proof. Let us display the matrix product $A(\operatorname{adj} A)$:

$$A(\operatorname{adj} A) = \begin{bmatrix} A_{11} & A_{12} & \cdots & A_{1n} \\ A_{21} & A_{22} & \cdots & A_{2n} \\ & \cdots & & \\ A_{n1} & A_{n2} & \cdots & A_{nn} \end{bmatrix} \begin{bmatrix} \operatorname{cof} A_{11} & \operatorname{cof} A_{21} & \cdots & \operatorname{cof} A_{n1} \\ \operatorname{cof} A_{12} & \operatorname{cof} A_{22} & \cdots & \operatorname{cof} A_{n2} \\ & \cdots & & \\ \operatorname{cof} A_{1n} & \operatorname{cof} A_{2n} & \cdots & \operatorname{cof} A_{nn} \end{bmatrix}$$

Now, using the definition of matrix product and the note following the definition of cofactor given above, we compute this product and get

$$A(\operatorname{adj} A) = \begin{bmatrix} \det A & 0 & 0 & \cdots & 0 \\ 0 & \det A & 0 & \cdots & 0 \\ 0 & 0 & \det A & \cdots & 0 \\ & & \cdots & & \\ 0 & 0 & 0 & \cdots & \det A \end{bmatrix} = (\det A)I$$

Corollary 2.12 If A is a nonsingular matrix, then

$$A^{-1} = \frac{1}{\det A} \operatorname{adj} A$$

Proof. If A is nonsingular, $\det A \neq 0$, and we multiply both sides of the equation in Theorem 2.12 by the scalar $1/(\det A)$ to get our result.

Example 1

We apply our new formula to Example 3 of Section 1.6, where

$$A = \begin{bmatrix} 2 & 3 & -2 \\ 0 & -1 & 5 \\ 0 & -2 & 4 \end{bmatrix}$$

We write the "cofactor matrix," cof A, which we get by replacing each entry with its cofactor.

$$\operatorname{cof} A_{11} = (-1)^{1+1}\det\begin{bmatrix} -1 & 5 \\ -2 & 4 \end{bmatrix} = (+1)(-4 + 10) = 6$$

$$\operatorname{cof} A_{12} = (-1)^{1+2}\det\begin{bmatrix} 0 & 5 \\ 0 & 4 \end{bmatrix} = (-1)(0 - 0) = 0$$

and so on. Thus,

$$\operatorname{cof} A = \begin{bmatrix} 6 & 0 & 0 \\ -8 & 8 & 4 \\ 13 & -10 & -2 \end{bmatrix}$$

To get the adjoint, we take the transpose:

$$\operatorname{adj} A = (\operatorname{cof} A)^t = \begin{bmatrix} 6 & -8 & 13 \\ 0 & 8 & -10 \\ 0 & 4 & -2 \end{bmatrix}$$

To get det A, we can match up any row (or column) with the corresponding row (or column) in our cofactor matrix. Using the first row involves less work and gives $\det A = (2)(6) + (3)(0) + (-2)(0) = 12$. Hence,

$$A^{-1} = \frac{1}{\det A} \operatorname{adj} A = \frac{1}{12}\begin{bmatrix} 6 & -8 & 13 \\ 0 & 8 & -10 \\ 0 & 4 & -2 \end{bmatrix}$$

which agrees with our result in Section 1.6. □

Another application of determinants is provided by *Cramer's Rule*, a formula for the solution of a system of equations, provided the coefficient matrix is nonsingular (in which case, by Corollary 1.9, we know there is a unique solution).

Theorem 2.13 *Cramer's Rule*. Let $AX = B$ by a system of n linear equations in n unknowns with nonsingular coefficient matrix A. Let

$$X = \begin{bmatrix} x_1 \\ x_2 \\ \vdots \\ x_n \end{bmatrix} \quad \text{and} \quad B = \begin{bmatrix} b_1 \\ b_2 \\ \vdots \\ b_n \end{bmatrix}$$

Then, the solution for the system is given by the quotient

$$x_i = \frac{\det \begin{bmatrix} A_{11} & \cdots & A_{1,i-1} & b_1 & A_{1,i+1} & \cdots & A_{1n} \\ A_{21} & \cdots & A_{2,i-1} & b_2 & A_{2,i+1} & \cdots & A_{2n} \\ & & & \cdots & & & \\ A_{n1} & \cdots & A_{n,i-1} & b_n & A_{n,i+1} & \cdots & A_{nn} \end{bmatrix}}{\det A}$$

for $i = 1, 2, \ldots, n$.

Proof. Since the coefficient matrix A is nonsingular, the solution for the system $AX = B$ is given by $X = A^{-1}B$. By our new formula for A^{-1}, this becomes

$$X = \frac{1}{\det A} (\text{adj } A)B$$

The i-th coordinate x_i of X, therefore, involves the scalar $1/(\det A)$ and, according to matrix multiplication, the i-th row of adj A and the column B. Computing the i-th coordinate of the matrix product (adj A) B, we get

$$(\text{cof } A_{1i})b_1 + (\text{cof } A_{2i})b_2 + \cdots + (\text{cof } A_{ni})b_n$$
$$= b_1(\text{cof } A_{1i}) + b_2(\text{cof } A_{2i}) + \cdots + b_n(\text{cof } A_{ni})$$

Upon close scrutiny, one can see that this is the determinant called for in the numerator of Cramer's Rule, *if* it is expanded about the i-th column containing the b's. Thus, if we attach to this the scalar factor $1/(\det A)$, we have the formula for x_i as stated in the theorem.

Cramer's Rule is not as formidable as it might appear. The numerator is just the determinant of the matrix obtained by replacing the i-th column of A by the numbers on the right of the equations in the system. For example, the 2×2 case translates to the following:

The solution for

$$ax + by = e$$
$$cx + dy = f$$

is given by

$$x = \frac{\det \begin{bmatrix} e & b \\ f & d \end{bmatrix}}{\det \begin{bmatrix} a & b \\ c & d \end{bmatrix}} \qquad y = \frac{\det \begin{bmatrix} a & e \\ c & f \end{bmatrix}}{\det \begin{bmatrix} a & b \\ c & d \end{bmatrix}}$$

Example 2
(Example 1, Section 1.4, revisited) We solve, by Cramer's Rule,

$$\begin{aligned} 2x \quad\quad\; + \; z &= 1 \\ x - 3y - 2z &= 3 \\ 3x + \; y + \; z &= 2 \end{aligned}$$

Here,

$$A = \begin{bmatrix} 2 & 0 & 1 \\ 1 & -3 & -2 \\ 3 & 1 & 1 \end{bmatrix}, \quad X = \begin{bmatrix} x \\ y \\ z \end{bmatrix}, \quad \text{and} \quad B = \begin{bmatrix} 1 \\ 3 \\ 2 \end{bmatrix}$$

Note that the solution will involve the calculation of four 3×3 determinants—a different numerator determinant for each unknown and the common denominator $\det A$. For the unknown x, we need

$$\det \begin{bmatrix} 1 & 0 & 1 \\ 3 & -3 & -2 \\ 2 & 1 & 1 \end{bmatrix} = (+1)(1)(-1) + (-1)(0)(7) + (+1)(1)(9) = 8$$

Now, $\det A$, which is needed for all unknowns, is

$$\det \begin{bmatrix} 2 & 0 & 1 \\ 1 & -3 & -2 \\ 3 & 1 & 1 \end{bmatrix} = (+1)(2)(-1) + (-1)(0)(7) + (+1)(1)(10) = 8$$

(We expanded about the first row in each case.) Thus $x = \frac{8}{8} = 1$. It can be verified that the determinant needed in the numerator for y has the value 0, and that the one needed for z has the value -8. So, $y = \frac{0}{8} = 0$ and $z = \frac{-8}{8} = -1$. The solution is therefore $(x, y, z) = (1, 0, -1)$, confirming what we obtained in Section 1.4 for the same system. □

In general, for systems with unique solutions, Cramer's Rule is not as efficient a method of solution as that of converting the augmented matrix to echelon form; finding A^{-1} by the adjoint method and then calculating the solution $X = A^{-1}B$ is likewise less efficient than converting to echelon form because more arithmetic operations are required in the former methods.

2.2 Problem Set

1. If

$$A = \begin{bmatrix} a & b \\ c & d \end{bmatrix}$$

and A is nonsingular, derive the formula

$$A^{-1} = \frac{1}{ad - bc}\begin{bmatrix} d & -b \\ -c & a \end{bmatrix}$$

Check this formula by computing $A\,A^{-1}$.

2. Use the adjoint to find the inverses of the given matrices.

a. $\begin{bmatrix} 1 & 1 \\ 3 & 2 \end{bmatrix}$

b. $\begin{bmatrix} 1 & 0 & 1 \\ 1 & 1 & 0 \\ 0 & 1 & 1 \end{bmatrix}$

c. $\begin{bmatrix} 0 & 5 & 0 \\ 1 & 0 & -4 \\ 0 & 0 & 1 \end{bmatrix}$

d. $\begin{bmatrix} 2 & 2 & 3 \\ 3 & 0 & 2 \\ 1 & 0 & 1 \end{bmatrix}$

3. Use Cramer's Rule to solve these systems.

a. $\begin{aligned} x + y - z &= 5 \\ 2x + y - 3z &= 10 \\ 3x + 4y + 5z &= 1 \end{aligned}$

b. $\begin{aligned} 2x - y + z &= 7 \\ 3x + 2y + z &= 3 \\ 2x - y + 3z &= 9 \end{aligned}$

c. $\begin{aligned} x - y + z - 2w &= 2 \\ 2x + y - 3z + 3w &= 1 \\ 3x + 2y + w &= 7 \\ 2x + y - z &= 2 \end{aligned}$

4. Use Cramer's Rule to solve these systems.

a. $\begin{aligned} 3x + 2y + 5z &= 2 \\ -x + y + 5z &= 1 \\ 2x - 2y + z &= 9 \end{aligned}$

b. $\begin{aligned} 2x + 3y + z - w &= 2 \\ x + 2y + 5z + 3w &= 5 \\ -x + 3z + w &= 1 \\ x - 2y + 3z &= -2 \end{aligned}$

c. $\begin{aligned} v + 3x + y + z &= 2 \\ w + 2x + 2y + z &= 1 \\ -v + 3w + 2x + 4y + 2z &= 1 \\ 2v - w + x - 4y + 3z &= 6 \\ 2v + 2w + 5x + 3y + 2z &= 4 \end{aligned}$

5. Explain why Cramer's Rule ensures that a *homogeneous* system of n equations in n unknowns with nonsingular coefficient matrix has only the trivial solution.

6. Let A be a nonsingular $n \times n$ matrix. Use Problem 6 of Section 2.1 (p. 66) and Corollary 2.12 to show that

$$\det(\text{adj } A) = (\det A)^{n-1}$$

7. Show that if A is a singular square matrix, then $A(\text{adj } A) = O$, the zero matrix.

8. We have seen that *determinant* associates with each square matrix a real number. Another number corresponding to a square matrix is the *trace* of that matrix; the *trace* of an $n \times n$ matrix A, written $\text{tr } A$, is the sum of the entries on the main diagonal of A:

$$\text{tr } A = \sum_{k=1}^{n} A_{kk}$$

For $n \times n$ matrices A, B, and C and any any scalar c, show that

a. $\text{tr}(A \pm B) = \text{tr } A \pm \text{tr } B$

b. $\text{tr}(cA) = c \text{ tr } A$

c. $\text{tr}(AB) = \text{tr}(BA)$

d. $\text{tr}(C^{-1}A\, C) = \text{tr } A$, if C is nonsingular

3

VECTORS

3.1 VECTORS IN $\mathbf{R}^2$ AND $\mathbf{R}^3$

Let us recall that an element in $\mathbf{R}^3$ is an ordered triple (x, y, z). We have observed that such a triple can be viewed as a 1×3 matrix

$$[x \quad y \quad z]$$

or as a 3×1 matrix

$$\begin{bmatrix} x \\ y \\ z \end{bmatrix}$$

In any case, what must be clear are the three coordinates and their order.

Elements of $\mathbf{R}^3$, since they can be represented as matrices, are subject to the operations as defined for matrices and the consequent rules accompanying these operations. We shall reiterate these rules here.

Addition: If $\mathbf{x} = (x_1, x_2, x_3)$ and $\mathbf{y} = (y_1, y_2, y_3)$, then the sum $\mathbf{x} + \mathbf{y}$ is given by

$$\mathbf{x} + \mathbf{y} = (x_1 + y_1, x_2 + y_2, x_3 + y_3)$$

Multiplication by scalars: If $\mathbf{x} = (x_1, x_2, x_3)$ and r is a scalar (that is, r is a real number), then $r\mathbf{x}$ is given by

$$r\mathbf{x} = (rx_1, rx_2, rx_3)$$

Properties: ($\mathbf{x}$, $\mathbf{y}$, and $\mathbf{z}$ are arbitrary members of $\mathbf{R}^3$ and r and s are arbitrary scalars.)

a. $\mathbf{x} + \mathbf{y} = \mathbf{y} + \mathbf{x}$

b. $(\mathbf{x} + \mathbf{y}) + \mathbf{z} = \mathbf{x} + (\mathbf{y} + \mathbf{z})$

c. There is a member $\mathbf{0} = (0, 0, 0)$ in $\mathbf{R}^3$ such that

$$\mathbf{x} + \mathbf{0} = \mathbf{x} \quad \text{for every } \mathbf{x} \text{ in } \mathbf{R}^3$$

d. For every $\mathbf{x}$ in $\mathbf{R}^3$ there is a member $-\mathbf{x}$ in $\mathbf{R}^3$ such that

$$\mathbf{x} + (-\mathbf{x}) = \mathbf{0}$$

e. $r(\mathbf{x} + \mathbf{y}) = r\mathbf{x} + r\mathbf{y}$

f. $(r + s)\mathbf{x} = r\mathbf{x} + s\mathbf{x}$

g. $(rs)\mathbf{x} = r(s\mathbf{x})$

h. $1 \cdot \mathbf{x} = \mathbf{x}$

Elements of $\mathbf{R}^3$ are frequently called *vectors*. (The discussion above would apply equally well to $\mathbf{R}^2$, except that the vectors in question would, of course, be ordered pairs instead of ordered triples.) Now, the term "vector" is widely used in science, especially in the physical sciences, and it is no coincidence that this word is also used for elements of $\mathbf{R}^3$ (or $\mathbf{R}^2$). Let us see how to make the connection between physical vectors and mathematical vectors.

In physics and other sciences, a "vector quantity" is something distinguished by two and only two characteristics, namely, direction and magnitude. Some examples of vector quantities are force, momentum, velocity, acceleration, and, perhaps the simplest of these, displacement. In $\mathbf{R}^3$ (or $\mathbf{R}^2$), such quantities are represented pictorially by directed line segments ("arrows"); the way the line points depicts the direction, and the length of the line depicts the magnitude. Such a directed line segment, in turn, is characterized by two points in a given order: an "initial" point $\mathbf{p}$ and a "terminal" point $\mathbf{q}$, $\mathbf{q}$ being where the "arrowhead" is.

Since vector quantities depend only on direction and magnitude, it is entirely possible that two different directed line segments represent the same vector quantity; all that is required is that they have the same length and point in the same direction. However, if we insist that a directed line segment emanate from the origin, then we shall have, for a given vector quantity, *exactly one* representative arrow (Figure 3.1).

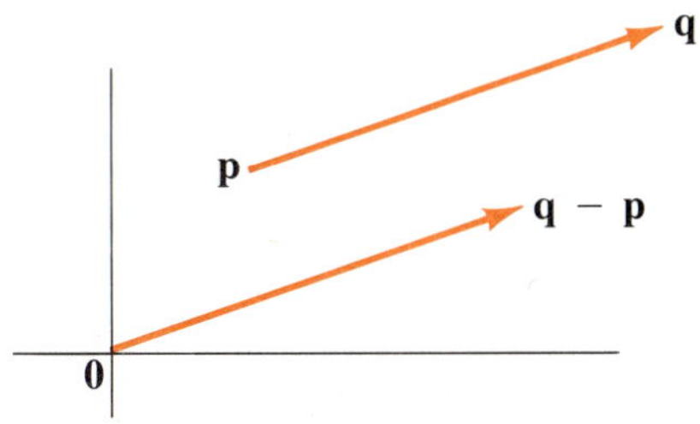

Figure 3.1

By elementary geometry, we can see that if a vector quantity is represented by a directed line segment from $\mathbf{p}$ to $\mathbf{q}$, then the unique representative arrow emanating from the origin will terminate at the point $\mathbf{q} - \mathbf{p}$, as examination of the coordinates of the points will show. Thus we see that it is possible to establish a one-to-one correspondence between "vector quantities" in science and points (which we have called "vectors") in $\mathbf{R}^3$ or $\mathbf{R}^2$. In dealing with vectors, we can allow ourselves to draw the representative arrows where it is convenient so long as we keep in mind this correspondence. It is essential to remember that when we identify a vector $\mathbf{x}$ by specifying coordinates (x_1, x_2) in the case of $\mathbf{R}^2$, or (x_1, x_2, x_3) in the case of $\mathbf{R}^3$, the arrowhead of the pictorial representation is at the point with those coordinates *only if* the arrow begins at the origin. Otherwise, the coordinates of $\mathbf{x}$ are obtained by subtracting the initial point from the terminal point. (We also must allow for the zero vector, that is, the origin, which as an "arrow" has *no* direction and *zero* length.)

There is a geometric way to describe our algebraic operations by using arrows. Let $\mathbf{x} = (x_1, x_2)$ and $\mathbf{y} = (y_1, y_2)$ in $\mathbf{R}^2$. The arrows $\mathbf{x}$ and $\mathbf{y}$, extending from the origin, determine a parallelogram, of which $\mathbf{x}$ and $\mathbf{y}$ are two adjacent sides. An arrow drawn from the origin to the opposite vertex of this parallelogram will be the vector $\mathbf{x} + \mathbf{y}$ (Figure 3.2).

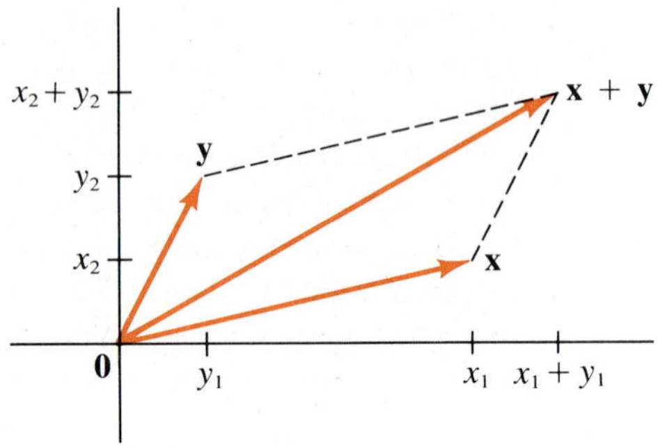

Figure 3.2

That this is the case is a consequence of elementary geometry; one need only verify that the coordinates of this fourth vertex are the required $(x_1 + y_1, x_2 + y_2)$. This method for obtaining $\mathbf{x} + \mathbf{y}$ pictorially is usually called the *parallelogram law,* which is valid even if the vectors emanate from a common point other than the origin.

The depiction of scalar multiplication of a vector is a similar consequence of elementary geometry. Let $\mathbf{x} = (x_1, x_2)$ and let r be a scalar. As before, the arrow representing $\mathbf{x}$ goes from the origin to the point (x_1, x_2). For $r > 0$, an arrow drawn from the origin in the same direction as $\mathbf{x}$ but with length r times the length of $\mathbf{x}$ will terminate at the point $r\mathbf{x}$ and hence represent $r\mathbf{x}$; for $r < 0$, the only difference is that the arrow must be drawn in the opposite direction. The geometric way of adding vectors or multiplying them by scalars is equally valid in $\mathbf{R}^3$, but is, of course, more difficult to visualize.

3.2 VECTORS IN $\mathbf{R}^n$

If $n > 3$, we do not have a convenient geometric representation for elements of $\mathbf{R}^n$; nevertheless, as in the case of $\mathbf{R}^2$ and $\mathbf{R}^3$, we may apply our matrix operations of addition and multiplication by scalars and all of the consequent properties we enumerated in the last section. We shall use our knowledge of the geometry of $\mathbf{R}^2$ and $\mathbf{R}^3$ to introduce certain new ideas in $\mathbf{R}^n$, where elements will also be called vectors.

Let $\mathbf{x} = (x_1, x_2)$ be a vector in $\mathbf{R}^2$. From the Pythagorean Theorem, the length $|\mathbf{x}|$ of $\mathbf{x}$ is given by $\sqrt{x_1^2 + x_2^2}$. In $\mathbf{R}^3$, the length of $\mathbf{x} = (x_1, x_2, x_3)$ is given by $|\mathbf{x}| = \sqrt{x_1^2 + x_2^2 + x_3^2}$ (Figure 3.3). So, we introduce the following in $\mathbf{R}^n$.

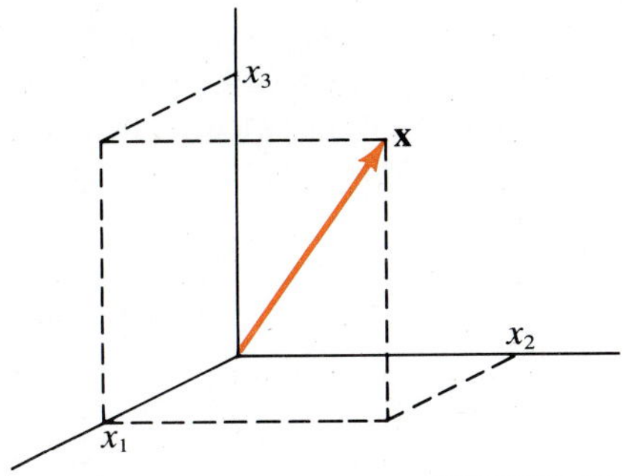

the length of $\mathbf{x} = |\mathbf{x}| = \sqrt{x_1^2 + x_2^2 + x_3^2}$

Figure 3.3

Definition

Let $\mathbf{x} = (x_1, x_2, \ldots, x_n)$ be a vector in $\mathbf{R}^n$. Then, the *length* $|\mathbf{x}|$ of $\mathbf{x}$ is given by

$$|\mathbf{x}| = \sqrt{x_1^2 + x_2^2 + \cdots + x_n^2}$$

Definition

If $\mathbf{x} = (x_1, x_2, \ldots, x_n)$ and $\mathbf{y} = (y_1, y_2, \ldots, y_n)$, then the *distance* between $\mathbf{x}$ and $\mathbf{y}$ is given by $|\mathbf{x} - \mathbf{y}|$.

This is the length of the vector $\mathbf{x} - \mathbf{y}$, which is as it should be, judging from $\mathbf{R}^2$ or $\mathbf{R}^3$. Explicitly, we have

$$|\mathbf{x} - \mathbf{y}| = \sqrt{(x_1 - y_1)^2 + (x_2 - y_2)^2 + \cdots + (x_n - y_n)^2}$$

which agrees with the distance formula from analytic geometry for $n = 2$ or 3.

Example 1

If $\mathbf{x} = (1, 0, 3)$, then

$$|\mathbf{x}| = \sqrt{1^2 + 0^2 + 3^2} = \sqrt{10}$$

The distance between the points $(1, 0, 3)$ and $(2, -1, 2)$ is

$$\sqrt{(1 - 2)^2 + (0 - (-1))^2 + (3 - 2)^2} = \sqrt{(-1)^2 + 1^2 + 1^2} = \sqrt{3}$$

□

Returning to $\mathbf{R}^2$, let us find the angle between two nonzero vectors $\mathbf{x} = (x_1, x_2)$ and $\mathbf{y} = (y_1, y_2)$. By this, we mean to find the angle α formed by the arrows emanating from the origin to the points $\mathbf{x}$ and $\mathbf{y}$ such that $0 \leq \alpha \leq 180°$ (Figure 3.4).

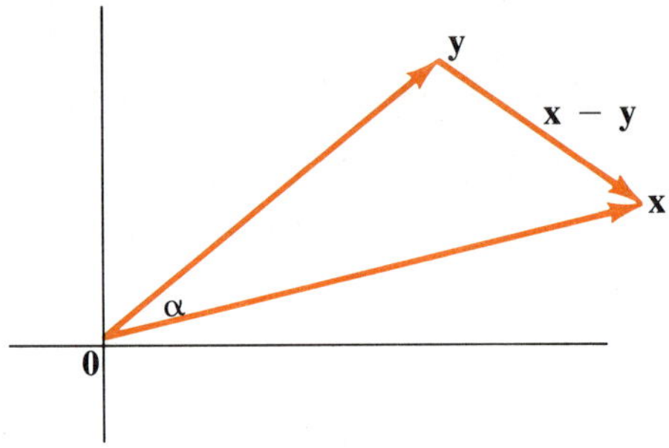

Figure 3.4

We shall employ the Law of Cosines from trigonometry to find this angle. Applying this law to the triangle with vertices $\mathbf{0}$, $\mathbf{x}$, and $\mathbf{y}$, we find

$$|\mathbf{x} - \mathbf{y}|^2 = |\mathbf{x}|^2 + |\mathbf{y}|^2 - 2|\mathbf{x}||\mathbf{y}| \cos \alpha$$

After substituting in coordinates, this equation becomes

$$(x_1 - y_1)^2 + (x_2 - y_2)^2 = x_1^2 + x_2^2 + y_1^2 + y_2^2 - 2\sqrt{x_1^2 + x_2^2}\sqrt{y_1^2 + y_2^2} \cos \alpha$$

Algebraic simplification of this equation will yield

$$\cos \alpha = \frac{x_1 y_1 + x_2 y_2}{\sqrt{x_1^2 + x_2^2}\sqrt{y_1^2 + y_2^2}}$$

This equation, together with the requirement that $0 \leq \alpha \leq 180°$, is enough to determine α.

Example 2

The angle α between $\mathbf{x} = (1, 0)$ and $\mathbf{y} = (2, -1)$ is determined by

$$\cos \alpha = \frac{2}{\sqrt{1}\sqrt{5}} = \frac{2}{\sqrt{5}}$$

Thus α is approximately 26.6° (by calculator). □

Example 3

The vectors $\mathbf{x} = (2, 1)$ and $\mathbf{y} = (2, -4)$ are perpendicular, since, in this case,

$$x_1y_1 + x_2y_2 = 2 \cdot 2 + 1(-4) = 0$$

Hence $\cos \alpha = 0$, which means that $\alpha = 90°$. □

Recall the expression $x_1y_1 + x_2y_2$, which appears in the numerator of the formula for the angle between two vectors. In case $\mathbf{x} = \mathbf{y}$, this would read $x_1^2 + x_2^2$, which we have seen in the formula for vector length. The recurrence of this expression leads us to the following definition.

Definition

Let $\mathbf{x} = (x_1, x_2, \ldots, x_n)$ and $\mathbf{y} = (y_1, y_2, \ldots, y_n)$. The *scalar* (or *dot*) *product* of $\mathbf{x}$ with $\mathbf{y}$ is given by

$$\mathbf{x} \cdot \mathbf{y} = x_1y_1 + x_2y_2 + \cdots + x_ny_n$$

Example 4

In $\mathbf{R}^5$, the dot product of $\mathbf{x} = (1, 2, 0, -3, 1)$ and $\mathbf{y} = (-2, 0, -1, 3, 1)$ is

$$\mathbf{x} \cdot \mathbf{y} = (1)(-2) + (2)(0) + (0)(-1) + (-3)(3) + (1)(1) = -10$$

□

Note that the value of the scalar product of two vectors is a scalar. Here are some immediate consequences of our definition.

Theorem 3.1

In $\mathbf{R}^n$,

a. $\mathbf{x} \cdot \mathbf{x} \geq 0$; $\mathbf{x} \cdot \mathbf{x} = 0$ only if $\mathbf{x} = \mathbf{0}$

b. $|\mathbf{x}| = \sqrt{\mathbf{x} \cdot \mathbf{x}}$; that is, $\mathbf{x} \cdot \mathbf{x} = |\mathbf{x}|^2$

c. $\mathbf{x} \cdot \mathbf{y} = \mathbf{y} \cdot \mathbf{x}$

d. $\mathbf{x} \cdot (\mathbf{y} + \mathbf{z}) = \mathbf{x} \cdot \mathbf{y} + \mathbf{x} \cdot \mathbf{z}$

e. $\mathbf{x} \cdot (r\mathbf{y}) = r(\mathbf{x} \cdot \mathbf{y}) = (r\mathbf{x}) \cdot \mathbf{y}$, for any scalar r

We leave the proof of this theorem and the subsequent corollary for the exercises.

Corollary 3.1 $|r\mathbf{x}| = |r||\mathbf{x}|$. ($|r|$ is the absolute value of the scalar r and $|r\mathbf{x}|$ and $|\mathbf{x}|$ are vector lengths.)

Definition A *unit vector* is a vector having length 1.

It is frequently desirable to obtain a unit vector in the same direction as a given nonzero vector $\mathbf{x}$. To accomplish this, one simply multiplies $\mathbf{x}$ by $1/|\mathbf{x}|$. According to Corollary 3.1, $(1/|\mathbf{x}|)\mathbf{x}$ will have length 1, for

$$\left|\frac{1}{|\mathbf{x}|}\mathbf{x}\right| = \frac{1}{|\mathbf{x}|}|\mathbf{x}| = 1$$

In $\mathbf{R}^2$, we have discovered that the angle α between two nonzero vectors x and y is determined by the formula

$$\cos \alpha = \frac{\mathbf{x} \cdot \mathbf{y}}{|\mathbf{x}||\mathbf{y}|}$$

and this formula can be derived in the same manner in $\mathbf{R}^3$. In $\mathbf{R}^n$, for $n > 3$, we have no concept of "angle" unless we introduce it ourselves.

Definition Let $\mathbf{x}$ and $\mathbf{y}$ be nonzero vectors in $\mathbf{R}^n$. The *angle* between $\mathbf{x}$ and $\mathbf{y}$ is that angle α such that $0 \leq \alpha \leq 180°$ and

$$\cos \alpha = \frac{\mathbf{x} \cdot \mathbf{y}}{|\mathbf{x}||\mathbf{y}|}$$

That is,

$$\alpha = \arccos \frac{\mathbf{x} \cdot \mathbf{y}}{|\mathbf{x}||\mathbf{y}|}$$

It is important to recognize that we have taken a formula *derived* in $\mathbf{R}^2$ and have made it the key element in a *definition* for $\mathbf{R}^n$ in general. However, it is conceivable that our definition, in certain instances, might not make sense. (A mathematician would say that the concept of angle between two vectors might not be "well-defined" by our definition.) Recall that the cosine of any angle must lie between -1 and 1, inclusive; that is, the absolute value of the cosine of an angle must not exceed 1 in value. We have, at this point, no immediate assurance that our angle formula will necessarily produce a legitimate cosine value. However, such assurance is provided by a famous inequality.

Theorem 3.2 *The Schwarz Inequality.** For any vectors $\mathbf{x}$ and $\mathbf{y}$ in $\mathbf{R}^n$,

$$|\mathbf{x} \cdot \mathbf{y}| \le |\mathbf{x}||\mathbf{y}|$$

Proof. Consider first a special case: assume that $\mathbf{y} = \mathbf{0}$. Then

$$\mathbf{x} \cdot \mathbf{y} = \mathbf{x} \cdot \mathbf{0} = 0$$

by direct computation, so the left-hand side of the inequality is just 0. On the right-hand side, we have a factor of $|\mathbf{y}| = |\mathbf{0}| = 0$, so that side is also 0. Thus, the inequality simply asserts that $0 \le 0$, which is certainly correct.

Now, we can assume that $\mathbf{y} \ne \mathbf{0}$. We begin with an inequality, the validity of which is assured by Theorem 3.1, part a:

$$\left(\mathbf{x} - \frac{\mathbf{x} \cdot \mathbf{y}}{\mathbf{y} \cdot \mathbf{y}}\mathbf{y}\right) \cdot \left(\mathbf{x} - \frac{\mathbf{x} \cdot \mathbf{y}}{\mathbf{y} \cdot \mathbf{y}}\mathbf{y}\right) \ge 0$$

We multiply out this dot product, utilizing the various parts of Theorem 3.1.

$$\mathbf{x} \cdot \mathbf{x} - \left(\frac{\mathbf{x} \cdot \mathbf{y}}{\mathbf{y} \cdot \mathbf{y}}\mathbf{y}\right) \cdot \mathbf{x} - \mathbf{x} \cdot \left(\frac{\mathbf{x} \cdot \mathbf{y}}{\mathbf{y} \cdot \mathbf{y}}\mathbf{y}\right) + \left(\frac{\mathbf{x} \cdot \mathbf{y}}{\mathbf{y} \cdot \mathbf{y}}\mathbf{y}\right) \cdot \left(\frac{\mathbf{x} \cdot \mathbf{y}}{\mathbf{y} \cdot \mathbf{y}}\mathbf{y}\right) \ge 0$$

$$\mathbf{x} \cdot \mathbf{x} - \left(\frac{\mathbf{x} \cdot \mathbf{y}}{\mathbf{y} \cdot \mathbf{y}}\right)(\mathbf{y} \cdot \mathbf{x}) - \left(\frac{\mathbf{x} \cdot \mathbf{y}}{\mathbf{y} \cdot \mathbf{y}}\right)(\mathbf{x} \cdot \mathbf{y}) + \left(\frac{\mathbf{x} \cdot \mathbf{y}}{\mathbf{y} \cdot \mathbf{y}}\right)^2(\mathbf{y} \cdot \mathbf{y}) \ge 0$$

$$\mathbf{x} \cdot \mathbf{x} - 2\frac{(\mathbf{x} \cdot \mathbf{y})^2}{\mathbf{y} \cdot \mathbf{y}} + \frac{(\mathbf{x} \cdot \mathbf{y})^2}{\mathbf{y} \cdot \mathbf{y}} \ge 0$$

$$\mathbf{x} \cdot \mathbf{x} - \frac{(\mathbf{x} \cdot \mathbf{y})^2}{\mathbf{y} \cdot \mathbf{y}} \ge 0$$

$$(\mathbf{x} \cdot \mathbf{x})(\mathbf{y} \cdot \mathbf{y}) \ge (\mathbf{x} \cdot \mathbf{y})^2$$

We take nonnegative square roots of both sides and get

$$|\mathbf{x}||\mathbf{y}| \ge |\mathbf{x} \cdot \mathbf{y}|$$

which was our goal. (This highly unmotivated proof was devised by working backwards from the desired conclusion to some valid starting point.)

---

*Called by others the Cauchy–Schwarz, or the Cauchy–Schwarz–Buniakovski, Inequality. If only one name is used, it is usually that of Schwarz.

Example 5

We use the vectors in Example 4 to illustrate Schwarz's Inequality. In that case, $|\mathbf{x}| = |\mathbf{y}| = \sqrt{15}$, and we found there that $\mathbf{x} \cdot \mathbf{y} = -10$, and thus the inequality becomes $|-10| \leq \sqrt{15}\sqrt{15}$, or $10 \leq 15$. □

Example 6

In $\mathbf{R}^3$ let $\mathbf{x} = (-2, 0, 1)$ and $\mathbf{y} = (6, 0, -3)$. Then $\mathbf{x} \cdot \mathbf{y} = -15$, while $|\mathbf{x}| = \sqrt{5}$ and $|\mathbf{y}| = \sqrt{45} = 3\sqrt{5}$. Thus, in this case, equality will in fact hold in Schwarz's Inequality: $15 = 15$. □

It is of interest to note that *equality* holds in Schwarz's Inequality exactly when either of **x** and **y** is a scalar multiple of the other. In $\mathbf{R}^2$ this is as it should be, for, in the case of nonzero **x** and **y**, this would mean that the cosine of the angle between **x** and **y** is 1 or -1, in which case **x** and **y** would have either the same or opposite directions (the angle being either 0° or 180°). In general, we have the following corollary.

Corollary 3.2a In $\mathbf{R}^n$, $|\mathbf{x} \cdot \mathbf{y}| = |\mathbf{x}||\mathbf{y}|$ if and only if either **x** or **y** is a scalar multiple of the other.

Proof. We prove the "only if" part and leave the "if" part as an exercise. Let $|\mathbf{x} \cdot \mathbf{y}| = |\mathbf{x}||\mathbf{y}|$; if $\mathbf{y} = \mathbf{0}$, then **y** is certainly a scalar multiple of **x** (since $\mathbf{y} = \mathbf{0} = 0\mathbf{x}$). Otherwise, we begin with the final line of the proof of the theorem, but replacing "$\geq$" with "$=$", and simply reverse the order of the steps. (All the operations involved are legitimate when reversed.) The outcome will be

$$\left(\mathbf{x} - \frac{\mathbf{x} \cdot \mathbf{y}}{\mathbf{y} \cdot \mathbf{y}}\mathbf{y}\right) \cdot \left(\mathbf{x} - \frac{\mathbf{x} \cdot \mathbf{y}}{\mathbf{y} \cdot \mathbf{y}}\mathbf{y}\right) = 0$$

which, by Theorem 3.1, part a, implies that

$$\mathbf{x} - \frac{\mathbf{x} \cdot \mathbf{y}}{\mathbf{y} \cdot \mathbf{y}}\mathbf{y} = \mathbf{0} \quad \text{or} \quad \mathbf{x} = \left(\frac{\mathbf{x} \cdot \mathbf{y}}{\mathbf{y} \cdot \mathbf{y}}\right)\mathbf{y}$$

Thus **x** is a scalar multiple of **y**.

Another important inequality, perhaps better known than Schwarz's itself, can be deduced as a consequence of the Schwarz Inequality. We present it in our next corollary.

Corollary 3.2b *The Triangle Inequality.*

$$|\mathbf{x} + \mathbf{y}| \le |\mathbf{x}| + |\mathbf{y}|$$

Proof. Once again, we employ the various parts of Theorem 3.1.

$$\begin{aligned}
|\mathbf{x} + \mathbf{y}|^2 &= (\mathbf{x} + \mathbf{y}) \cdot (\mathbf{x} + \mathbf{y}) \\
&= \mathbf{x} \cdot \mathbf{x} + \mathbf{x} \cdot \mathbf{y} + \mathbf{y} \cdot \mathbf{x} + \mathbf{y} \cdot \mathbf{y} \\
&= \mathbf{x} \cdot \mathbf{x} + 2(\mathbf{x} \cdot \mathbf{y}) + \mathbf{y} \cdot \mathbf{y} \\
&\le \mathbf{x} \cdot \mathbf{x} + 2|\mathbf{x} \cdot \mathbf{y}| + \mathbf{y} \cdot \mathbf{y} && \text{(because } \mathbf{x} \cdot \mathbf{y} \text{ is a number, so it is less than or equal to its own absolute value)} \\
&\le \mathbf{x} \cdot \mathbf{x} + 2|\mathbf{x}||\mathbf{y}| + \mathbf{y} \cdot \mathbf{y} && \text{(by Schwarz)} \\
&\le |\mathbf{x}|^2 + 2|\mathbf{x}||\mathbf{y}| + |\mathbf{y}|^2 \\
&\le (|\mathbf{x}| + |\mathbf{y}|)^2
\end{aligned}$$

We take nonnegative square roots and get

$$|\mathbf{x} + \mathbf{y}| \le |\mathbf{x}| + |\mathbf{y}|$$

A sketch in $\mathbf{R}^2$ shows why this inequality is called the Triangle Inequality: after all, the length of one side of a triangle cannot exceed the sum of the lengths of the other two sides. (Put another way, the shortest distance between two points is along a straight line.) See Figure 3.5.

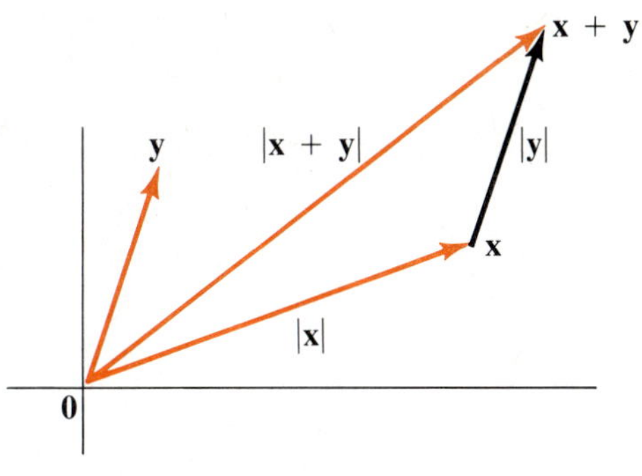

Figure 3.5

In much of the above, the expression

$$\frac{\mathbf{x} \cdot \mathbf{y}}{\mathbf{y} \cdot \mathbf{y}}\mathbf{y}$$

appears, and this vector has additional significance as well.

Definition

For vectors $\mathbf{x}$ and $\mathbf{y}$, with $\mathbf{y} \neq \mathbf{0}$, the vector

$$\frac{\mathbf{x} \cdot \mathbf{y}}{\mathbf{y} \cdot \mathbf{y}}\mathbf{y}$$

is called the *projection of* $\mathbf{x}$ *onto* $\mathbf{y}$. We shall write

$$\text{proj}_{\mathbf{y}}\mathbf{x} = \frac{\mathbf{x} \cdot \mathbf{y}}{\mathbf{y} \cdot \mathbf{y}}\mathbf{y}$$

We justify this terminology in our usual way: we see that it is geometrically appropriate in $\mathbf{R}^2$.

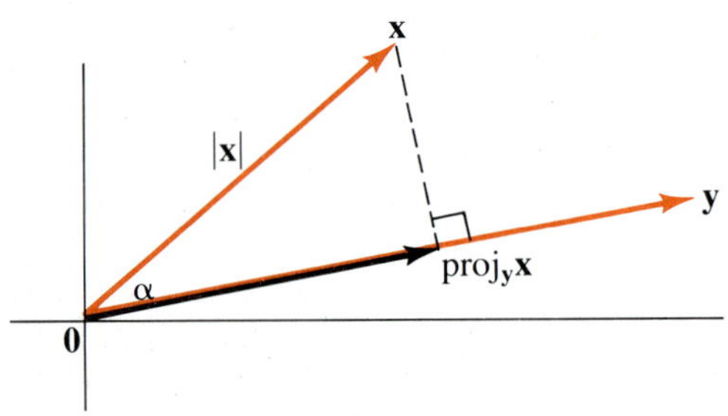

Figure 3.6

The *length,* as Figure 3.6 shows, of the projection of $\mathbf{x}$ onto $\mathbf{y}$ should be $|\mathbf{x}| \cos \alpha$, since in a right triangle, the cosine of an acute angle is the ratio of the length of the adjacent side to that of the hypotenuse. But,

$$|\mathbf{x}| \cos \alpha = \frac{\mathbf{x} \cdot \mathbf{y}}{|\mathbf{y}|}$$

So, we are looking for a vector of length

$$\frac{\mathbf{x} \cdot \mathbf{y}}{|\mathbf{y}|}$$

in the direction of $\mathbf{y}$. A vector of length 1 in the direction of $\mathbf{y}$ is

$$\frac{1}{|\mathbf{y}|}\mathbf{y}$$

so the desired projection of $\mathbf{x}$ onto $\mathbf{y}$ is

$$\left(\frac{\mathbf{x} \cdot \mathbf{y}}{|\mathbf{y}|}\right)\left(\frac{1}{|\mathbf{y}|}\mathbf{y}\right) = \frac{\mathbf{x} \cdot \mathbf{y}}{|\mathbf{y}||\mathbf{y}|}\mathbf{y} = \frac{\mathbf{x} \cdot \mathbf{y}}{\mathbf{y} \cdot \mathbf{y}}\mathbf{y}$$

(This analysis is valid for acute α; a slight modification will make it work for obtuse α as well.)

Example 7

In $\mathbf{R}^3$ let $\mathbf{x} = (-1, 0, -2)$ and $\mathbf{y} = (3, 1, 1)$. Then

$$\text{proj}_{\mathbf{y}}\mathbf{x} = \frac{\mathbf{x} \cdot \mathbf{y}}{\mathbf{y} \cdot \mathbf{y}}\mathbf{y} = \frac{-5}{11}(3, 1, 1)$$

Since this projection turned out to be a negative multiple of $\mathbf{y}$, it must be a vector pointing in the direction opposite to that of $\mathbf{y}$. That means that the angle between $\mathbf{x}$ and $\mathbf{y}$ must be obtuse. This is confirmed by the observation that it is $\mathbf{x} \cdot \mathbf{y}$ which determines the sign of the cosine of the angle between the two vectors and in this case it is negative. □

Example 8

In $\mathbf{R}^3$ let $\mathbf{x} = (1, 1, 2)$ and let $\mathbf{y} = (2, 0, -1)$. Here, $\mathbf{x} \cdot \mathbf{y} = 0$, so $\text{proj}_{\mathbf{y}}\mathbf{x} = 0\mathbf{y} = \mathbf{0}$, the zero vector. A glance at Figure 3.6 will indicate that this must be the result of having selected perpendicular vectors, a situation which we shall now explore further. □

In $\mathbf{R}^2$ (or $\mathbf{R}^3$) two nonzero vectors $\mathbf{x}$ and $\mathbf{y}$ are perpendicular if the angle α between them is 90°; that is, if $\cos \alpha = 0$. According to our formula expressing $\cos \alpha$ in terms of $\mathbf{x} \cdot \mathbf{y}$, this situation would require that $\mathbf{x} \cdot \mathbf{y} = 0$. Extrapolating once more from this setting, we formulate the following definition.

Definition

Vectors $\mathbf{x}$ and $\mathbf{y}$ in $\mathbf{R}^n$ are said to be *orthogonal* provided that $\mathbf{x} \cdot \mathbf{y} = 0$.

Note that, under this definition, the zero vector $\mathbf{0}$ is orthogonal to every vector. On the other hand, it is entirely possible that $\mathbf{x} \cdot \mathbf{y} = 0$ even if neither $\mathbf{x}$ nor $\mathbf{y}$ is $\mathbf{0}$. (See Example 3 or 8, for instance.)

The vector methods just discussed have interesting applications to some classical theorems of plane geometry. The following example uses them to prove the Pythagorean Theorem; other applications will be found in the Problem Set.

Example 9

(*The Pythagorean Theorem*) A triangle is a right triangle if and only if the square of the length of one side is equal to the sum of the squares of the lengths of the other two sides.

To prove this using vector methods, we let the triangle in question be formed by the vectors $\mathbf{x}$ and $\mathbf{y}$, as shown in Figure 3.7, so that the lengths of the sides are $|\mathbf{x}|$, $|\mathbf{y}|$, and $|\mathbf{x} + \mathbf{y}|$.

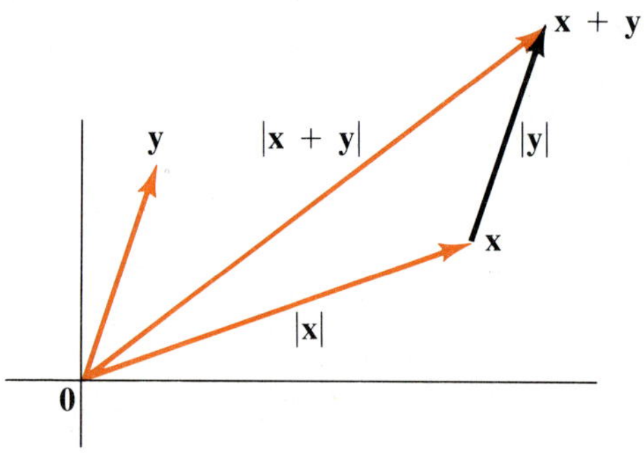

Figure 3.7

If the triangle is a right triangle, we can assume that $\mathbf{x} \cdot \mathbf{y} = 0$. Then

$$\begin{aligned} |\mathbf{x} + \mathbf{y}|^2 &= (\mathbf{x} + \mathbf{y}) \cdot (\mathbf{x} + \mathbf{y}) \\ &= \mathbf{x} \cdot \mathbf{x} + 2\mathbf{x} \cdot \mathbf{y} + \mathbf{y} \cdot \mathbf{y} = |\mathbf{x}|^2 + 0 + |\mathbf{y}|^2 \end{aligned}$$

so the square of the length of one side is the sum of the squares of the lengths of the other two. The converse can be proved by starting with $|\mathbf{x} + \mathbf{y}|^2 = |\mathbf{x}|^2 + |\mathbf{y}|^2$ and deducing $\mathbf{x} \cdot \mathbf{y} = 0$ (see Problem 10). □

3.2

Problem Set

1. For each of the following pairs of vectors, find the lengths $|\mathbf{x}|$ and $|\mathbf{y}|$; find the distance between the points $\mathbf{x}$ and $\mathbf{y}$; and find the angle between the vectors $\mathbf{x}$ and $\mathbf{y}$.
   a. $\mathbf{x} = (1, -1, 0, 1)$, $\mathbf{y} = (1, -1, 1, 1)$
   b. $\mathbf{x} = (-1, 0, 1)$, $\mathbf{y} = (-2, 2, 0)$
   c. $\mathbf{x} = (1, 1, 0)$, $\mathbf{y} = (2, 0, 0)$
2. Find two vectors in $\mathbf{R}^3$ which are not scalar multiples of each other and each of which is perpendicular to $(1, -1, 2)$.
3. Find a vector in $\mathbf{R}^3$ which is perpendicular to both $(1, -1, 0)$ and $(2, 0, 1)$.
4. Prove Theorem 3.1.
5. Prove Corollary 3.1.
6. For each pair of vectors in Problem 1,
   a. Verify Schwarz's Inequality.
   b. Verify the Triangle Inequality.
   c. Find $\text{proj}_{\mathbf{y}}\mathbf{x}$.
7. Prove: If equality holds in the triangle inequality (that is, if $|\mathbf{x} + \mathbf{y}| = |\mathbf{x}| + |\mathbf{y}|$), then the vectors $\mathbf{x}$ and $\mathbf{y}$ have the same direction. (*Hint:* Square both sides of the equation $|\mathbf{x} + \mathbf{y}| = |\mathbf{x}| + |\mathbf{y}|$. Simplify and conclude that the cosine of the angle between $\mathbf{x}$ and $\mathbf{y}$ is 1.) Interpret geometrically.
8. For any vectors $\mathbf{x}$ and $\mathbf{y}$ ($\mathbf{y} \neq \mathbf{0}$) in $\mathbf{R}^n$, show that the vectors $\mathbf{x} - (\text{proj}_{\mathbf{y}}\mathbf{x})$ and $\mathbf{y}$ are perpendicular by computing the dot product of
   $$\mathbf{x} - \frac{\mathbf{x} \cdot \mathbf{y}}{\mathbf{y} \cdot \mathbf{y}}\mathbf{y} \quad \text{and} \quad \mathbf{y}$$
   and simplifying by means of Theorem 3.1. Check this with the vectors in Problem 1, part a.
9. For each of the vectors $\mathbf{x}$ and $\mathbf{y}$ in Problem 1, part b, write a unit vector having the same direction as the given vector.
10. Complete the proof of the Pythagorean Theorem (Example 9).
11. Prove, by vector methods: In a parallelogram, the sum of the squares of the lengths of the diagonals is equal to the sum of the squares of the lengths of all four sides (Figure 3.8).

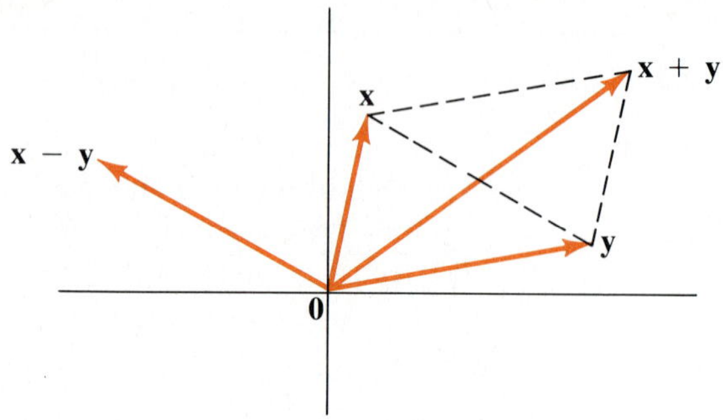

Figure 3.8

12. Prove, by vector methods: A parallelogram is a rectangle if and only if its diagonals are of equal length.
13. Prove, by vector methods: A parallelogram is a rhombus (has all sides of equal length) if and only if its diagonals are perpendicular.
14. Prove, by vector methods: A triangle inscribed in a semicircle with one side lying along the diameter and the opposite vertex lying on the arc must be a right triangle. (*Hint:* See Figure 3.9.)

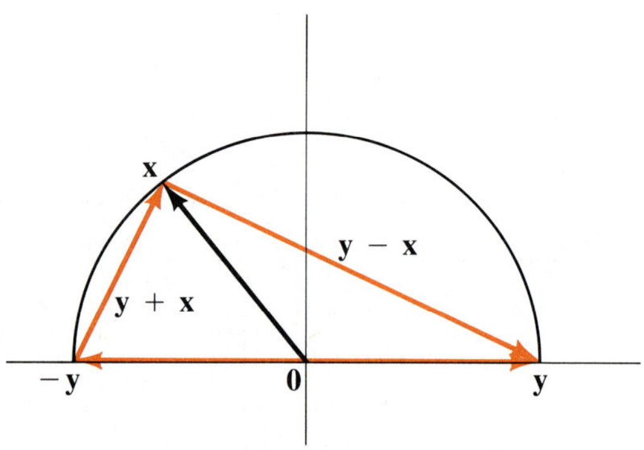

Show: $(\mathbf{y} - \mathbf{x}) \cdot (\mathbf{y} + \mathbf{x}) = 0$

Figure 3.9

15. Complete the proof of Corollary 3.2a.
16. Let $\mathbf{y}$ and $\mathbf{z}$ be *orthogonal* vectors in $\mathbf{R}^n$. Show that if $\mathbf{x}$ is in $\mathbf{R}^n$, then $\mathbf{x} - \text{proj}_{\mathbf{y}}\mathbf{x} - \text{proj}_{\mathbf{z}}\mathbf{x}$ is orthogonal to each of $\mathbf{y}$ and $\mathbf{z}$.

3.3

VECTOR GEOMETRY OF LINES AND PLANES

In this section, we shall apply vector methods to the study of lines and planes. As in the preceding section, we will analyze lines and planes in the familiar, concrete settings of $\mathbf{R}^2$ and $\mathbf{R}^3$ and then use our findings to motivate definitions for the abstract setting of $\mathbf{R}^n$. We begin with lines.

Let $\mathbf{u}$ be a nonzero vector in $\mathbf{R}^2$ or $\mathbf{R}^3$. We want to describe the line through the origin that has the same direction as $\mathbf{u}$. We visualize $\mathbf{u}$ as an arrow emanating from the origin. Then a point $\mathbf{x}$ will lie on our line if and only if there is a scalar t such that $\mathbf{x} = t\mathbf{u}$, for such a point $\mathbf{x}$ will correspond to an arrow emanating from the origin in a direction the same as or opposite to that of $\mathbf{u}$ (Figure 3.10).

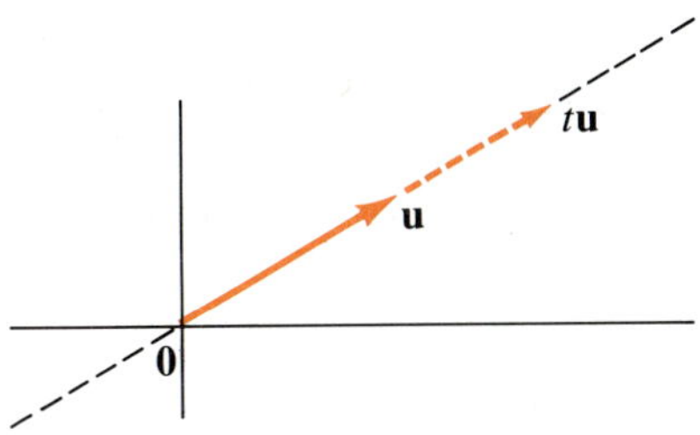

Figure 3.10

Definition

In $\mathbf{R}^n$ the *line through the origin in the direction of* $\mathbf{u}$ ($\mathbf{u} \neq \mathbf{0}$) is the set of all points $\mathbf{x}$ satisfying

$$\mathbf{x} = t\mathbf{u} \qquad t \text{ arbitrary}$$

The equation above is called a *parametric* equation for the line, and the variable t is called a *parameter*.

Example 1

In $\mathbf{R}^2$ the line through the origin in the direction of $(-1, 2)$ is given by $(x, y) = t(-1, 2)$. Note that, when separated by coordinates, the description of this line can be given by a pair of equations:

$$x = -t$$
$$y = 2t$$

(See Figure 3.11.) □

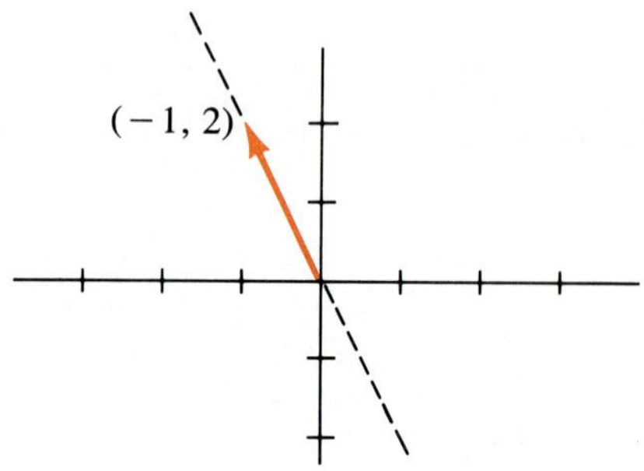

the line $\mathbf{x} = t(-1, 2)$, *or* $2x + y = 0$

Figure 3.11

In $\mathbf{R}^2$ it is possible to convert a parametric equation of a line to the more familiar form $ax + by = c$, and vice versa.

Example 2

For the line given in Example 1, we can solve the two coordinate equations $x = -t$ and $y = 2t$ to write t in terms of x and in terms of y. We find

$$t = -x \quad \text{and} \quad t = \left(\frac{1}{2}\right)y$$

Thus $-x = (\frac{1}{2})y$, which yields $2x + y = 0$ as an equation for the line. □

Definition In $\mathbf{R}^n$ the line through the point $\mathbf{x}_0$ in the direction of $\mathbf{u}$ ($\mathbf{u} \neq \mathbf{0}$) is given by the parametric equation

$$\mathbf{x} = \mathbf{x}_0 + t\mathbf{u} \qquad t \text{ arbitrary}$$

(See Figure 3.12.)

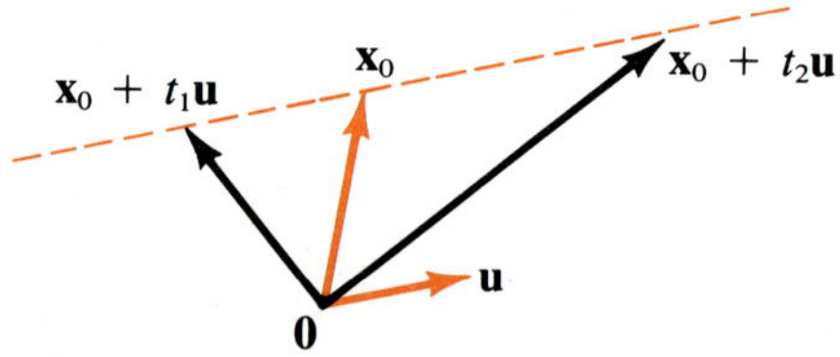

Figure 3.12

Example 3

According to our definition, the line in $\mathbf{R}^3$ through the point $(1, 0, -1)$ in the direction of the vector $(2, 1, 3)$ is given by the parametric equation

$$(x, y, z) = (1, 0, -1) + t(2, 1, 3)$$ □

It would be natural to conjecture that a line in $\mathbf{R}^3$ could be described by an equation of the form $ax + by + cz = d$. However, this is not the case. For, if we proceed as in Example 2, we obtain a *trio* of equations giving t in terms of x, y, and z. For instance, the parametric equation of Example 3 would yield

$$x = 1 + 2t$$
$$y = t$$
$$z = -1 + 3t$$

which, in turn, gives us

$$t = \frac{x-1}{2}, \qquad t = y, \quad \text{and} \quad t = \frac{z+1}{3}$$

Equating these three expressions for t will not result in a single equation relating x, y, and z. Instead, the result is a combination of equations

$$\frac{x-1}{2} = y = \frac{z+1}{3}$$

This is called the "symmetric equation" for the line in question.

In general, the *symmetric equation* of a line given parametrically by

$$(x, y, z) = (x_0, y_0, z_0) + t(a, b, c)$$

will be

$$\frac{x-x_0}{a} = \frac{y-y_0}{b} = \frac{z-z_0}{c}$$

as is readily seen. It is written in this form even if one or more of a, b, and c is zero. Now, we can see that all points on a line must satisfy the *three* equations

$$\frac{x-x_0}{a} = \frac{y-y_0}{b}, \qquad \frac{y-y_0}{b} = \frac{z-z_0}{c}, \quad \text{and} \quad \frac{x-x_0}{a} = \frac{z-z_0}{c}$$

We shall see that any one of these equations represents a *plane* in $\mathbf{R}^3$. Thus the symmetric equation provides us with three planes, all of which intersect in the line in question. We shall return to this fact when we focus our attention on planes. Meanwhile, our next example leads to another piece of information the symmetric equation displays.

Example 4
We find both a parametric equation and a symmetric equation for the line in $\mathbf{R}^3$ through $\mathbf{x}_0 = (1, 3, 5)$ in the direction of $\mathbf{u} = \frac{1}{3}(1, 2, -2)$. Note that we have chosen a *unit vector* to give the direction. In parametric form, we have the equation

$$(x, y, z) = (1, 3, 5) + t\left(\frac{1}{3}, \frac{2}{3}, -\frac{2}{3}\right)$$

In symmetric form, the equation is

$$\frac{x - 1}{\frac{1}{3}} = \frac{y - 3}{\frac{2}{3}} = \frac{z - 5}{-\frac{2}{3}}$$

□

The denominators of the components of the symmetric form are, of course, the coordinates of the direction vector. However, if the direction vector is a *unit vector*, as in Example 4, it can be shown that these denominators are also the cosines of the angles made by $\mathbf{u}$ with the three coordinate axes. These angles determine the direction of the line, and their cosines are called the *direction cosines* of the line. In Example 4, we have $\cos\alpha = \frac{1}{3}$, $\cos\beta = \frac{2}{3}$, and $\cos\gamma = -\frac{2}{3}$; thus the angles formed by $\mathbf{u}$ and the x, y, and z axes are, respectively, $\alpha = 70.53°$, $\beta = 48.19°$, $\gamma = 131.81°$.

"Two points determine a line" is a statement oft-repeated in high-school geometry. Because this statement is true, if two points in $\mathbf{R}^n$ are given, it ought to be possible to determine a parametric equation for the line through them. A parametric line equation requires one point on the line and a "direction vector" for the line. Suppose that the given points are $\mathbf{x}_1$ and $\mathbf{x}_2$. Then either can be used for the required point, and the difference $\mathbf{x}_2 - \mathbf{x}_1$ (or any scalar multiple of it) can be the required vector. Thus *the line through* $\mathbf{x}_1$ *and* $\mathbf{x}_2$ is given by

$$\mathbf{x} = \mathbf{x}_1 + t(\mathbf{x}_2 - \mathbf{x}_1)$$

(Figure 3.13). Note that $t = 0$ produces $\mathbf{x}_1$ and $t = 1$ produces $\mathbf{x}_2$.

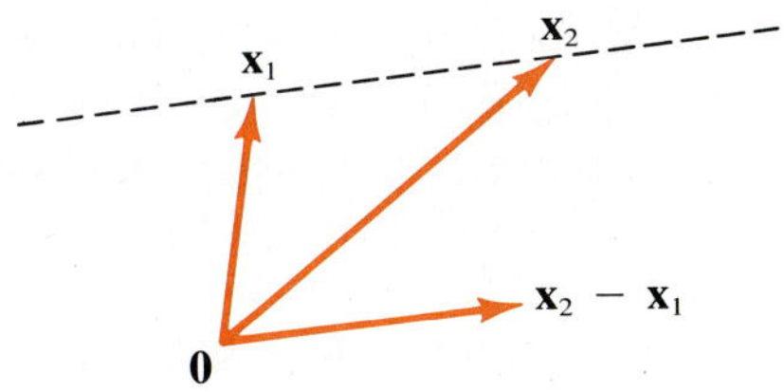

Figure 3.13

Example 5

We find a parametric equation for the line in $\mathbf{R}^4$ through $(1, 2, 0, -3)$ and $(2, 0, -1, -1)$. One such equation is

$$(x, y, z, w) = (1, 2, 0, -3) + t(1, -2, -1, 2)$$

although

$$(x, y, z, w) = (2, 0, -1, -1) + s(-1, 2, 1, -2)$$

would be equally correct. (As t and s range over all possible scalar values, the set of points produced will be the same for both equations.) □

We now consider how pairs of lines might be oriented.

Definition

Let $\mathbf{x} = \mathbf{x}_1 + t\mathbf{u}$ and $\mathbf{x} = \mathbf{x}_2 + s\mathbf{v}$ be parametric equations for two lines. Then,

a. The lines are *parallel* provided that $\mathbf{u}$ and $\mathbf{v}$ are scalar multiples of each other.

b. The lines *intersect* provided that there exist scalar values for s and t such that $\mathbf{x}_1 - \mathbf{x}_2 = s\mathbf{v} - t\mathbf{u}$.

c. The lines are *perpendicular* provided that $\mathbf{u} \cdot \mathbf{v} = 0$.

d. The lines are *skew* provided that they neither intersect nor are parallel.

Some remarks are in order about the appropriateness of these definitions. The criterion for parallelism simply assures that the direction vectors for the lines are parallel (have the same direction or opposite directions). It should be noted, however, that the intersection of parallel lines is not precluded; should parallel lines intersect, the two parametric equations in fact represent the same line. For lines to intersect, they must share a common point and hence, for some s and t, $\mathbf{x}_1 + t\mathbf{u} = \mathbf{x}_2 + s\mathbf{v}$, which leads to the intersection criterion in part b. Perpendicularity results from orthogonality of the direction vectors. In $\mathbf{R}^n$ for $n > 2$, nonparallel lines need not intersect, so skewness may occur.

Example 6

We determine the orientation of the lines in $\mathbf{R}^3$ given by

$$\mathbf{x} = (2, 3, 4) + t(-2, 1, 1) \quad \text{and} \quad \mathbf{x} = (0, 1, 3) + s(-1, 3, 1)$$

These lines are not parallel, since the direction vectors $(-2, 1, 1)$ and $(-1, 3, 1)$ are not parallel. (This also rules out the possibility of having two representations of the same line.) The lines are also not perpendicular, since the dot product of the direction vectors is not 0. To determine if they intersect, we must consider the possibility of being able to solve for s and t in the equation $(2, 3, 4) - (0, 1, 3) = s(-1, 3, 1) - t(-2, 1, 1)$. This gives the system

$$\begin{aligned} -s + 2t &= 2 \\ 3s - t &= 2 \\ s - t &= 1 \end{aligned}$$

Operating on the augmented matrix

$$\left[\begin{array}{rr|r} -1 & 2 & 2 \\ 3 & -1 & 2 \\ 1 & -1 & 1 \end{array}\right]$$

results in the echelon form

$$\left[\begin{array}{rr|r} 1 & 0 & 0 \\ 0 & 1 & 0 \\ 0 & 0 & 1 \end{array}\right]$$

which, as an *augmented* matrix, tells us there is no solution. Therefore, the lines do not intersect and, in fact, are skew. □

Before turning our attention to planes, we shall apply the ideas we have developed to the solution of some geometric problems involving lines. It is important in our applications to allow vectors to "float"; that is, we must be able to position them where appropriate for the problem, even if that means that they are not "anchored" at the origin. We also note that we will rely heavily on the concept of projection (Section 3.2).

Example 7

We find the distance between the point $\mathbf{x}_1 = (1, 0, 2)$ and the line $\mathbf{x} = (5, 1, -2) + t(2, 1, 3)$. (By "the distance between a point and a line," we mean the distance measured at a right angle to the line.)

Since $\mathbf{u} = (2, 1, 3)$ is the direction vector for the line, we visualize it as lying along the line, emanating from $\mathbf{x}_0 = (5, 1, -2)$, a point on the line. Next, we consider the vector from $\mathbf{x}_0$ to $\mathbf{x}_1$, which is $\mathbf{x}_1 - \mathbf{x}_0$ (Figure 3.14).

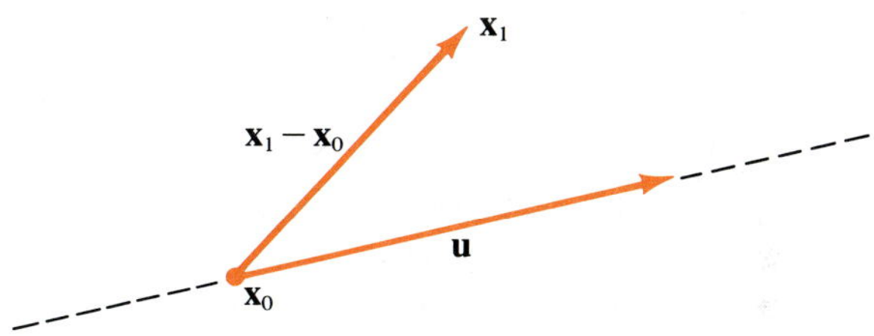

Figure 3.14

We find the projection of $\mathbf{x}_1 - \mathbf{x}_0$ onto $\mathbf{u}$:

$$\begin{aligned}\text{proj}_{\mathbf{u}}(\mathbf{x}_1 - \mathbf{x}_0) &= \frac{(\mathbf{x}_1 - \mathbf{x}_0) \cdot \mathbf{u}}{\mathbf{u} \cdot \mathbf{u}}\mathbf{u} \\ &= \frac{(-4, -1, 4) \cdot (2, 1, 3)}{(2, 1, 3) \cdot (2, 1, 3)}\mathbf{u} \\ &= \frac{3}{14}(2, 1, 3)\end{aligned}$$

The difference $(\mathbf{x}_1 - \mathbf{x}_0) - \text{proj}_{\mathbf{u}}(\mathbf{x}_1 - \mathbf{x}_0)$ is a vector perpendicular to $\mathbf{u}$ (see Problem 8 of Section 3.2), and, if suitably positioned, will stretch exactly from the line to the point $\mathbf{x}_1$. Thus, the *length* of this difference will be our answer (Figure 3.15).

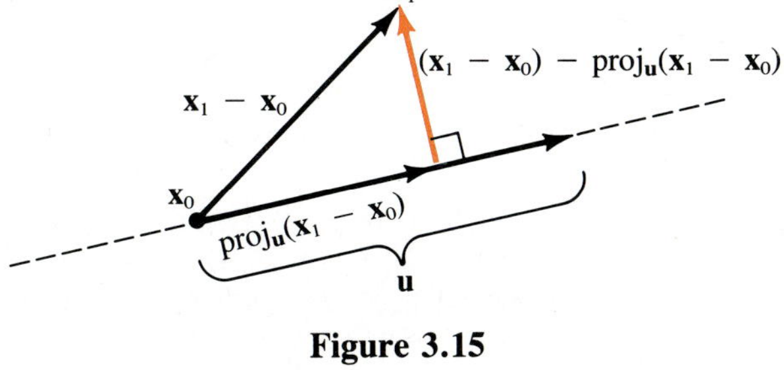

Figure 3.15

$$(\mathbf{x}_1 - \mathbf{x}_0) - \text{proj}_{\mathbf{u}}(\mathbf{x}_1 - \mathbf{x}_0) = (-4, -1, 4) - \frac{3}{14}(2, 1, 3)$$

$$= \left(-\frac{62}{14}, -\frac{17}{14}, \frac{47}{14}\right) = \frac{1}{14}(-62, -17, 47)$$

The length of this vector is

$$\frac{1}{14}\sqrt{(-62)^2 + (-17)^2 + (47)^2} = \frac{1}{14}\sqrt{6342} = 5.69 \text{ (rounded)}$$

so the point $\mathbf{x}_1$ is 5.69 units away from the given line. (A calculus solution is asked for in Problem 7.) □

Note that in Example 7 the point $\mathbf{x}_1$ was indeed off the given line. If $\mathbf{x}_1$ had been *on* the line, however, the length of $(\mathbf{x}_1 - \mathbf{x}_0) - \text{proj}_{\mathbf{u}}(\mathbf{x}_1 - \mathbf{x}_0)$ would turn out to be zero. Obtaining a zero distance between a point and a line is, in fact, one way of determining that a given point is on a given line.

The next example shows how to find the distance between two parallel lines.

Example 8

We find the distance between the two parallel lines,

$$\mathbf{x} = (1, -2, 3) + s(2, 2, -1) \quad \text{and} \quad \mathbf{x} = (-2, 3, 4) + t(4, 4, -2)$$

This distance is the same as the distance between one of the lines and any point on the other line, since the lines are parallel. Hence, we can proceed as in Example 7. We take $\mathbf{x}_1 = (-2, 3, 4)$, $\mathbf{x}_0 = (1, -2, 3)$, and $\mathbf{u} = (2, 2, -1)$ (Figure 3.16) and get

$$(\mathbf{x}_1 - \mathbf{x}_0) - \text{proj}_{\mathbf{u}}(\mathbf{x}_1 - \mathbf{x}_0) = (-3, 5, 1) - \frac{3}{9}(2, 2, -1)$$

$$= \frac{1}{9}(-33, 39, 12)$$

$$= \frac{1}{3}(-11, 13, 4)$$

The length of this vector is

$$\frac{1}{3}\sqrt{(-11)^2 + (13)^2 + (4)^2} = \sqrt{34}$$

so the distance between the lines is $\sqrt{34}$ units. □

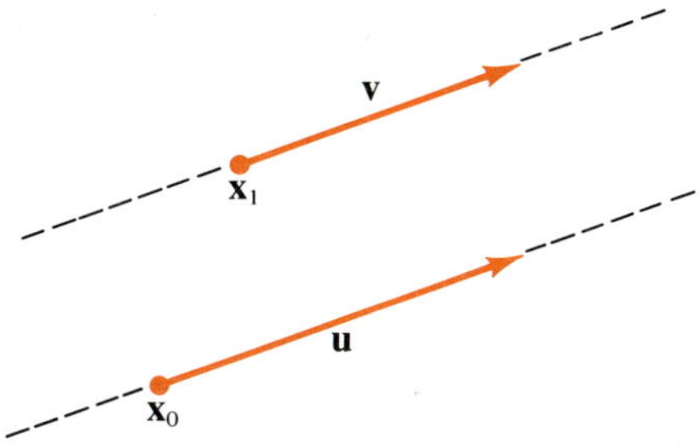

Figure 3.16

In $\mathbf{R}^3$ we can deal with planes at the outset in a manner similar to the way we dealt with lines in $\mathbf{R}^2$ or $\mathbf{R}^3$.

Let $\mathbf{u}$ and $\mathbf{v}$ be *linearly independent* vectors in $\mathbf{R}^3$; that is, neither $\mathbf{u}$ nor $\mathbf{v}$ is a scalar multiple of the other (or, put another way, the points $\mathbf{u}$ and $\mathbf{v}$ are not collinear with the origin). Then these vectors will determine the "orientation" of certain planes in $\mathbf{R}^3$, and if we also specify a certain point $\mathbf{x}_0$ through which a plane must pass, we shall have described one and only one plane.

Definition

Let $\mathbf{u}$ and $\mathbf{v}$ be linearly independent vectors in $\mathbf{R}^3$ and let $\mathbf{x}_0$ be a point in $\mathbf{R}^3$. Then the *plane through* $\mathbf{x}_0$ *parallel to* $\mathbf{u}$ *and* $\mathbf{v}$ is the set of all points $\mathbf{x}$ satisfying the parametric equation

$$\mathbf{x} = \mathbf{x}_0 + s\mathbf{u} + t\mathbf{v}$$

(Figure 3.17).

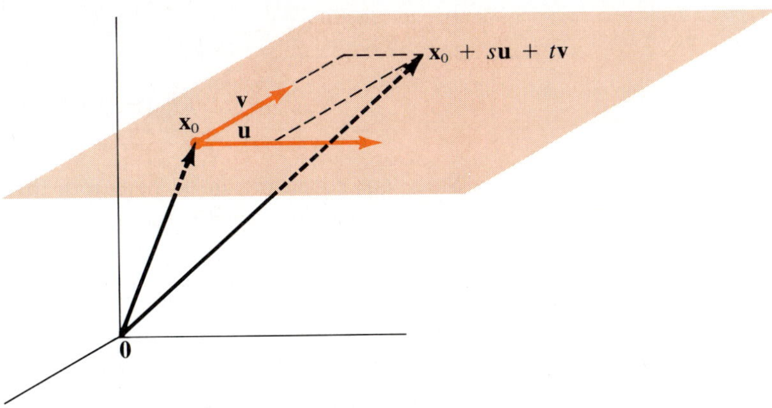

Figure 3.17

A more familiar form for an equation of a plane in $\mathbf{R}^3$ might be the coordinate equation $ax + by + cz = d$. It is possible, as our examples show, to convert from one form to the other. However, it is more satisfactory to take a different view of what determines a plane, a view which will lead directly to the coordinate form of its equation. First, we let $\mathbf{n}$ be a nonzero vector in $\mathbf{R}^3$. If we consider the planes in $\mathbf{R}^3$ that are *perpendicular* to $\mathbf{n}$, we get a family of parallel planes, and thus $\mathbf{n}$ determines the orientation. A certain *one* of these planes is singled out if we require that the plane pass through a given point $\mathbf{x}_0$. Now, let $\mathbf{x}$ be any point on the plane through $\mathbf{x}_0$ perpendicular to $\mathbf{n}$. What condition must $\mathbf{x}$ satisfy? The vector $\mathbf{x} - \mathbf{x}_0$, as an arrow, can be drawn from $\mathbf{x}_0$ to $\mathbf{x}$; thus $\mathbf{x} - \mathbf{x}_0$ is in the plane and hence must be perpendicular to $\mathbf{n}$. Therefore, we have the following definition.

Definition

In $\mathbf{R}^n$, let $\mathbf{n}$ be a nonzero vector and let $\mathbf{x}_0$ be a point. Then, the *hyperplane through* $\mathbf{x}_0$ *perpendicular to* $\mathbf{n}$ is given by the equation

$$(\mathbf{x} - \mathbf{x}_0) \cdot \mathbf{n} = 0$$

or, equivalently,

$$\mathbf{x} \cdot \mathbf{n} = \mathbf{x}_0 \cdot \mathbf{n}$$

The vector $\mathbf{n}$ is called a *normal* vector for the hyperplane (Figure 3.18).

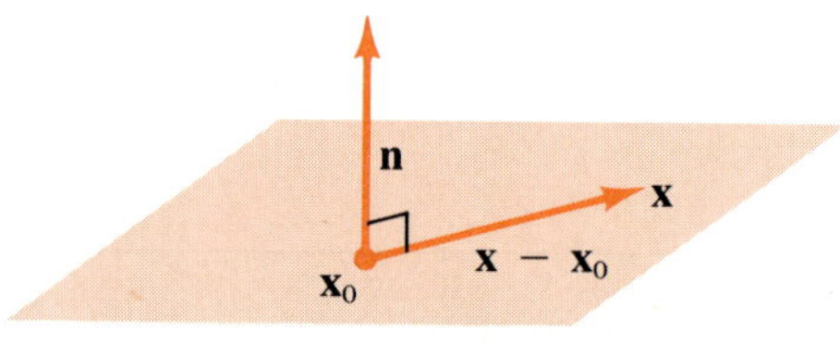

Figure 3.18

Since our equation can be written in the form $\mathbf{x} \cdot \mathbf{n} = \mathbf{x}_0 \cdot \mathbf{n}$, if everything is written out with coordinates we will get for a plane in $\mathbf{R}^3$ an equation of the type $ax + by + cz = d$. For if we let $\mathbf{x} = (x, y, z)$, $\mathbf{n} = (a, b, c)$, and $\mathbf{x}_0 = (x_0, y_0, z_0)$, we get

$$(x, y, z) \cdot (a, b, c) = (x_0, y_0, z_0) \cdot (a, b, c)$$

or

$$ax + by + cz = ax_0 + by_0 + cz_0$$

Example 9

We consider the plane in $\mathbf{R}^3$ through (1, 0, 1) parallel to the vectors (2, −1, 3) and (−1, 1, 0). Its parametric equation, by virtue of the definition we gave of parametric equation of a plane in $\mathbf{R}^3$, is $\mathbf{x} = (1, 0, 1) + s(2, -1, 3) + t(-1, 1, 0)$. To convert to the coordinate equation, we break apart this parametric equation into its coordinate components:

$$\begin{aligned} x &= 1 + 2s - t \\ y &= \quad\; - s + t \\ z &= 1 + 3s \end{aligned}$$

Now, the idea is to eliminate s and t and get one relationship among x, y, and z. Though there is no set procedure for this, we choose to add the first two equations, and we get $x + y = 1 + s$, which says $s = x + y - 1$. On the other hand, the third equation says that $s = \frac{1}{3}(z - 1)$. Equating these expressions for s gives us

$$x + y - 1 = \frac{1}{3}(z - 1)$$

or

$$3x + 3y - z = 2$$

We can confirm that this result is indeed the coordinate equation for our plane by observing that the given point (1, 0, 1) does satisfy the equation; we also observe that the vector $\mathbf{n} = (3, 3, -1)$ (consisting of the coefficients of x, y, and z) is perpendicular to the plane, according to the discussion earlier, and hence ought to be perpendicular to each of (2, −1, 3) and (−1, 1, 0), which indeed is the case. □

Example 10

We find the plane in $\mathbf{R}^3$ perpendicular to $\mathbf{n} = (5, 3, -2)$ and containing the point $\mathbf{x}_0 = (2, -1, 0)$. By definition, we merely write

$$\mathbf{x} \cdot \mathbf{n} = \mathbf{x}_0 \cdot \mathbf{n}$$

or

$$(x, y, z) \cdot (5, 3, -2) = (2, -1, 0) \cdot (5, 3, -2)$$

which gives the coordinate equation

$$5x + 3y - 2z = 7$$

To obtain a parametric form for this plane, we simply let $s = y$ and $t = z$ and solve the equation for x to get $x = \frac{1}{5}(7 - 3y + 2z)$. Thus we can write

$$\begin{aligned} x &= \frac{7}{5} - \frac{3}{5}s + \frac{2}{5}t \\ y &= \phantom{\frac{7}{5} - \frac{3}{5}}s \\ z &= \phantom{\frac{7}{5} - \frac{3}{5}s + \frac{2}{5}}t \end{aligned}$$

or

$$(x, y, z) = \left(\frac{7}{5}, 0, 0\right) + s\left(-\frac{3}{5}, 1, 0\right) + t\left(\frac{2}{5}, 0, 1\right)$$

It is easy to verify that $(\frac{7}{5}, 0, 0)$ is in fact a point on the plane and that each "direction vector" $(-\frac{3}{5}, 1, 0)$ and $(\frac{2}{5}, 0, 1)$ is perpendicular to $\mathbf{n} = (5, 3, -2)$. □

At this point, we can offer a geometric interpretation of certain types of results we obtained when solving systems of equations. Suppose, for instance, we have a system of two linear equations in three unknowns. As we have seen in previous examples, this system can have one of three types of solutions: $(x, y, z) = \mathbf{x}_0 + r\mathbf{u} + t\mathbf{v}$ (with two

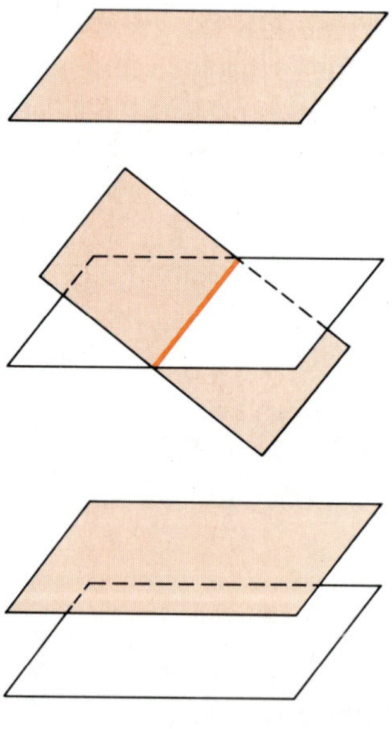

two equations in three unknowns

Figure 3.19

arbitrary unknowns); or $(x, y, z) = \mathbf{x}_0 + t\mathbf{u}$ (with one arbitrary unknown); or we have no solution at all. Because a linear equation in three unknowns represents a plane in $\mathbf{R}^3$, the solutions for a system of two such equations actually represent those points which two planes have in common. Respective to the order listed above, we have that two planes in $\mathbf{R}^3$ can intersect in a plane (that is, they coincide), in a line, or not at all (Figure 3.19).

Similarly, the solution set for a system of three equations in three unknowns represents the points of intersection of three planes in $\mathbf{R}^3$. We may obtain a plane (if two unknowns may be declared arbitrary), a line (if one unknown is arbitrary), a single point (if there is a unique solution), or no solution (which may occur in a variety of ways) (Figure 3.20).

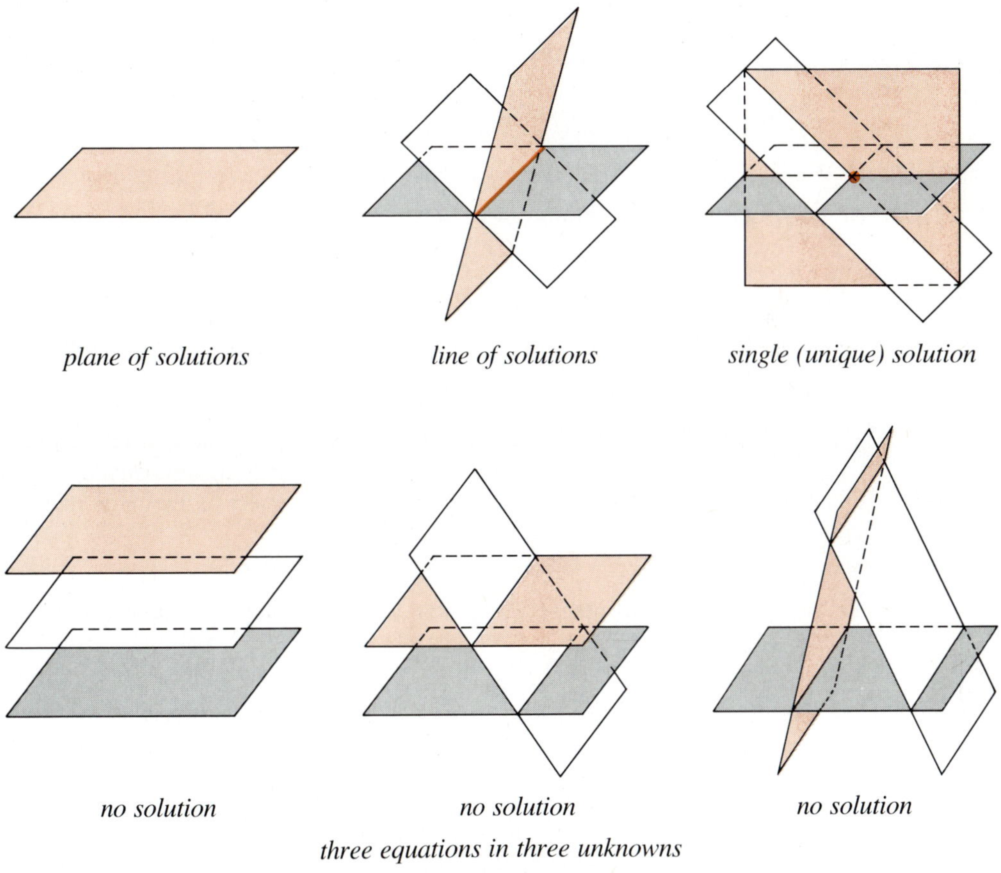

Figure 3.20

Our definition of *hyperplane* applies in $\mathbf{R}^n$ for any n. However, only in $\mathbf{R}^3$ does it give what we are accustomed to thinking of as a plane. Of course, in $\mathbf{R}^n$ for $n > 3$ it does not give a "real" object at all; on the other hand, in $\mathbf{R}^2$ a hyperplane by our definition will in fact be a line.

Example 11

We find the line (hyperplane) in $\mathbf{R}^2$ perpendicular to (2, 1) and passing through the origin. The solution is simply

$$(x, y) \cdot (2, 1) = (0, 0) \cdot (2, 1)$$

or

$$2x + y = 0 \qquad \square$$

Now, let us recall our discussion earlier in this section of certain consequences of the symmetric equation.

$$\frac{x - x_0}{a} = \frac{y - y_0}{b} = \frac{z - z_0}{c}$$

of a line in $\mathbf{R}^3$. We found that the points on a line with the given symmetric equation must satisfy these three equations:

$$\frac{x - x_0}{a} = \frac{y - y_0}{b}, \quad \frac{y - y_0}{b} = \frac{z - z_0}{c}, \quad \text{and} \quad \frac{x - x_0}{a} = \frac{z - z_0}{c}$$

It is easy to see that each of these three equations represents a plane in $\mathbf{R}^3$. Moreover, the first is a plane parallel to the z axis, the second is a plane parallel to the x axis, and the third is a plane parallel to the y axis. Any two of the equations are sufficient to determine the line in question, so the symmetric equation gives us an immediate way of writing equations for two planes which intersect in the given line.

Example 12

We find two planes whose line of intersection is the line given parametrically by

$$(x, y, z) = (-1, 2, 5) + t(3, -2, 4)$$

The symmetric equation for this line is

$$\frac{x + 1}{3} = \frac{y - 2}{-2} = \frac{z - 5}{4}$$

If we wish to view this line as the line of intersection of a plane parallel to the x axis and another plane parallel to the y axis, we take from the symmetric equation the following two plane equations:

$$\frac{y-2}{-2} = \frac{z-5}{4} \quad \text{and} \quad \frac{x+1}{3} = \frac{z-5}{4}$$

These can be written in the more customary form

$$4y + 2z = 18 \quad \text{and} \quad 4x - 3z = -19$$

Since the normal vector for the first of these planes is (0,4,2), the normal vector is perpendicular to the x axis, so the plane itself is parallel to this axis. Similarly, the second plane is parallel to the y axis. □

In a manner similar to what we have done with lines, we can apply our projection techniques to solve certain geometric problems involving planes.

Example 13

We find the distance between the point $(2, 1, -3)$ and the plane $x - 2y + z = 2$.

We know from the equation for the plane that $\mathbf{n} = (1, -2, 1)$ is a normal vector for the plane. We let $\mathbf{x}_0$ be *any* point on the plane; for instance, let $\mathbf{x}_0 = (2, 0, 0)$ (the coordinates need only satisfy the equation). Now, we project $\mathbf{x}_1 - \mathbf{x}_0$ onto $\mathbf{n}$, where $\mathbf{x}_1 = (2, 1, -3)$ (Figure 3.21):

$$\begin{aligned}\text{proj}_{\mathbf{n}}(\mathbf{x}_1 - \mathbf{x}_0) &= \frac{(0, 1, -3) \cdot (1, -2, 1)}{(1, -2, 1) \cdot (1, -2, 1)}(1, -2, 1) \\ &= \frac{-5}{6}(1, -2, 1)\end{aligned}$$

The *length* of this projection is our answer: $|\frac{-5}{6}(1, -2, 1)| = \frac{5}{6}\sqrt{6}$. □

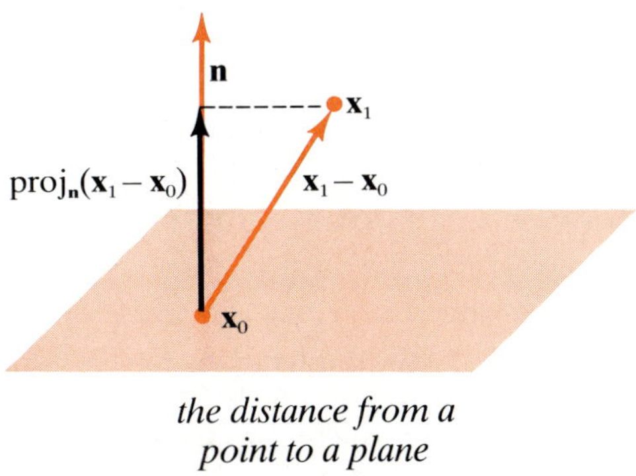

the distance from a point to a plane

Figure 3.21

Recall the notion that, in $\mathbf{R}^3$, three noncollinear points determine a plane. If we have three such points, say $\mathbf{x}_1$, $\mathbf{x}_2$, $\mathbf{x}_3$, it should be possible to produce an equation for the plane they determine. To find such a coordinate equation directly—without going by way of a parametric equation—we would, of course, need a normal vector $\mathbf{n}$ and a point on the plane, for which any of the three can serve. To discover this normal vector, we first note that the difference vectors $\mathbf{x}_2 - \mathbf{x}_1$ and $\mathbf{x}_3 - \mathbf{x}_1$ will be parallel to the plane in question. Intuitively we can perceive that any vector perpendicular to these two vectors will also be perpendicular to the plane. We must, then, discover a formula that will produce a vector normal to two given vectors. That is, for two vectors (u_1, u_2, u_3) and (v_1, v_2, v_3), we must find a vector (x, y, z) satisfying

$$\begin{aligned} u_1x + u_2y + u_3z &= 0 \\ v_1x + v_2y + v_3z &= 0 \end{aligned}$$

We shall assume that such a vector exists and we shall also assume that the two given vectors (u_1, u_2, u_3) and (v_1, v_2, v_3) are not parallel to each other. It can be shown that the coefficients of *some two* of the unknowns in this system must constitute a *nonsingular* 2×2 matrix, so let us suppose that the coefficients of x and y in our pair of equations constitute the nonsingular 2×2 matrix

$$\begin{bmatrix} u_1 & u_2 \\ v_1 & v_2 \end{bmatrix}$$

Because of our assumption about the coefficients of x and y, it can be argued geometrically that the coordinate z of the vector we seek is not zero. Thus we may divide all coordinates of the desired vectors by z, if necessary. Such division by a scalar will not affect the perpendicularity conditions imposed and will let us assume that the vector in question is of the form $(x, y, 1)$. Now, our pair of equations translates to

$$\begin{aligned} u_1x + u_2y &= -u_3 \\ v_1x + v_2y &= -v_3 \end{aligned}$$

where we have only two unknowns, so we can apply Cramer's Rule to obtain

$$x = \frac{-u_3v_2 + u_2v_3}{u_1v_2 - u_2v_1}$$

$$y = \frac{-u_1v_3 + u_3v_1}{u_1v_2 - u_2v_1}$$

These values for x and y, together with $z = 1$, should give us our perpendicular vector. Since we won't affect the perpendicularity if we alter the length of the vector, we simplify our result by multiplying all coordinates by the denominator $u_1v_2 - u_2v_1$. Our analysis prompts the following definition.

Definition

Let $\mathbf{u} = (u_1, u_2, u_3)$ and $\mathbf{v} = (v_1, v_2, v_3)$. Then the *cross product* (some say "vector product") *of* **u** *with* **v** is given by

$$\mathbf{u} \times \mathbf{v} = (u_2v_3 - u_3v_2, u_3v_1 - u_1v_3, u_1v_2 - u_2v_1)$$

Note that the cross product of two vectors is a *vector* (unlike the dot product, which is a number) and that it is an operation valid only in $\mathbf{R}^3$.*

Example 14

We compute a cross product and confirm its perpendicularity to the two given vectors. Let $\mathbf{u} = (1, -1, 3)$ and $\mathbf{v} = (0, 4, 2)$. We find that

$$\begin{aligned}\mathbf{u} \times \mathbf{v} &= ((-1)(2) - (3)(4), (3)(0) - (1)(2), (1)(4) - (-1)(0)) \\ &= (-14, -2, 4)\end{aligned}$$

Now, $\mathbf{u} \cdot (\mathbf{u} \times \mathbf{v}) = (1)(-14) + (-1)(-2) + (3)(4) = -14 + 2 + 12 = 0$ and $\mathbf{v} \cdot (\mathbf{u} \times \mathbf{v}) = (0)(-14) + (4)(-2) + (2)(4) = -8 + 8 = 0$. Thus $\mathbf{u} \times \mathbf{v}$ is orthogonal to each of **u** and **v**, since the appropriate dot products are each zero. □

Now that we have exhibited an explicit formula for the cross product of two vectors in $\mathbf{R}^3$, it is an easy matter to verify that our motivating discussion did, in fact, lead us to a vector perpendicular to each of two given vectors.

Theorem 3.3

In $\mathbf{R}^3$, $\mathbf{u} \times \mathbf{v}$ is perpendicular to each of **u** and **v** (Figure 3.22).

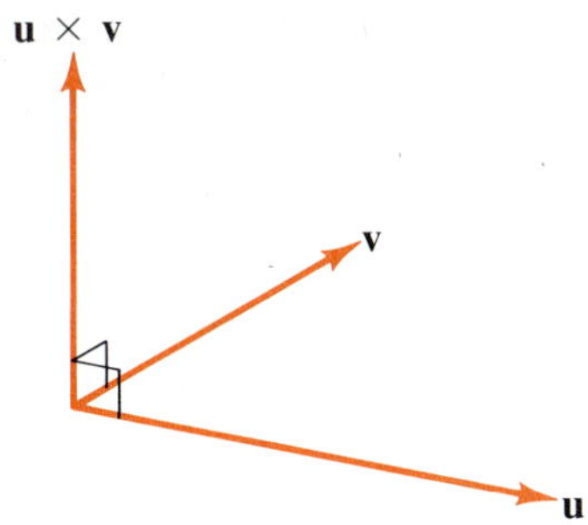

Figure 3.22

*Memorizing this cross-product formula is not required. Our next section provides a more easily retained formula.

Proof. It must be verified that $\mathbf{u} \cdot (\mathbf{u} \times \mathbf{v}) = 0$ and that $\mathbf{v} \cdot (\mathbf{u} \times \mathbf{v}) = 0$. Since these involve only direct computation using the coordinates of the vectors involved, we leave the rest of the proof as an exercise.

We will examine the cross product further in the next section, but first let us recall what first led us to consider the cross product: namely, the desire to have a convenient method for producing a normal vector for a plane in $\mathbf{R}^3$ containing three given noncollinear points $\mathbf{x}_1$, $\mathbf{x}_2$, and $\mathbf{x}_3$. We noted that the vectors $\mathbf{u} = \mathbf{x}_2 - \mathbf{x}_1$ and $\mathbf{v} = \mathbf{x}_3 - \mathbf{x}_1$ would be parallel to the plane, so the desired normal vector would be perpendicular to each of $\mathbf{x}_2 - \mathbf{x}_1$ and $\mathbf{x}_3 - \mathbf{x}_1$. We can now obtain the normal vector by computing a cross product. Any of the three given points can be used as $\mathbf{x}_0$. Hence we can readily find an equation $\mathbf{x} \cdot \mathbf{n} = \mathbf{x}_0 \cdot \mathbf{n}$ for a plane through three points (Figure 3.23).

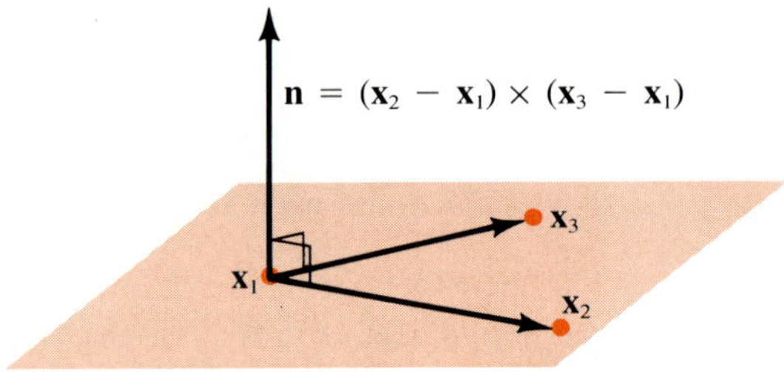

the plane through $\mathbf{x}_1, \mathbf{x}_2, \mathbf{x}_3$

Figure 3.23

Example 15

We find an equation for the plane passing through the three points (1, 0, 2), (1, 3, 2), and (2, 1, 1). According to our discussion, the difference vectors $\mathbf{u} = (1, 3, 2) - (1, 0, 2) = (0, 3, 0)$ and $\mathbf{v} = (2, 1, 1) - (1, 0, 2) = (1, 1, -1)$ are parallel to the plane in question, so the cross product $\mathbf{u} \times \mathbf{v}$ will serve as a normal vector $\mathbf{n}$. By definition,

$$\begin{aligned} \mathbf{u} \times \mathbf{v} &= ((3)(-1) - (0)(1), (0)(1) - (0)(-1), (0)(1) - (3)(1)) \\ &= (-3, 0, -3) \end{aligned}$$

Thus, using (1, 0, 2) as the point on the plane, we obtain:

$$\begin{aligned} (x, y, z) \cdot (-3, 0, -3) &= (1, 0, 2) \cdot (-3, 0, -3) \\ -3x - 3z &= -9 \\ x + z &= 3 \end{aligned}$$

To verify the answer, we can check that the three given points satisfy this equation. □

3.3 Problem Set

1. For each of the following, write a parametric equation for the line through $\mathbf{x}_0$ in the direction of $\mathbf{u}$. Convert it to an equation of the form $ax + by = c$.

   a. $\mathbf{x}_0 = (-1, 2)$, $\mathbf{u} = (2, 3)$

   b. $\mathbf{x}_0 = (2, 3)$, $\mathbf{u} = (-1, 2)$

   c. $\mathbf{x}_0 = (0, 2)$, $\mathbf{u} = (1, 3)$

   d. $\mathbf{x}_0 = (0, 0)$, $\mathbf{u} = (1, -7)$

2. For each of the following, write a parametric equation for the line through $\mathbf{x}$ in the direction of $\mathbf{u}$. Convert it to a symmetric equation and find the direction cosines. Find planes parallel to the x axis and to the y axis, respectively, whose line of intersection is the line in question.

   a. $\mathbf{x}_0 = (1, -2, 1)$, $\mathbf{u} = (1, 0, 3)$

   b. $\mathbf{x}_0 = (1, 0, 3)$, $\mathbf{u} = (1, -2, 1)$

   c. $\mathbf{x}_0 = (2, 1, 3)$, $\mathbf{u} = (5, -1, -3)$

3. For each of the following, write a parametric equation for the line through the given points.

   a. $(5, 2), (-6, 0)$

   b. $(3, 3, -3), (5, 2, -1)$

   c. $(2, 1, 0, 3), (0, -1, 2, 0)$

   d. $(3, -1, 0, 2, 0), (-4, 2, -1, 0, 3)$

4. Consider the lines given by $\mathbf{x} = (-2, 1, 4) + t(-3, 1, 2)$ and $\mathbf{x} = (1, 0, 2) + s(1, 0, 0)$.

   a. Verify that these lines intersect and find the point of intersection.

   b. Find the cosine of the angle of intersection of these lines.

   c. Find an equation in parametric form for the line perpendicular to both of these lines at the point of intersection.

5. Show that the point $(1, -5, 7)$ is on the line given by $\mathbf{x} = (1, -2, 1) + t(0, -1, 2)$:

   a. by finding a value of the parameter t which produces the point.

b. by showing that the distance between the point and the line is 0.

6. In each of the following, find the distance from the first point to the line through the second and third points.

   a. (1, 2); (1, 1), (3, 5)

   b. (1, 0, 3); (3, 1, −2), (2, 1, −3)

   c. (3, 1, −2); (1, 0, 3), (2, 1, −3)

   d. (−14, −2, 11); (−2, 0, 3), (4, 1, −1)

7. Use calculus to solve Example 7 as follows:

   a. Write the distance between the given points and an arbitrary point on the line by applying the distance formula to (1, 0, 2) and $(5 + 2t,\ 1 + t,\ -2 + 3t)$.

   b. Find the value of t which produces a minimum for this distance by finding the critical point of the *square* of the distance as a function of t. (It is easier to work with the square of the distance.)

   c. Evaluate the distance for the critical value of t.

8. Follow the method of Problem 7 to rework all parts of Problem 6.

9. In each of the following, find a line through the first point parallel to the line through the second and third points. Find a line through the first point perpendicular to the line through the second and third points.

   a. (1, 2); (1, 1), (3, 5)

   b. (1, 0, 3); (3, 1, −2), (2, 1, 3)

10. In each of the following, tell whether the lines intersect, are parallel, or are skew; find the distance between the lines.

    a. $\mathbf{x} = (1, 0, 2) + t(-3, 1, 2)$, $\mathbf{x} = (1, 0, 0) + s(6, -2, -4)$

    b. $\mathbf{x} = (-2, 1, 4) + t(-3, 1, 2)$, $\mathbf{x} = (0, 0, 3) + s(1, 0, 0)$

    c. $\mathbf{x} = (1, 0, 2) + t(1, 0, -1)$, $\mathbf{x} = (2, -1, 3) + s(3, 2, 0)$

    d. $\mathbf{x} = (1, -1, 2) + t(2, 0, 0)$, $\mathbf{x} = (2, 1, -2) + s(5, 2, 0)$

11. Find an equation for the plane containing the intersecting lines in Problem 4.

12. For each plane described below, find a parametric equation and a coordinate equation of the form $ax + by + cz = d$.

    a. the plane through the point (1, −1, 3) with normal vector (3, 5, −2)

    b. the plane through the point (1, 1, 1) with normal vector (2, 1, −3)

c. the plane through the point $(2, 1, -3)$ with normal vector $(1, 1, 1)$

d. the plane parallel to the plane in part a through the point $(1, 0, 0)$

e. the plane parallel to the plane in part b through the point $(2, 6, -1)$

f. the plane parallel to the plane in part c through the point $(2, 6, -1)$

13. In Problem 12, find the distance between the planes in

a. parts a and d

b. parts b and e

c. parts c and f

14. In each of the following, find a coordinate equation of the plane containing the given points.

a. $(0, -1, 0), (1, 1, 0), (1, 3, 1)$

b. $(1, 0, 0), (0, 1, 0), (0, 0, 1)$

c. $(3, -2, 1), (5, 2, 3), (0, 0, 0)$

d. $(5, -2, -1), (4, 2, 6), (-1, 1, 0)$

15. Find equations for the lines through $(0, -1, 0)$ perpendicular to the planes in Problem 14.

16. Find an equation for the plane containing the given point and the given line.

a. $(1, -1, 2)$; $\mathbf{x} = (4, -1, 2) + t(0, 2, 1)$

b. $(2, 3, 4)$; $\mathbf{x} = (1, -1, 0) + t(3, 0, -2)$

c. $(0, 0, 0)$; $\mathbf{x} = (1, -7, 3) + t(5, 3, -2)$

17. Find the distance between the given point and the given plane.

a. $(0, 0, 0)$; $3x + 2y + z = 4$

b. $(1, 3, -1)$; $x - 3y + 2z = -1$

c. $(2, 3, -1)$; $4y - z = 3$

18. If $\mathbf{u} = (u_1, u_2, u_3)$ is a *unit* vector, show that the cosines of the angles made by $\mathbf{u}$ with the x, y, and z axes are, in order, u_1, u_2, and u_3. (*Hint:* Find the cosine of the angle between $\mathbf{u}$ and $(1, 0, 0)$, and so forth.)

19. Derive this general formula for the distance in $\mathbf{R}^2$ between a point (x_0, y_0) and a line $ax + by + c = 0$:

$$\text{distance} = \frac{|ax_0 + by_0 + c|}{\sqrt{a^2 + b^2}}$$

(*Hint:* (a, b) is the normal vector for the line. Let (x_1, y_1) be a point on the line; project $(x_0 - x_1, y_0 - y_1)$ onto (a, b) and find the length of this projection. Sketch.)

20. Show that in $\mathbf{R}^3$, the distance between the point (x_0, y_0, z_0) and the plane $ax + by + cz + d = 0$ is given by

$$\frac{|ax_0 + by_0 + cz_0 + d|}{\sqrt{a^2 + b^2 + c^2}}$$

21. Prove Theorem 3.3.
22. Show that the line in $\mathbf{R}^2$ through the points (x_1, y_1) and (x_2, y_2) is given by the determinant equation

$$\begin{vmatrix} x & y & 1 \\ x_1 & y_1 & 1 \\ x_2 & y_2 & 1 \end{vmatrix} = 0$$

Generalize to a plane in $\mathbf{R}^3$.

3.4 Cross Product and Further Applications

The cross product, as defined in the previous section, possesses some interesting algebraic properties, which we shall discover in this section. The cross product can also be effectively employed in a wider variety of geometric applications than we have seen. Our work in both of these regards will be simplified by deriving a new way to compute cross product. To discover this new formula for cross product, we need to consider three vectors in $\mathbf{R}^3$ which play an especially significant role there. We are referring to the three vectors (1, 0, 0), (0, 1, 0), and (0, 0, 1). These vectors are called the "natural basis" vectors for $\mathbf{R}^3$ because any vector in $\mathbf{R}^3$ can be expressed in a *natural and unique* way in terms of these three specific vectors. (In Chapter 4, the idea of "basis" vector will take on added significance in a more general context.) To see that a vector in $\mathbf{R}^3$ has a natural and unique expression in terms of our three natural basis vectors, consider, for instance, the fact that

$$(-2, 0, 3) = (-2)(1, 0, 0) + (0)(0, 1, 0) + (3)(0, 0, 1)$$

and, in general,

$$(a, b, c) = a(1, 0, 0) + b(0, 1, 0) + c(0, 0, 1)$$

for any (a, b, c) in $\mathbf{R}^3$.

It is customary to denote the natural basis vectors in $\mathbf{R}^3$ by the symbols $\mathbf{i}$, $\mathbf{j}$, and $\mathbf{k}$, as follows:

$$\mathbf{i} = (1, 0, 0) \qquad \mathbf{j} = (0, 1, 0) \qquad \mathbf{k} = (0, 0, 1)$$

As a consequence, we can write any vector (a, b, c) in terms of the natural basis vectors as

$$(a, b, c) = a\mathbf{i} + b\mathbf{j} + c\mathbf{k}$$

Although we will explore further applications of these natural basis vectors later, right now we can use them to present a handy device for computing the cross product of two vectors. This device makes memorizing the cross product formula from Section 3.3 unnecessary.

Let $\mathbf{u} = (u_1, u_2, u_3)$ and $\mathbf{v} = (v_1, v_2, v_3)$. Using the natural basis vectors, we can get $\mathbf{u} \times \mathbf{v}$ by writing

$$\mathbf{u} \times \mathbf{v} = \det\begin{bmatrix} \mathbf{i} & \mathbf{j} & \mathbf{k} \\ u_1 & u_2 & u_3 \\ v_1 & v_2 & v_3 \end{bmatrix}$$

and expanding this "determinant" in the usual manner about the first row, as in the definition of determinant. Performing the computations indeed results in the cross product as given in Section 3.3:

$$\begin{aligned} \mathbf{u} \times \mathbf{v} &= \mathbf{i} \det\begin{bmatrix} u_2 & u_3 \\ v_2 & v_3 \end{bmatrix} - \mathbf{j} \det\begin{bmatrix} u_1 & u_3 \\ v_1 & v_3 \end{bmatrix} + \mathbf{k} \det\begin{bmatrix} u_1 & u_2 \\ v_1 & v_2 \end{bmatrix} \\ &= (u_2v_3 - u_3v_2)\mathbf{i} - (u_1v_3 - u_3v_1)\mathbf{j} + (u_1v_2 - u_2v_1)\mathbf{k} \end{aligned}$$

Example 1

We compute $(1, -2, 1) \times (3, 0, 2)$ by our new device:

$$\begin{aligned} (1, -2, 1) \times (3, 0, 2) &= \det\begin{bmatrix} \mathbf{i} & \mathbf{j} & \mathbf{k} \\ 1 & -2 & 1 \\ 3 & 0 & 2 \end{bmatrix} \\ &= ((-2)(2) - (1)(0))\mathbf{i} \\ &\quad - ((1)(2) - (1)(3))\mathbf{j} \\ &\quad + ((1)(0) - (-2)(3))\mathbf{k} \\ &= -4\mathbf{i} + \mathbf{j} + 6\mathbf{k} \\ &= (-4, 1, 6) \end{aligned}$$

□

Our new formula does not actually produce a "legitimate" determinant, since some entries are not real numbers. Nevertheless, this device works and, moreover, makes it relatively simple to deduce certain basic properties of the cross product operation.

Theorem 3.4 For vectors $\mathbf{u}$, $\mathbf{v}$, $\mathbf{w}$ in $\mathbf{R}^3$, and any scalar r,

a. *Anti-commutativity*:

$$\mathbf{u} \times \mathbf{v} = -(\mathbf{v} \times \mathbf{u})$$

b. $\mathbf{u} \cdot (\mathbf{u} \times \mathbf{v}) = \mathbf{v} \cdot (\mathbf{u} \times \mathbf{v}) = 0$

c. *Left distributive law*:

$$\mathbf{u} \times (\mathbf{v} + \mathbf{w}) = (\mathbf{u} \times \mathbf{v}) + (\mathbf{u} \times \mathbf{w})$$

d. *Right distributive law*:

$$(\mathbf{v} + \mathbf{w}) \times \mathbf{u} = (\mathbf{v} \times \mathbf{u}) + (\mathbf{w} \times \mathbf{u})$$

e. $\mathbf{u} \times (r\mathbf{v}) = (r\mathbf{u}) \times \mathbf{v} = r(\mathbf{u} \times \mathbf{v})$

f. $\mathbf{u} \times \mathbf{u} = \mathbf{0}$

Proof. Part a follows from the effect on a determinant of interchanging two rows; to obtain part b, observe that we obtain $\mathbf{u} \cdot (\mathbf{u} \times \mathbf{v})$ if we replace in our determinant formula the symbols $\mathbf{i}, \mathbf{j}, \mathbf{k}$ by the coordinates of $\mathbf{u}$, which, in turn, produces a determinant of a matrix having equal second and third rows and hence the value 0. (Note that part b was recorded earlier in Theorem 3.3.) Part c can be obtained by direct computation and then part d by combining parts a and c. Part e follows from a determinant property (the effect of multiplying a row by a scalar) as does part f (a determinant with two equal rows).

Henceforth, if we are required to produce a vector perpendicular to two given vectors, we shall use this determinant method to get their cross product.

Displaying the coordinates, as we have done for cross product, is one way of identifying a vector in $\mathbf{R}^3$; another way is to specify the vector's length and its direction. We shall give the length of a cross product in our next theorem and describe its direction after that.

Theorem 3.5 The length $|\mathbf{u} \times \mathbf{v}|$ of the cross product of $\mathbf{u}$ with $\mathbf{v}$ is given by

$$|\mathbf{u} \times \mathbf{v}| = |\mathbf{u}||\mathbf{v}| \sin \alpha$$

where α, as usual, is the angle between $\mathbf{u}$ and $\mathbf{v}$. (We are assuming here that neither $\mathbf{u}$ nor $\mathbf{v}$ is the zero vector.)

Proof. Let $\mathbf{u} = (u_1, u_2, u_3)$ and $\mathbf{v} = (v_1, v_2, v_3)$. It is a straightforward—albeit tedious and uninteresting—exercise in algebra to verify the formula

$$\begin{aligned}(u_2v_3 - u_3v_2)^2 &+ (u_3v_1 - u_1v_3)^2 + (u_1v_2 - u_2v_1)^2 \\ &= (u_1^2 + u_2^2 + u_3^2)(v_1^2 + v_2^2 + v_3^2) - (u_1v_1 + u_2v_2 + u_3v_3)^2\end{aligned}$$

This translates directly to

$$\begin{aligned}|\mathbf{u} \times \mathbf{v}|^2 &= |\mathbf{u}|^2|\mathbf{v}|^2 - (\mathbf{u} \cdot \mathbf{v})^2 \\ &= |\mathbf{u}|^2|\mathbf{v}|^2\left(1 - \frac{(\mathbf{u} \cdot \mathbf{v})^2}{|\mathbf{u}|^2|\mathbf{v}|^2}\right) \qquad \text{(by algebra)} \\ &= |\mathbf{u}|^2|\mathbf{v}|^2(1 - \cos^2\alpha) \qquad \text{(see Section 3.2)}\end{aligned}$$

By the trigonometric identity $\sin^2\alpha + \cos^2\alpha = 1$, we now get

$$|\mathbf{u} \times \mathbf{v}|^2 = |\mathbf{u}|^2|\mathbf{v}|^2\sin^2\alpha$$

We take nonnegative square roots and get

$$|\mathbf{u} \times \mathbf{v}| = |\mathbf{u}||\mathbf{v}| \sin \alpha$$

($\sqrt{\sin^2\alpha} = +\sin \alpha$, since by definition $0° \leq \alpha \leq 180°$, where $\sin \alpha$ is nonnegative.)

Example 2

We calculate, for two vectors $\mathbf{u}$ and $\mathbf{v}$, the length of $\mathbf{u} \times \mathbf{v}$ by means of the formula of Theorem 3.5. Let $\mathbf{u} = (1, 2, -1)$ and $\mathbf{v} = (1, 3, 1)$. We shall need $\sin \alpha$, where α is the angle between $\mathbf{u}$ and $\mathbf{v}$. Although this angle is not readily available by inspection, we can find its sine indirectly by finding its cosine, using the formula

$$\cos \alpha = \frac{\mathbf{u} \cdot \mathbf{v}}{|\mathbf{u}||\mathbf{v}|}$$

from Section 3.2. We obtain

$$\begin{aligned}\cos \alpha &= \frac{(1, 2, -1) \cdot (1, 3, 1)}{\sqrt{1 + 4 + 1}\sqrt{1 + 9 + 1}} \\ &= \frac{6}{\sqrt{6}\sqrt{11}} \\ &= \frac{6}{\sqrt{66}}\end{aligned}$$

Now,

$$\begin{aligned}\sin\alpha &= \sqrt{1-\cos^2\alpha}\\ &= \sqrt{1-\left(\frac{36}{66}\right)}\\ &= \sqrt{\frac{30}{66}}\end{aligned}$$

From Theorem 3.5, we find

$$\begin{aligned}|\mathbf{u}\times\mathbf{v}| &= |\mathbf{u}||\mathbf{v}|\sin\alpha\\ &= \sqrt{6}\sqrt{11}\sqrt{\frac{30}{66}}\\ &= \sqrt{66}\sqrt{\frac{30}{66}}\\ &= \sqrt{30}\end{aligned}$$

To confirm this result, we may compute $\mathbf{u}\times\mathbf{v}$ by our earlier method and calculate its length directly.

$$\begin{aligned}\mathbf{u}\times\mathbf{v} &= \det\begin{bmatrix}\mathbf{i} & \mathbf{j} & \mathbf{k}\\ 1 & 2 & -1\\ 1 & 3 & 1\end{bmatrix}\\ &= 5\mathbf{i}-2\mathbf{j}+\mathbf{k}\\ &= (5, -2, 1)\end{aligned}$$

Therefore,

$$|\mathbf{u}\times\mathbf{v}| = \sqrt{25+4+1} = \sqrt{30}$$ □

So much for the *length* of $\mathbf{u}\times\mathbf{v}$. What about its *direction*? We already know that $\mathbf{u}\times\mathbf{v}$ is perpendicular to each of $\mathbf{u}$ and $\mathbf{v}$, which, assuming that $\mathbf{u}$ and $\mathbf{v}$ are not parallel, narrows things down to two possibilities: it is perpendicular to the plane parallel to $\mathbf{u}$ and $\mathbf{v}$ either in one direction or in the opposite direction. Which of these two directions is correct is determined by the so-called "right-hand rule." This rule, while easy to visualize in one's mind, is not so simple to describe on paper. When we draw on paper a depiction of three-dimensional coordinate space, we normally represent three mutually perpendicular axes (an x axis, a y axis, and a z axis) (Figure 3.24).

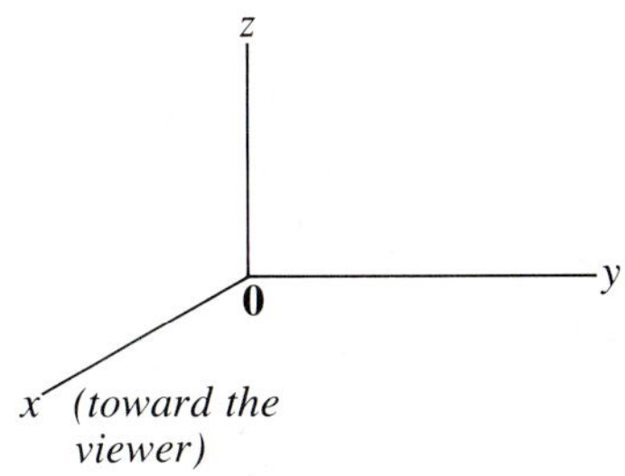

Figure 3.24

It is assumed that, in the mind's eye of the reader, the y and z axes are seen to be in the plane of the page and that the x axis is protruding *toward* the viewer. The axes—x, y, and z, *in that order*—are said, in this case, to determine a "right-handed" coordinate system. The reason is that if a normal ("right-handed") screw is aligned so that it points in the positive z direction, then turning it *from* the positive x direction *toward* the positive y direction (taking the shorter route) will cause the screw to *advance* in the positive z direction. If this explanation seems a bit complicated, perhaps a "hands-on" approach will be helpful: in a right-handed system, if the *right hand* is wrapped about the z axis so that the fingers curl in the direction *from* the positive x axis *toward* the positive y axis, then the thumb will point in the positive z direction.

Now, we can say that the direction of $\mathbf{u} \times \mathbf{v}$ is such that the three vectors $\mathbf{u}$, $\mathbf{v}$, and $\mathbf{u} \times \mathbf{v}$, when drawn emanating from the origin (or any other *common* point) must form a right-handed system, just as did the x, y, and z axes (Figure 3.25).

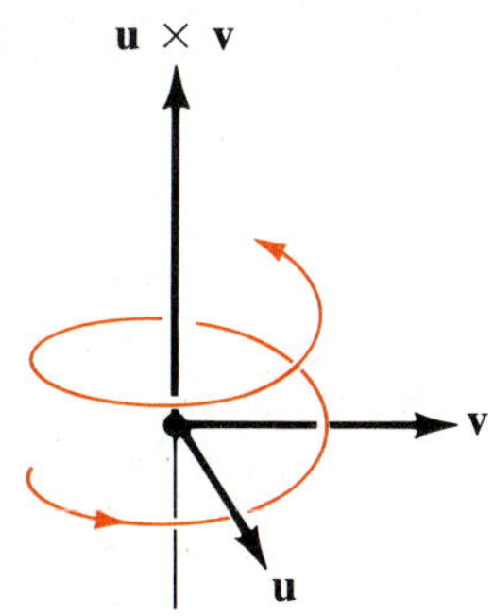

the right-hand rule

Figure 3.25

To confirm this, we could compute $\mathbf{i} \times \mathbf{j}$—that is, $(1, 0, 0) \times (0, 1, 0)$—and find that the result is $\mathbf{i} \times \mathbf{j} = \mathbf{k}$. But, of course, $\mathbf{i}$, $\mathbf{j}$, and $\mathbf{k}$, when drawn from the origin, point respectively in the positive x, y, and z directions (Figure 3.26).

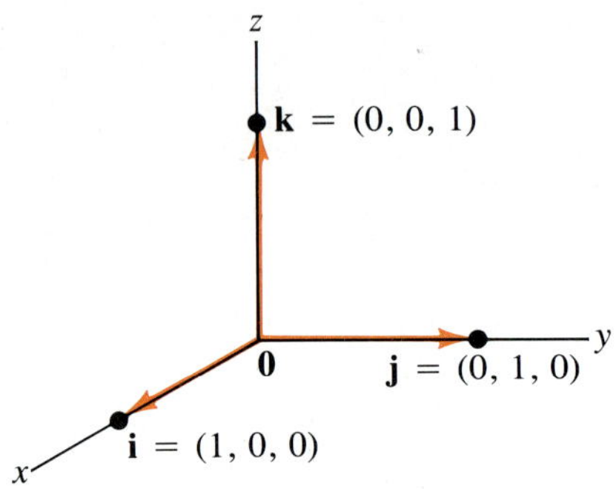

Figure 3.26

We close this section with some final applications of the cross product to geometry.

Let $\mathbf{x}_1$, $\mathbf{x}_2$, and $\mathbf{x}_3$ be three noncollinear points in $\mathbf{R}^3$. The vectors from $\mathbf{x}_1$ to $\mathbf{x}_2$ and from $\mathbf{x}_1$ to $\mathbf{x}_3$ constitute two sides of a parallelogram, and since they are adjacent sides, the other two sides are automatically determined. Then, the vertices of the parallelogram are $\mathbf{x}_1$, $\mathbf{x}_2$, $\mathbf{x}_3$, and $\mathbf{x}_2 + \mathbf{x}_3 - \mathbf{x}_1$, as can be verified.

We determine the area of this parallelogram (Figure 3.27). For the base, we can use the vector from $\mathbf{x}_1$ to $\mathbf{x}_2$: its length is $|\mathbf{x}_2 - \mathbf{x}_1|$. Now, by trigonometry, the height of the parallelogram is given by $|\mathbf{x}_3 - \mathbf{x}_1| \sin \alpha$, where α is the angle between the vectors $\mathbf{x}_2 - \mathbf{x}_1$ and

$$b = |\mathbf{x}_2 - \mathbf{x}_1|$$
$$h = |\mathbf{x}_3 - \mathbf{x}_1| \sin \alpha$$
$$A = bh$$

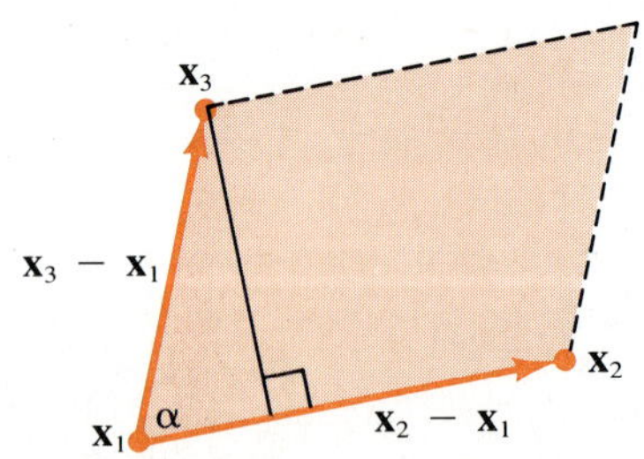

Figure 3.27

$\mathbf{x}_3 - \mathbf{x}_1$. Thus, the area is

$$|\mathbf{x}_2 - \mathbf{x}_1||\mathbf{x}_3 - \mathbf{x}_1| \sin \alpha$$

which, by Theorem 3.5, is the length of the cross product

$$|(\mathbf{x}_2 - \mathbf{x}_1) \times (\mathbf{x}_3 - \mathbf{x}_1)|$$

Example 3

Use the cross product to find the area of the parallelogram with vertices (1, 2, 3), (2, −1, −3), (0, 1, 2), and (1, −2, −4). To solve this problem, we let the first three points be $\mathbf{x}_1$, $\mathbf{x}_2$, and $\mathbf{x}_3$, respectively. Then $\mathbf{x}_2 - \mathbf{x}_1 = (1, -3, -6)$ and $\mathbf{x}_3 - \mathbf{x}_1 = (-1, -1, -1)$. Therefore

$$(\mathbf{x}_2 - \mathbf{x}_1) \times (\mathbf{x}_3 - \mathbf{x}_1) = (1, -3, -6) \times (-1, -1, -1)$$

$$= \det\begin{bmatrix} \mathbf{i} & \mathbf{j} & \mathbf{k} \\ 1 & -3 & -6 \\ -1 & -1 & -1 \end{bmatrix} = -3\mathbf{i} + 7\mathbf{j} - 4\mathbf{k}$$

The area of the parallelogram is the length of this vector,

$$|-3\mathbf{i} + 7\mathbf{j} - 4\mathbf{k}| = \sqrt{9 + 49 + 16} = \sqrt{74}$$ □

Suppose we wish to apply our cross product formula to the problem of finding the area of a parallelogram in $\mathbf{R}^2$. Because the cross product applies only in $\mathbf{R}^3$, we must find a way to "translate" a parallelogram from $\mathbf{R}^2$ into $\mathbf{R}^3$. To do so, let two sides of a parallelogram in $\mathbf{R}^2$ be determined by line segments from (a, b) to (c, d) and from (a, b) to (e, f). The three points (a, b), (c, d), and (e, f) are thus three of the vertices of the parallelogram (Figure 3.28).

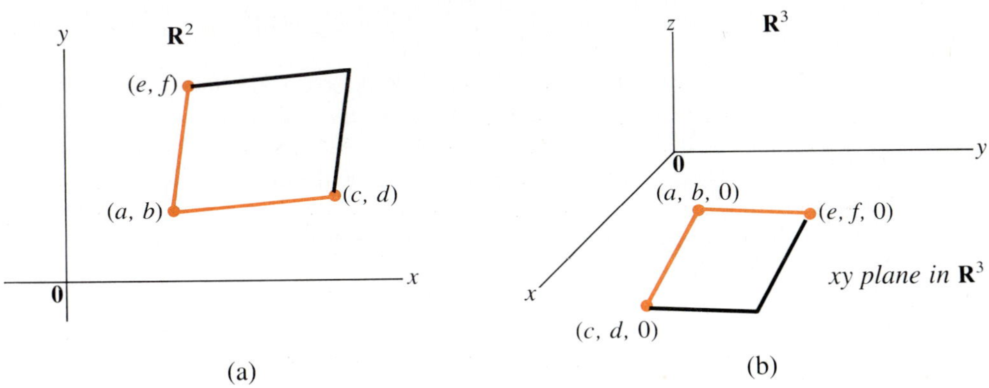

Figure 3.28

The idea here is to identify points in $\mathbf{R}^2$ with the points in $\mathbf{R}^3$ which lie in the xy plane; that is, we want to translate them into the points in $\mathbf{R}^3$ which have third coordinate 0 (Figure 3.28b). We find in the xy plane of $\mathbf{R}^3$ an exact geometric replica of $\mathbf{R}^2$. Thus, under this correspondence, we can transfer our problem of finding the area of the parallelogram in $\mathbf{R}^2$ to that of finding the area of the parallelogram in $\mathbf{R}^3$ with adjacent sides determined by the vectors from $(a, b, 0)$ to $(c, d, 0)$ and from $(a, b, 0)$ to $(e, f, 0)$. According to our preceding discussion, we must obtain the length of the cross product

$$\begin{aligned}((c, d, 0) - (a, b, 0)) \times ((e, f, 0) - (a, b, 0))\\ &= \det\begin{bmatrix} \mathbf{i} & \mathbf{j} & \mathbf{k} \\ c-a & d-b & 0 \\ e-a & f-b & 0 \end{bmatrix}\\ &= [(c-a)(f-b) - (d-b)(e-a)]\mathbf{k}\end{aligned}$$

For the length of this cross product, we get

$$\begin{aligned}|[(c-a)(f-b) - (d-b)(e-a)]\mathbf{k}|\\ &= |(c-a)(f-b) - (d-b)(e-a)||\mathbf{k}|\\ &= |(c-a)(f-b) - (d-b)(e-a)|\end{aligned}$$

since $|\mathbf{k}| = 1$.

Now, we observe that this result is simply the absolute value of a certain 2×2 determinant, namely

$$\left|\det\begin{bmatrix} c-a & d-b \\ e-a & f-b \end{bmatrix}\right|$$

Therefore, the area of the parallelogram in question could have been found directly from the given points without ever having detoured through $\mathbf{R}^3$. In particular, suppose that one vertex of a parallelogram in $\mathbf{R}^2$ is at the origin (which would be the case with *any* parallelogram upon translation, if necessary), and that two sides extend from the origin to points (p, q) and (r, s) (Figure 3.29). Then, the area of the parallelogram is simply

$$\begin{aligned} A &= \left|\det\begin{bmatrix} p & q \\ r & s \end{bmatrix}\right| \\ &= |ps - qr| \end{aligned}$$

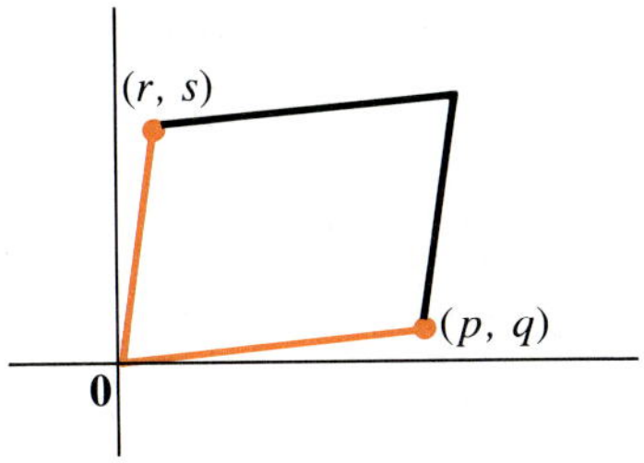

Figure 3.29

Example 4

Find the area of the parallelogram in $\mathbf{R}^2$ with vertices (1, 2), (0, 4), (3, 5), and (2, 7). We can translate this parallelogram one unit to the left and two units down, relocating the vertices at (0, 0), (−1, 2), (2, 3), and (1, 5). The area is

$$\left|\det\begin{bmatrix} -1 & 2 \\ 2 & 3 \end{bmatrix}\right| = |-3 - 4| = |-7| = 7 \qquad \square$$

A *parallelepiped* is a three-dimensional figure with six faces, all parallelograms. The six faces are in opposing pairs, which are parallel. A parallelepiped is the three-dimensional analog of a parallelogram (Figure 3.30).

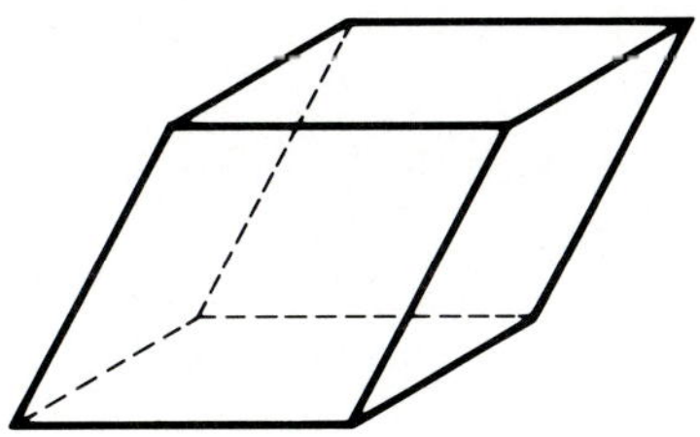

Figure 3.30

If thc *area* of a parallelogram in $\mathbf{R}^2$ is given by the absolute value of a certain 2×2 determinant, one might hope that the *volume* of a parallelepiped in $\mathbf{R}^3$ would be given by the absolute value of a certain 3×3 determinant. We shall show that this is indeed the case. First, we need a formula which might be considered a corollary to part of Theorem 3.4.

Theorem 3.6 For vectors $\mathbf{u} = (u_1, u_2\ u_3)$, $\mathbf{v} = (v_1, v_2, v_3)$, $\mathbf{w} = (w_1, w_2, w_3)$ in $\mathbf{R}^3$,

$$\mathbf{u} \cdot (\mathbf{v} \times \mathbf{w}) = \det \begin{bmatrix} u_1 & u_2 & u_3 \\ v_1 & v_2 & v_3 \\ w_1 & w_2 & w_3 \end{bmatrix}$$

Proof. As in the proof of part b of Theorem 3.4, we can obtain the desired result merely by replacing **i**, **j**, and **k** in our determinant formula for a cross product with the coordinates u_1, u_2, and u_3.

Now, let us consider a parallelepiped determined by four noncoplanar points $\mathbf{x}_1, \mathbf{x}_2, \mathbf{x}_3, \mathbf{x}_4$ so that $\mathbf{x}_2 - \mathbf{x}_1$, $\mathbf{x}_3 - \mathbf{x}_1$, and $\mathbf{x}_4 - \mathbf{x}_1$ form three edges with the common vertex $\mathbf{x}_1$. We take the viewpoint that the base of this parallelepiped is the parallelogram determined by $\mathbf{x}_1$, $\mathbf{x}_2$, $\mathbf{x}_3$. Then, by our parallelogram formula, the area of the base is given by $|(\mathbf{x}_2 - \mathbf{x}_1) \times (\mathbf{x}_3 - \mathbf{x}_1)|$. But the *vector* $(\mathbf{x}_2 - \mathbf{x}_1) \times (\mathbf{x}_3 - \mathbf{x}_1)$ is perpendicular to the base and the height of the figure will be measured in this direction (Figure 3.31). If β is the angle between $\mathbf{x}_4 - \mathbf{x}_1$ and $(\mathbf{x}_2 - \mathbf{x}_1) \times (\mathbf{x}_3 - \mathbf{x}_1)$, then the height is $|\mathbf{x}_4 - \mathbf{x}_1||\cos\beta|$. Thus we find

$$\begin{aligned} \text{volume} &= \text{(area of base) times (height)} \\ &= |(\mathbf{x}_2 - \mathbf{x}_1) \times (\mathbf{x}_3 - \mathbf{x}_1)||\mathbf{x}_4 - \mathbf{x}_1||\cos\beta| \\ &= |\{(\mathbf{x}_2 - \mathbf{x}_1) \times (\mathbf{x}_3 - \mathbf{x}_1)\} \cdot (\mathbf{x}_4 - \mathbf{x}_1)| \\ &= |(\mathbf{x}_4 - \mathbf{x}_1) \cdot \{(\mathbf{x}_2 - \mathbf{x}_1) \times (\mathbf{x}_3 - \mathbf{x}_1)\}| \end{aligned}$$

which, by Theorem 3.6, can be put in determinant form.

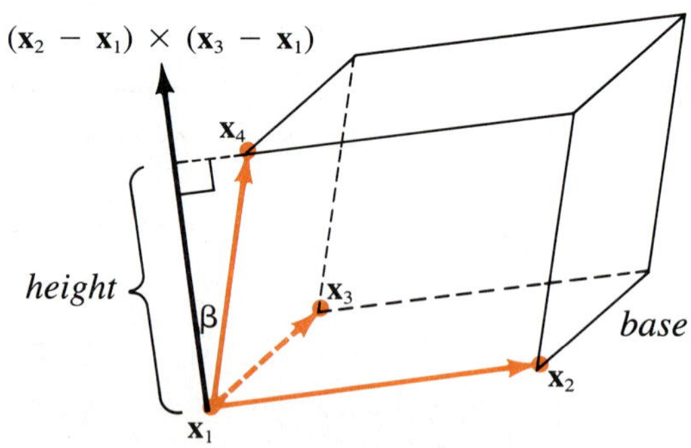

Figure 3.31

If $\mathbf{x}_1$ is at the origin and the other three determining vertices are $\mathbf{u} = (u_1, u_2, u_3)$, $\mathbf{v} = (v_1, v_2, v_3)$, and $\mathbf{w} = (w_1, w_2, w_3)$, then the volume is given by

$$V = \left| \det \begin{bmatrix} u_1 & u_2 & u_3 \\ v_1 & v_2 & v_3 \\ w_1 & w_2 & w_3 \end{bmatrix} \right|$$

which compares with our formula for the area of a parallelogram.

Example 5

Let a parallelepiped have one vertex at the origin and three adjacent edges extending from the origin to (1, 0, 0), (0, 1, 0), and (1, 1, 1). Then the volume is given by

$$V = \left| \det \begin{bmatrix} 1 & 0 & 0 \\ 0 & 1 & 0 \\ 1 & 1 & 1 \end{bmatrix} \right| = 1$$

This result is easily confirmed by elementary geometry, since the parallelepiped has the unit square for its base and its height is 1 (Figure 3.32). □

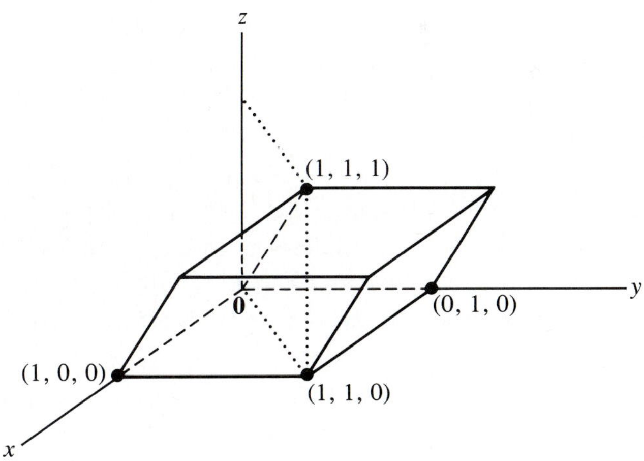

Figure 3.32

Example 6

Let four adjacent vertices of a parallelepiped be the points $\mathbf{x}_1 = (2, -1, 0)$, $\mathbf{x}_2 = (4, 0, 0)$, $\mathbf{x}_3 = (2, 0, 2)$, and $\mathbf{x}_4 = (4, 1, 5)$, so that three adjacent edges are given by $\mathbf{x}_2 - \mathbf{x}_1$, $\mathbf{x}_3 - \mathbf{x}_1$, and $\mathbf{x}_4 - \mathbf{x}_1$. We have $\mathbf{x}_2 - \mathbf{x}_1 = (2, 1, 0)$, $\mathbf{x}_3 - \mathbf{x}_1 = (0, 1, 2)$, and $\mathbf{x}_4 - \mathbf{x}_1 = (2, 2, 5)$. Therefore

$$\begin{aligned} V &= |(2, 2, 5) \cdot [(2, 1, 0) \times (0, 1, 2)]| \\ &= \left| \det \begin{bmatrix} 2 & 2 & 5 \\ 2 & 1 & 0 \\ 0 & 1 & 2 \end{bmatrix} \right| \\ &= 6 \end{aligned}$$

Note that the order in which the edges are used as rows in the determinant is immaterial since absolute value is applied to the determinant value. □

3.4 Problem Set

1. Find a coordinate equation for the plane passing through (1, 2, 1), (2, −1, 3), and (0, 4, 3).
2. Find an equation for the plane containing the line $\mathbf{x} = (4, -1, 2) + t(0, 2, 1)$ and also the point (1, −1, 2).
3. Find an equation for the plane perpendicular to the given plane and containing the given line.
   a. $2x + 3y + 3z = 0$; $\mathbf{x} = t(1, -1, 2)$
   b. $5x + 2y - 6z = 3$; $\mathbf{x} = t(1, -1, 4)$
   c. $x + y - 4z = -1$; $\mathbf{x} = (0, 2, 1) + t(3, 1, 2)$
4. For each pair of planes, find the cosine of the angle between them (by finding the cosine of the angle between their normal vectors); find an equation for the line of intersection of the two planes.
   a. $2x - 3y + 3z = 15$, $x + 2y - z = 4$
   b. $x + 4y - 2z = -1$, $4x - 3y + z = 2$
5. Show that $\mathbf{i} \times \mathbf{j} = \mathbf{k}$; $\mathbf{j} \times \mathbf{k} = \mathbf{i}$; $\mathbf{k} \times \mathbf{i} = \mathbf{j}$.
6. Show that if $\mathbf{x}_1$, $\mathbf{x}_2$, and $\mathbf{x}_3$ are three noncollinear points in $\mathbf{R}^3$ and the vectors $\mathbf{x}_2 - \mathbf{x}_1$ and $\mathbf{x}_3 - \mathbf{x}_1$ are two adjacent sides of a certain parallelogram, then the other vertex must be $\mathbf{x}_2 + \mathbf{x}_3 - \mathbf{x}_1$. (*Hint:* Consider $\mathbf{x}_1 + (\mathbf{x}_2 - \mathbf{x}_1) + (\mathbf{x}_3 - \mathbf{x}_1)$.)
7. Show that the four points in Example 3 are coplanar and that they are the vertices of a parallelogram.

8. Use Theorem 3.5 to show that if $\mathbf{u}$ and $\mathbf{v}$ are nonzero vectors but $\mathbf{u} \times \mathbf{v} = \mathbf{0}$, then $\mathbf{u}$ and $\mathbf{v}$ are parallel. (*Hint:* Show that the angle between $\mathbf{u}$ and $\mathbf{v}$ is either 0° or 180°.)

9. Find the area of the parallelogram with the given vertices.

   a. $(-3, 1), (-2, 2), (1, -2), (2, -1)$

   b. $(1, -1, 2), (2, 1, 3), (1, 0, 1), (2, 2, 2)$

   c. $(5, 3, -1), (4, 2, 3), (-1, -3, 0), (-2, -4, 2)$

   d. $(3, -1, 2), (2, 1, 2), (5, 6, 2), (4, 8, 2)$

10. Find the area of the triangle with the given vertices.

    a. $(1, 0), (2, 3), (-1, 2)$

    b. $(1, 1, -1), (2, 0, 3), (2, 1, 3)$

    c. $(5, 2, 9), (3, 5, 7), (-2, -4, 6)$

    d. $(1, 0, 0), (0, 0, 1), (3, 2, 4)$

11. In each of the following, $\mathbf{x}_1$, $\mathbf{x}_2$, $\mathbf{x}_3$, and $\mathbf{x}_4$ are four adjacent vertices of a parallelepiped so that $\mathbf{x}_2 - \mathbf{x}_1$, $\mathbf{x}_3 - \mathbf{x}_1$, and $\mathbf{x}_4 - \mathbf{x}_1$ form three adjacent edges. Find the volume of each parallelepiped.

    a. $\mathbf{x}_1 = (2, -1, 1)$, $\mathbf{x}_2 = (3, 0, 0)$, $\mathbf{x}_3 = (0, 1, 1)$, $\mathbf{x}_4 = (1, 2, 1)$

    b. $\mathbf{x}_1 = (0, 0, 0)$, $\mathbf{x}_2 = (1, 0, -1)$, $\mathbf{x}_3 = (2, 1, 1)$, $\mathbf{x}_4 = (3, 1, 5)$

    c. $\mathbf{x}_1 = (1, 2, 1)$, $\mathbf{x}_2 = (-1, 3, 2)$, $\mathbf{x}_3 = (2, 0, 2)$, $\mathbf{x}_4 = (1, 3, 2)$

12. Show that the four points $(0, -1, 0)$, $(0, 0, 2)$, $(1, 1, 1)$, and $(2, 1, -2)$ are coplanar by showing that the volume of the "parallelepiped" of which the given points are four adjacent vertices is zero.

4

Vector Spaces

4.1 Abstract Vector Spaces

In Section 1.3 we established (in Theorems 1.1 and 1.2) seven properties of matrices concerning addition and scalar multiplication. The properties given there, together with an eighth property not listed but readily established, distinguish an important class of mathematical systems called *vector spaces*.

Definition A *vector space* is a set V of objects called *vectors* which may be added together and multiplied by scalars so as to produce other vectors of V, subject to the following rules. (Here $\mathbf{u}, \mathbf{v}, \mathbf{w}$ in V and a, b in $\mathbf{R}$ are arbitrary.)

VS1 *Commutativity*:

$$\mathbf{u} + \mathbf{v} = \mathbf{v} + \mathbf{u}$$

VS2 *Associativity*:

$$(\mathbf{u} + \mathbf{v}) + \mathbf{w} = \mathbf{u} + (\mathbf{v} + \mathbf{w})$$

VS3 *Existence of a zero vector*: there is a vector $\mathbf{0}$ in V such that

$$\mathbf{u} + \mathbf{0} = \mathbf{u} \qquad \text{for each } \mathbf{u} \text{ in } V$$

VS4 *Existence of inverses*: for each $\mathbf{u}$ in V there is a vector, denoted $-\mathbf{u}$, such that

$$\mathbf{u} + (-\mathbf{u}) = \mathbf{0}$$

VS5 *Distributivity*:

$$a(\mathbf{u} + \mathbf{v}) = a\mathbf{u} + a\mathbf{v}$$

VS6 *Distributivity*:

$$(a + b)\mathbf{u} = a\mathbf{u} + b\mathbf{u}$$

VS7 *Scalar associativity*:

$$a(b\mathbf{u}) = (ab)\mathbf{u}$$

VS8 *Significance of the scalar 1*:

$$1 \cdot \mathbf{u} = \mathbf{u}$$

Note that in a vector space V, the *sum* of any two vectors in V must also be *a vector in* V, and a *scalar multiple* of any vector in V must also be *a vector in* V. When these requirements are met, V is said to be *closed under addition* and *closed under scalar multiplication*. Thus, to confirm that a system is a vector space, we must verify that these closure properties, as well as rules VS1–8, hold true for the system.

Example 1

The set M_{mn} of all $m \times n$ matrices, with the usual matrix operations, is an example of a vector space. It is closed under the operations of addition and scalar multiplication and satisfies VS1–8, as we have already mentioned. For instance,

$$\begin{bmatrix} 1 & 3 & 4 \\ 2 & 0 & -1 \end{bmatrix}$$

is a vector in the vector space M_{23}. In particular, if $m = 1$ or $n = 1$, M_{mn} reduces essentially to $\mathbf{R}^n$ or $\mathbf{R}^m$, respectively, very important special cases. □

Example 2

Let $\mathcal{F}$ be the set of all real-valued functions with domain **R** (or at least some common domain such as an interval in **R**). Recall that, for functions, sum is defined by the rule $(f + g)(x) = f(x) + g(x)$ and scalar multiple by the rule $(cf)(x) = c\,f(x)$. Thus $\mathcal{F}$ satisfies both closures. The "zero vector" of $\mathcal{F}$ is the function Θ given by $\Theta(x) = 0$ for all x in **R**. For f in $\mathcal{F}$, define $-f$ by the rule $(-f)(x) = -f(x)$. It is routine to check that all eight vector-space axioms are satisfied by $\mathcal{F}$. □

Example 3

a. Let $\mathcal{C}$ be the set of all real-valued functions continuous on some common domain in **R**.

b. Let $\mathcal{D}$ be the set of all real-valued functions differentiable on some common domain in **R**.

c. Let $\mathcal{P}_n$ be the set of all polynomials of degree $\leq n$. (*Note:* The zero polynomial is not usually said to have degree 0, or any specific *finite* degree, for that matter. This is to avoid violating the formula $\deg(p(x)q(x)) = \deg p(x) + \deg q(x)$. If we simply declare the zero polynomial to have degree $-\infty$, then most such problems are overcome, and the zero polynomial will indeed belong to $\mathcal{P}_n$, which is the membership needed if $\mathcal{P}_n$ is to satisfy VS3.)

Each of $\mathcal{C}$, $\mathcal{D}$, and $\mathcal{P}_n$ constitutes a vector space, although closures in the case of $\mathcal{C}$ and $\mathcal{D}$ depend on certain facts from calculus. Moreover, if we use the same domains for $\mathcal{D}$, $\mathcal{C}$, and $\mathcal{F}$, we have

$$\mathbf{R} = \mathcal{P}_0 \subset \mathcal{P}_1 \subset \mathcal{P}_2 \subset \cdots \subset \mathcal{D} \subset \mathcal{C} \subset \mathcal{F}$$

which illustrates a key concept—the notion of *subspace*—in vector spaces. □

Example 4

Let $\mathcal{N}$ be the set of all real-valued functions f such that, for every x in **R**, $f(x)$ is an even integer. If we use the ordinary operations as in the vector space $\mathcal{F}$, we find that $\mathcal{N}$ *fails* to be a vector space only because it is not closed under scalar multiplication. □

Several basic properties of vector spaces are outlined in the following theorem.

Theorem 4.1 If V is a vector space, then

a. There is only one zero vector in V.

b. Each vector has only one inverse.

c. For each $\mathbf{u}$ in V, $0\mathbf{u} = \mathbf{0}$.

d. For each c in $\mathbf{R}$, $c\mathbf{0} = \mathbf{0}$.

e. If $c\mathbf{u} = \mathbf{0}$, then either $c = 0$ or $\mathbf{u} = \mathbf{0}$.

f. For each $\mathbf{u}$ in V, $(-1)\mathbf{u} = -\mathbf{u}$.

Proof. We prove parts a, c, e, and f.

a. Suppose $\mathbf{0}$ and $\mathbf{0}'$ are each zeros of V. Then

$$\begin{aligned} \mathbf{0}' &= \mathbf{0}' + \mathbf{0} && \text{(since } \mathbf{0} \text{ is a zero of } V) \\ &= \mathbf{0} + \mathbf{0}' && \text{(by VS1)} \\ &= \mathbf{0} && \text{(since } \mathbf{0}' \text{ is a zero of } V) \end{aligned}$$

Thus any two zeros must be equal; that is, there can be only one zero in V.

c. We use the fact that $0 = 0 + 0$, so $0\mathbf{u} = (0 + 0)\mathbf{u} = 0\mathbf{u} + 0\mathbf{u}$ (by VS6). Now,

$$\begin{aligned} \mathbf{0} &= 0\mathbf{u} + (-0\mathbf{u}) \\ &= (0\mathbf{u} + 0\mathbf{u}) + (-0\mathbf{u}) \\ &= 0\mathbf{u} + (0\mathbf{u} + (-0\mathbf{u})) \\ &= 0\mathbf{u} + \mathbf{0} \\ &= 0\mathbf{u} \end{aligned}$$

as desired.

e. Suppose $c\mathbf{u} = \mathbf{0}$. If $c \neq 0$, we can multiply each side by the reciprocal of c:

$$\frac{1}{c}(c\mathbf{u}) = \frac{1}{c}\mathbf{0}$$

By VS7 and 8,

$$\frac{1}{c}(c\mathbf{u}) = \left(\frac{1}{c}c\right)\mathbf{u} = 1 \cdot \mathbf{u} = \mathbf{u}$$

and by part d above,

$$\frac{1}{c}\mathbf{0} = \mathbf{0}$$

That is, if $c \neq 0$, then $\mathbf{u} = \mathbf{0}$.

f. We have

$$\begin{aligned} \mathbf{0} &= 0\mathbf{u} && \text{(by part c above)} \\ &= (1 + (-1))\mathbf{u} && \text{(since } 1 + (-1) = 0\text{)} \\ &= 1 \cdot \mathbf{u} + (-1)\mathbf{u} && \text{(VS6)} \\ &= \mathbf{u} + (-1)\mathbf{u} && \text{(VS8)} \end{aligned}$$

This equation says that $(-1)\mathbf{u}$ is the inverse of $\mathbf{u}$ (there is exactly one such by part b). That is, $(-1)\mathbf{u} = -\mathbf{u}$.

Definition

A *subspace* of a vector space V is a subset of V which is itself a vector space (with addition and scalar multiplication as in V).

In Example 3 the vector spaces listed are subspaces of those vector spaces which contain them as subsets. For instance, $\mathcal{P}_3$ is a subspace of $\mathcal{C}$. However, not every sub*set* of a vector space is a sub*space*.

Example 5

Let $\mathcal{S}$ be the set of all polynomials of degree *exactly* 3. Then $\mathcal{S}$ is a subset of $\mathcal{P}_3$ but is not a vector space, because, among other reasons, it does not contain the zero polynomial and thus fails to satisfy VS3. It also fails the closure tests (Problem 8). □

There is a simple test to determine whether subsets of vector spaces are subspaces.

Theorem 4.2

A nonempty subset W of a vector space V is a subspace of V if and only if it is closed under addition and scalar multiplication; that is, for $\mathbf{u}$, $\mathbf{v}$ in W and c in $\mathbf{R}$, we have $\mathbf{u} + \mathbf{v}$ and $c\mathbf{u}$ in W.

Proof. Suppose W is closed and nonempty in V. As a subset, W inherits from V all vector-space properties except VS3 and 4. Let $\mathbf{w}$ belong to W. Then by scalar closure $0\mathbf{w}$ and $(-1)\mathbf{w}$ each belong to W. But $0\mathbf{w} = \mathbf{0}$ and $(-1)\mathbf{w} = -\mathbf{w}$ by Theorem 4.1, when parts c and f are applied to V. Thus VS3 and 4 also hold, and W is a subspace of V.

Conversely, if W is a subspace, it is closed and nonempty by definition; the proof is now complete.

It can be seen geometrically that a plane in $\mathbf{R}^3$ that passes through the origin is itself a subspace of $\mathbf{R}^3$. This is because addition and scalar multiplication, when applied to vectors in a plane through the origin, will yield vectors in the same plane (Figure 4.1). A nongeometric argument will be called for in the exercises.

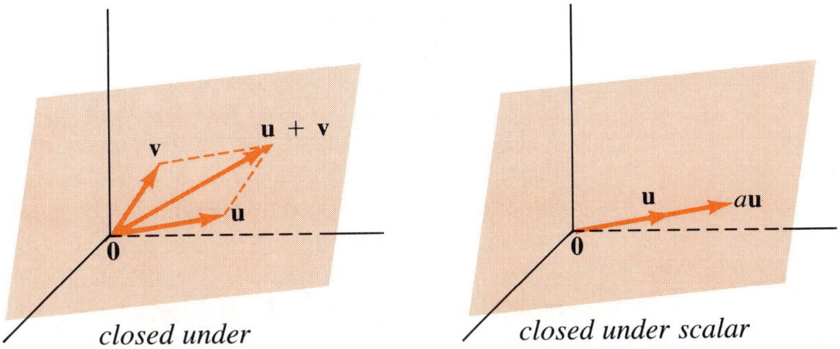

closed under addition

closed under scalar multiplication

a plane through the origin is a subspace of $\mathbf{R}^3$

Figure 4.1

Example 6

Let W be the subset of $\mathbf{R}^3$ consisting of all triples with middle coordinate 0. (This is the plane whose equation is $y = 0$; that is, it is the xz plane. Since all sums and scalar multiples of such vectors also have middle coordinate 0, W is a subspace of $\mathbf{R}^3$ (Figure 4.2). (W is clearly nonempty.) □

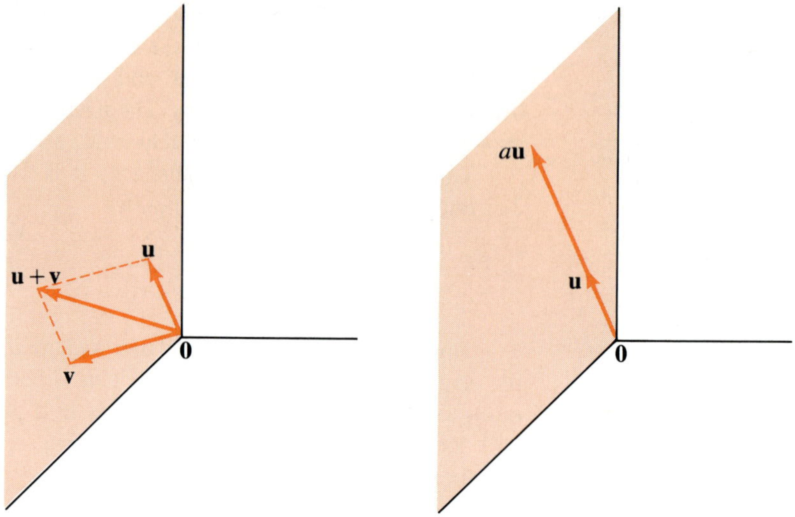

The xz plane is closed under addition and scalar multiplication.

Figure 4.2

Example 7

Let U be the subset of $\mathbf{R}^3$ consisting of all triples with middle coordinate 1 (Figure 4.3). U fails to be a subspace for any one of several reasons. First, $\mathbf{0} = (0, 0, 0)$ is not in U because its middle coordinate is not 1. Second, U is not closed under addition because, to cite a specific illustration, $(2, 1, 4)$ and $(4, 1, -3)$ belong to U, but their sum does not, since its middle coordinate is 2, not 1. Neither is U closed under scalar multiplication, as is easily verified. □

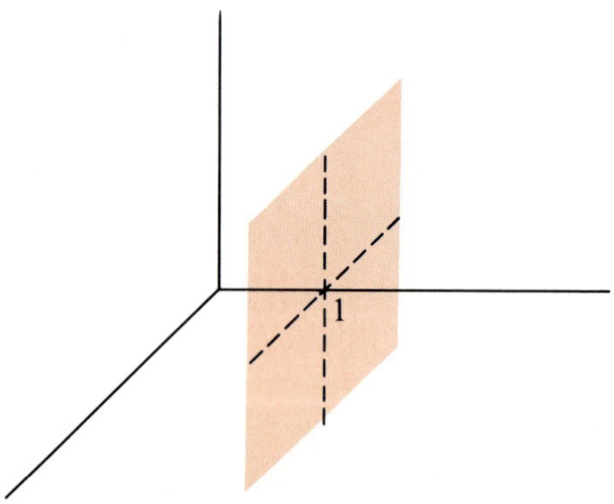

U is not a subspace,
but it is an affine subspace.

Figure 4.3

Although U in Example 7 is not a subspace, it can be obtained naturally from the subspace W of Example 6 by simply adding the vector $(0, 1, 0)$ to each vector of W. (U is therefore called the *translate of* W *by* $(0, 1, 0)$.) Subsets such as U are described in the following definition.

Definition

A subset U of a vector space V is called an *affine subspace* of V provided that, for some subspace W of V and some fixed vector $\mathbf{b}$ in V, $U = \{\mathbf{w} + \mathbf{b} | \mathbf{w} \in W\}$.

Note that an affine subspace is *not* a subspace (despite the name) *unless* $\mathbf{b}$ is in W. (Verify.)

Example 8

The line $y = mx + b$ (technically, the set $U = \{(x, mx + b) | x \in \mathbf{R}\}$ where m and b are constants) is an affine subspace of $\mathbf{R}^2$. Here W is the subspace given as the line $y = mx$ and $\mathbf{b}$ is the vector $(0, b)$. (See Figure 4.4.) □

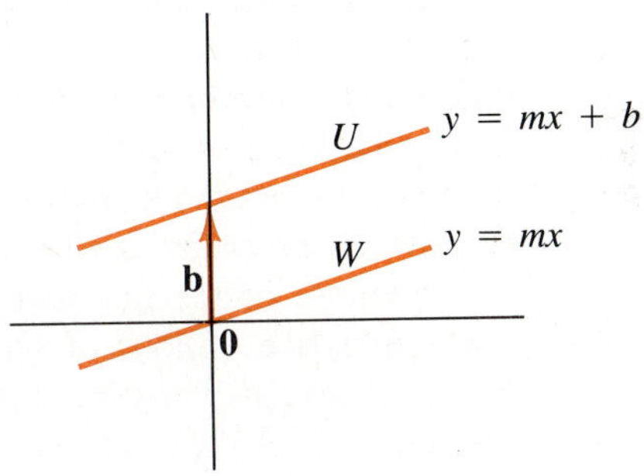

$U = W + \mathbf{b}$

Figure 4.4

Theorem 4.3

If W_1 and W_2 are subspaces of a vector space V, then so is $W_1 \cap W_2$.

Proof. Since $\mathbf{0}$ is in both W_1 and W_2, we have $\mathbf{0}$ in $W_1 \cap W_2$; thus $W_1 \cap W_2$ is nonempty. If we have $\mathbf{u}, \mathbf{v}$ in $W_1 \cap W_2$, then, in particular, we have $\mathbf{u}, \mathbf{v}$ in W_1, and since W_1 is a subspace, by its closure we get $\mathbf{u} + \mathbf{v}$ in W_1. Similar reasoning gives us $\mathbf{u} + \mathbf{v}$ in W_2. Thus $\mathbf{u} + \mathbf{v}$ is in $W_1 \cap W_2$, so $W_1 \cap W_2$ is closed under addition. That $W_1 \cap W_2$ is closed under scalar multiplication is left as an exercise.

Example 9
In $\mathbf{R}^3$, if two planes through the origin do not coincide, their intersection will be a line through the origin, and that line of intersection must, by Theorem 4.3, be a subspace of $\mathbf{R}^3$. Verification that a line through the origin constitutes a subspace of $\mathbf{R}^3$ can be done without using this theorem and is left as an exercise. □

Example 10
Let W_1 be the subset $\{(x, 0, z) | x, z \in \mathbf{R}\}$ of $\mathbf{R}^3$ (this is the subspace of Example 6). Let W_2 be the subset $\{(x, y, 2x) | x, y \in \mathbf{R}\}$ (this is the set of all triples having third coordinate twice the first). It can be shown that W_2 is a subspace of $\mathbf{R}^3$. The intersection of W_1 and W_2 is that set of all triples having middle coordinate 0 *and* third coordinate twice the first. By Theorem 4.3, this intersection set is a subspace of $\mathbf{R}^3$. □

Let W_1 and W_2 be as in Example 10. Then the *union* $W_1 \cup W_2$—the set of all triples having middle coordinate 0 *or* having third coordinate twice the first—is *not* a subspace of $\mathbf{R}^3$; sufficient evidence is that while (1, 0, 4) and (4, 3, 8) both belong to $W_1 \cup W_2$, their sum does not. In fact, a union of two subspaces cannot be a subspace *unless* one of the subspaces is a subspace of the other. (See exercises.)

Because of the fact that the union of two subspaces is normally not a subspace, unions of sets play almost no role in linear algebra. Instead, we consider what are called *sums* of subspaces. As shown in Problem 15, the sum of two subspaces is a subspace containing each of the two subspaces being added together. This is the sort of result we might have expected—but do not obtain—from the union of sets.

To close our introduction to vector spaces, we remark that a more general definition of a vector space does not preclude the use of something other than the real numbers as scalars. All that is required is that the scalars form a mathematical system closed under addition and multiplication and satisfying properties N1–9 in Section 1.1. The *rational numbers*, for instance, meet this requirement. Later in this book, we shall introduce the *complex numbers* and consider some of the consequences of using them as scalars.

4.1

Problem Set

1. a. Show that $W = \{(x, y, 2x) | x, y \in \mathbf{R}\}$ is a subspace of $\mathbf{R}^3$.
   b. Find a parametric equation for the line of intersection of the two planes in Example 10.
2. Show that $W = \{(x, y, x + y, x - y) | x, y \in \mathbf{R}\}$ is a subspace of $\mathbf{R}^4$.
3. Show that $W = \{(x, y, x + 2) | x, y \in \mathbf{R}\}$ is not a subspace of $\mathbf{R}^3$. Is it an affine subspace?
4. Does the set of all vectors of the given form constitute a subspace of $\mathbf{R}^3$? If not, does it constitute an affine subspace?
   a. $(x, y, x + y)$ b. $(z + 2y, y, z)$
   c. $(z + 2, y, z)$ d. (x, y, x^2)
5. Follow the instructions of Problem 4, but for $\mathbf{R}^2$.
   a. $(x, 2x)$ b. $(x, |x|)$ c. $(x, \sin x)$
6. a. Show that $W = \{f | f \text{ in } \mathscr{C}, f(2) = 0\}$ is a subspace of $\mathscr{C}$, the vector space defined in Example 3.
   b. Show that $U = \{f | f \text{ in } \mathscr{D}, f'(x) + f(x) = 0 \text{ for all } x\}$ is a subspace of the vector space $\mathscr{D}$ of Example 3.
7. Verify that $\mathscr{N}$, in Example 4, is not closed under scalar multiplication.
8. Show that $\mathscr{S}$, in Example 5, fails the closure tests.
9. Show that, in an arbitrary vector space V, $\{\mathbf{0}\}$ and V itself are each subspaces of V.
10. Recall that M_{33} is the vector space of all 3×3 matrices with real entries. The following describe certain subsets of M_{33}. Determine whether each constitutes a subspace of M_{33}. If it does not, does it constitute an affine subspace?
    a. all symmetric matrices
    b. all skew-symmetric matrices (where $A^t = -A$)
    c. all singular matrices
    d. all nonsingular matrices
    e. all upper-triangular matrices
    f. all matrices whose main-diagonal entries sum to 0 (where $\operatorname{tr} A = 0$)

g. all matrices whose main-diagonal entries sum to 1 (where $\text{tr } A = 1$)

h. all matrices having all entries even integers

11. Assuming domain equal to **R**, show that $\mathscr{C}$, $\mathscr{D}$, and $\mathscr{P}_n$ as defined in Example 3 are each subspaces of $\mathscr{F}$, as defined in Example 2.

12. a. Show that a plane through the origin (given by $\mathbf{x} \cdot \mathbf{n} = 0$) is a subspace of $\mathbf{R}^3$.

    b. Show that a line through the origin (given by $\mathbf{x} = t\mathbf{u}$) is a subspace of $\mathbf{R}^3$.

13. Establish this criterion for subspaces: A nonempty subset W of a vector space V is a subspace of V if and only if, for all scalars a and b and all vectors $\mathbf{u}$ and $\mathbf{v}$ in W, $a\mathbf{u} + b\mathbf{v}$ is in W.

14. Prove that, for subspaces W_1 and W_2 of a vector space V, $W_1 \cup W_2$ is a subspace of V if and only if either $W_1 \subset W_2$ or $W_2 \subset W_1$.

15. Let each of W_1 and W_2 be a subspace of a vector space V. Let $W_1 + W_2$ be the set of all possible sums of the form $\mathbf{w}_1 + \mathbf{w}_2$, where $\mathbf{w}_1 \in W_1$ and $\mathbf{w}_2 \in W_2$. Show that $W_1 + W_2$ is a subspace of V which contains each of W_1 and W_2.

16. Let V be the set of real numbers with the usual operation of addition. Use the *rational* numbers as scalars and let $a\mathbf{v}$ be the usual product of a rational number a and a real number $\mathbf{v}$. Does this give a vector space? Justify.

17. Repeat Problem 16, but reverse the roles of the reals and the rationals: let the rationals be the "vectors" and the reals be the "scalars."

18. For a subspace W of a vector space V, prove that the affine subspace $U = \{\mathbf{w} + \mathbf{b} | \mathbf{w} \in W\}$ is a subspace of V if and only if $\mathbf{b}$ is in W, in which case U is the same subspace as W.

19. Consider the set of all infinite, real-valued sequences $(a_1, a_2, \ldots)$. Define $(a_1, a_2, \ldots) + (b_1, b_2, \ldots) = (a_1 + b_1, a_2 + b_2, \ldots)$ and $c(a_1, a_2, \ldots) = (ca_1, ca_2, \ldots)$. We may call this system $\mathbf{R}^\infty$ (although that is not standard). Show that $\mathbf{R}^\infty$ is a vector space.

20. Let $\mathbf{R}^\infty$ be as in Problem 19.

    a. Let W_1 be the subset of $\mathbf{R}^\infty$ consisting of all sequences $(a_1, a_2, \ldots)$ such that the sum of the infinite series $\Sigma_{k=1}^{\infty} a_k$ is finite. Show that W_1 is a subspace of $\mathbf{R}^\infty$.

    b. Let W_2 be the subset of $\mathbf{R}^\infty$ consisting of all sequences $(a_1, a_2, \ldots)$ such that, for some N, $a_n = 0$ when $n > N$ (all values are 0 from some term on). Show that W_2 is a subspace of W_1.

c. Let W_3 be the subset of $\mathbf{R}^\infty$ consisting of all sequences $(a_1, a_2, \ldots)$ such that the sum of the infinite series $\Sigma_{k=1}^{\infty} a_k^2$ is finite. Show that W_3 is a subspace of $\mathbf{R}^\infty$. (*Hint:* The hard part is closure under addition. For any $n \geq 1$, we can write

$$\begin{aligned}\sum_{k=1}^{n} (a_k + b_k)^2 &= \sum_{k=1}^{n} a_k^2 + 2 \sum_{k=1}^{n} a_k b_k + \sum_{k=1}^{n} b_k^2 \\ &\leq \sum_{k=1}^{n} a_k^2 + 2\left|\sum_{k=1}^{n} a_k b_k\right| + \sum_{k=1}^{n} b_k^2 \\ &\leq \sum_{k=1}^{n} a_k^2 + 2\left(\sum_{k=1}^{n} a_k^2\right)^{1/2}\left(\sum_{k=1}^{n} b_k^2\right)^{1/2} + \sum_{k=1}^{n} b_k^2\end{aligned}$$

Justify all these steps and let $n \to \infty$.)

d. Show that W_1 is a subset of W_3 (and hence a subspace). (*Hint:* This is a calculus problem and requires a knowledge of tests for convergence of infinite series; you must apply a "comparison test" and the fact that in a convergent series the terms must tend toward 0.)

21. Finish the proofs of Theorems 4.1 and 4.3.

22. Let V be the set of all *positive* real numbers; these will be the "vectors." Let "sum" mean the usual *product* of real numbers. Let the "scalars" be the system of *all* real numbers and let the "scalar product" of a real number c and a positive real number v be v^c. Show that this gives a vector space. (*Hint:* After checking for closures, one must check VS1–8 after re-interpreting them correctly. For instance, VS5 becomes $(uv)^a = u^a v^a$.)

4.2 Basis of a Vector Space

The linear structure of a vector space makes it possible to focus on only a relatively few of its elements from which all characteristics of the entire vector space can be inferred. We shall now proceed toward the description of such important subsets.

Definition

A *linear combination* of vectors $\mathbf{u}_1, \mathbf{u}_2, \ldots, \mathbf{u}_r$ in a vector space is a vector that can be expressed in the form

$$c_1\mathbf{u}_1 + c_2\mathbf{u}_2 + \cdots + c_r\mathbf{u}_r$$

where the c's are scalars.

Example 1

Let $\mathbf{u}_1 = (2, 1, 3)$, $\mathbf{u}_2 = (1, 1, -2)$, and $\mathbf{u}_3 = (0, 3, 1)$. Then $2\mathbf{u}_1 + 3\mathbf{u}_2 - 5\mathbf{u}_3 = (7, -10, -5)$, so we have expressed $(7, -10, -5)$ as a linear combination of $\mathbf{u}_1, \mathbf{u}_2, \mathbf{u}_3$. □

Definition

A set of vectors $A = \{\mathbf{u}_1, \mathbf{u}_2, \ldots, \mathbf{u}_r\}$ is *linearly independent* provided that the only linear combination of $\mathbf{u}_1, \mathbf{u}_2, \ldots, \mathbf{u}_r$ giving the zero vector $\mathbf{0}$ is the trivial one $0\mathbf{u}_1 + 0\mathbf{u}_2 + \cdots + 0\mathbf{u}_r$. In other words, A is linearly independent if the only solution to the equation

$$c_1\mathbf{u}_1 + c_2\mathbf{u}_2 + \cdots + c_r\mathbf{u}_r = \mathbf{0}$$

is

$$c_1 = c_2 = \cdots = c_r = 0$$

If a set A is not linearly independent, then it is *linearly dependent*, which means that there are nontrivial solutions to the equation in question, as well as the trivial one.

Example 2

Let $A = \{\mathbf{u}_1, \mathbf{u}_2, \mathbf{u}_3\}$ where $\mathbf{u}_1, \mathbf{u}_2, \mathbf{u}_3$ are the same as in Example 1. We determine whether or not A is linearly independent by solving for c_1, c_2, c_3 in:

$$c_1\mathbf{u}_1 + c_2\mathbf{u}_2 + c_3\mathbf{u}_3 = \mathbf{0}$$

or

$$(2c_1, c_1, 3c_1) + (c_2, c_2, -2c_2) + (0, 3c_3, c_3) = (0, 0, 0)$$

which gives the system

$$\begin{aligned} 2c_1 + c_2 \phantom{{}+ 3c_3} &= 0 \\ c_1 + c_2 + 3c_3 &= 0 \\ 3c_1 - 2c_2 + c_3 &= 0 \end{aligned}$$

Solving this system in the usual way (that is, reducing to echelon form) gives the single solution $c_1 = c_2 = c_3 = 0$. We conclude that, since the only solution to $c_1\mathbf{u}_1 + c_2\mathbf{u}_2 + c_3\mathbf{u}_3 = \mathbf{0}$ is the trivial one, the set A is linearly independent. □

Example 3

Let $\mathbf{u}_1 = (2, 1, 3)$, $\mathbf{u}_2 = (1, 1, -2)$, and $\mathbf{u}_3 = (8, 5, 5)$. To test for linear independence, we proceed as in Example 2 and obtain the homogeneous system

$$\begin{aligned} 2c_1 + c_2 + 8c_3 &= 0 \\ c_1 + c_2 + 5c_3 &= 0 \\ 3c_1 - 2c_2 + 5c_3 &= 0 \end{aligned}$$

Now, the determinant of the coefficient matrix is 0; thus there must be a nontrivial solution which guarantees linear dependence. If we wish to produce a specific nontrivial linear combination of $\mathbf{u}_1$, $\mathbf{u}_2$, $\mathbf{u}_3$ equal to $\mathbf{0}$, we can actually solve the system and find the solutions given by $(-3t, -2t, t)$, t arbitrary. For $t = 1$, for instance, we obtain $c_1 = -3$, $c_2 = -2$, $c_3 = 1$. Hence $-3\mathbf{u}_1 - 2\mathbf{u}_2 + \mathbf{u}_3$ is a specific nontrivial linear combination of $\mathbf{u}_1$, $\mathbf{u}_2$, $\mathbf{u}_3$ equal to $\mathbf{0}$, which may be verified. □

Our next theorem provides an alternative view of the nature of linear dependence.

☆ **Theorem 4.4** The set $A = \{\mathbf{u}_1, \mathbf{u}_2, \ldots, \mathbf{u}_r\}$ is linearly dependent if and only if at least one of the vectors in A can be expressed as a linear combination of the others.

Proof. Suppose A is linearly dependent. Then there are coefficients $c_1, c_2, \ldots, c_r$, not all of which are zero, such that $c_1\mathbf{u}_1 + c_2\mathbf{u}_2 + \cdots + c_r\mathbf{u}_r = \mathbf{0}$. Without loss of generality we may assume that $c_1 \neq 0$. Using properties VS1–8 we may solve the equation for $\mathbf{u}_1$:

$$\mathbf{u}_1 = -\frac{c_2}{c_1}\mathbf{u}_2 - \frac{c_3}{c_1}\mathbf{u}_3 - \cdots - \frac{c_r}{c_1}\mathbf{u}_r$$

Thus $\mathbf{u}_1$ is a linear combination of the remaining vectors.

Conversely, if one vector, say $\mathbf{u}_1$, is a linear combination of the others, we have $\mathbf{u}_1 = d_2\mathbf{u}_2 + d_3\mathbf{u}_3 + \cdots + d_r\mathbf{u}_r$ for some coefficients $d_2, d_3, \ldots, d_r$. We may rewrite this equation as $\mathbf{u}_1 - d_2\mathbf{u}_2 - d_3\mathbf{u}_3 - \cdots - d_r\mathbf{u}_r = \mathbf{0}$, which expresses $\mathbf{0}$ as a nontrivial linear combination of $\mathbf{u}_1, \mathbf{u}_2, \ldots, \mathbf{u}_r$ (nontrivial because the coefficient of $\mathbf{u}_1$ is 1, not 0.) Hence A is linearly dependent.

If zero vector in set, then dependent

Because of this theorem, linear dependence is easily visualized in the case of either two or three vectors. For two vectors, linear dependence occurs precisely when one of the vectors is a scalar multiple of the other, that is, when the vectors lie along the same line (Figure 4.5).

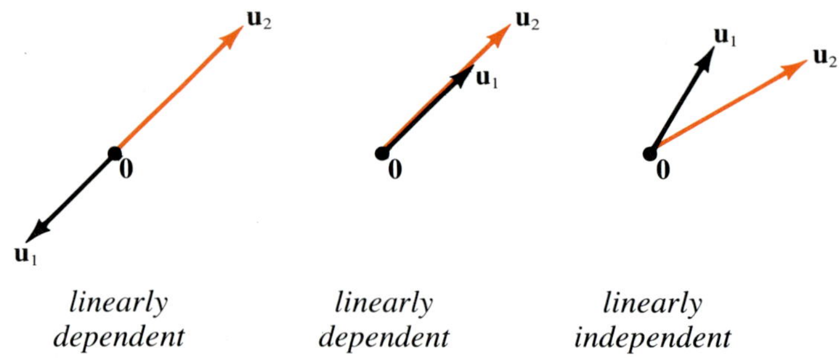

Figure 4.5

For three vectors, linear dependence occurs when one vector is in the plane of the other two, that is, when all three vectors are coplanar (Figure 4.6).

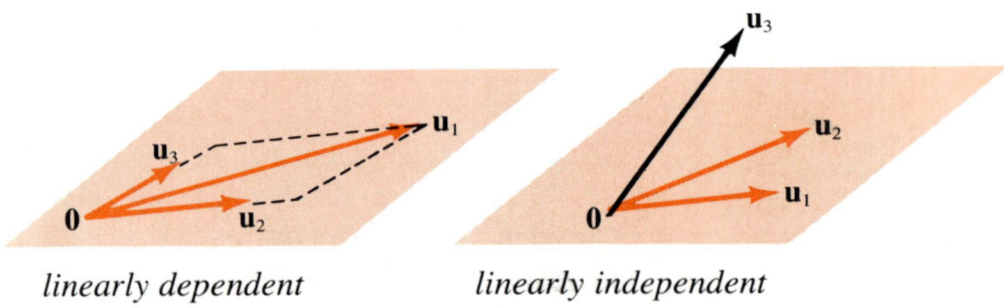

Figure 4.6

Accordingly, having more than two vectors in a linearly independent set in $\mathbf{R}^2$ would seem impossible, as would having more than three vectors in an independent set $\mathbf{R}^3$, and so on. We shall confirm this formally in Section 4.4.

☆ **Theorem 4.5** If a set $A = \{\mathbf{u}_1, \mathbf{u}_2, \ldots, \mathbf{u}_r\}$ of vectors in a vector space V is linearly dependent, then so is any finite superset of A (that is, any finite subset of V containing A). Equivalently, if a set A of vectors in V is linearly independent, then so is any subset of A.

Proof. Suppose A is linearly dependent and let $B = \{\mathbf{u}_1, \mathbf{u}_2, \ldots, \mathbf{u}_r, \mathbf{u}_{r+1}, \ldots, \mathbf{u}_s\}$. There is a nontrivial combination of the vectors of A equal to $\mathbf{0}$:

$$c_1\mathbf{u}_1 + c_2\mathbf{u}_2 + \cdots + c_r\mathbf{u}_r = \mathbf{0}$$

where at least one scalar coefficient is not 0. This equation can be expanded to

$$c_1\mathbf{u}_1 + c_2\mathbf{u}_2 + \cdots + c_r\mathbf{u}_r + 0\mathbf{u}_{r+1} + \cdots + 0\mathbf{u}_s = \mathbf{0}$$

expressing $\mathbf{0}$ as a nontrivial linear combination of the vectors of B; hence B is linearly dependent.

Example 4

The set $A = \{(1, 2, 1), (0, -1, 4), (-1, -3, 3)\}$ is linearly dependent because $(1, 2, 1) - (0, -1, 4) + (-1, -3, 3) = (0, 0, 0)$. From this fact we may deduce, by Theorem 4.5, that the set $\{(4, 0, 2), (1, 2, 1), (-1, -3, 3), (2, 1, 1), (0, -1, 4)\}$, a superset of A, must also be linearly dependent. □

Definition

Let A be a finite, nonempty subset of a vector space V. The *span* of A, denoted Sp A, is the set of all linear combinations of the vectors in A. That is, if $A = \{\mathbf{u}_1, \mathbf{u}_2, \ldots, \mathbf{u}_r\}$, then

$$\text{Sp } A = \{c_1\mathbf{u}_1 + c_2\mathbf{u}_2 + \cdots + c_r\mathbf{u}_r | c_1, c_2, \ldots, c_r \in \mathbf{R}\}$$

A is said to *span* (or *generate*) Sp A.

Example 5

Let $A = \{(1, 0, 0), (0, 0, 1)\}$. Then

$$\begin{aligned} \text{Sp } A &= \{c_1(1, 0, 0) + c_2(0, 0, 1) | c_1, c_2 \in \mathbf{R}\} \\ &= \{(c_1, 0, c_2) | c_1, c_2 \in \mathbf{R}\} \end{aligned}$$

which is just the set of all elements of $\mathbf{R}^3$ with middle coordinate 0. □

Recall that we argued in Example 6 of Section 4.1 that the set from Example 5 above is a subspace of $\mathbf{R}^3$. This fact is also a consequence of the next theorem.

Theorem 4.6

The span of a nonempty subset $A = \{\mathbf{u}_1, \mathbf{u}_2, \ldots, \mathbf{u}_r\}$ of a vector space V is a subspace of V (and not merely a subset of V).

Proof. First, Sp A is nonempty, since it contains A. Second, the sum of two vectors $a_1\mathbf{u}_1 + a_2\mathbf{u}_2 + \cdots + a_r\mathbf{u}_r$ and $b_1\mathbf{u}_1 + b_2\mathbf{u}_2 + \cdots + b_r\mathbf{u}_r$ in Sp A is $(a_1 + b_1)\mathbf{u}_1 + (a_2 + b_2)\mathbf{u}_2 + \cdots + (a_r + b_r)\mathbf{u}_r$ which is also in Sp A, so Sp A is closed under addition. Finally, for $a_1\mathbf{u}_1 + a_2\mathbf{u}_2 + \cdots + a_r\mathbf{u}_r$ in Sp A and any scalar c, we have $c(a_1\mathbf{u}_1 + a_2\mathbf{u}_2 + \cdots + a_r\mathbf{u}_r) = (ca_1)\mathbf{u}_1 + (ca_2)\mathbf{u}_2 + \cdots + (ca_r)\mathbf{u}_r$ which is also in Sp A, so Sp A is closed under scalar multiplication. Hence Sp A is a subspace of V.

If one begins with some finite set of vectors in a vector space V and freely applies the vector space operations, the resulting collection of vectors will be the span of the original set.

Each vector space has a "smallest" collection of vectors in terms of which all other vectors in the space can be described. This core subset, called a *basis* of the space, supplies all the necessary information about the vector space as a whole. Its characterization combines the notions of linear independence and spanning.

Definition A *basis* of a vector space V is a subset B of V such that

a. B is linearly independent, and

b. B spans V.

Example 6

Once again, consider the subspace W of $\mathbf{R}^3$ consisting of all vectors with middle coordinate 0. We want to show that the set $A = \{(1, 0, 0), (0, 0, 1)\}$ is a basis for W. We have already seen in Example 5 that Sp $A = W$; it remains only to show that A is linearly independent. Now, $c_1(1, 0, 0) + c_2(0, 0, 1) = (0, 0, 0)$ simplifies to $(c_1, 0, c_2) = (0, 0, 0)$ which has the single, trivial solution $c_1 = c_2 = 0$. Thus by definition A is linearly independent. Hence A is a basis for W. □

Example 7

We show here that a second basis for W, the subspace of Example 6, is the set $B = \{(1, 0, 2), (3, 0, -1)\}$. To show $W =$ Sp B, we prove that Sp $B \subset W$ and that $W \subset$ Sp B. For the former, let $\mathbf{u} \in$ Sp B. Then we can write

$$\mathbf{u} = c_1(1, 0, 2) + c_2(3, 0, -1) = (c_1 + 3c_2, 0, 2c_1 - c_2)$$

for certain scalars c_1 and c_2. Since the middle coordinate of $\mathbf{u}$ is 0, $\mathbf{u}$ is in W, so $\text{Sp } B \subset W$. On the other hand, let $\mathbf{w} = (a, 0, b)$, $a, b \in \mathbf{R}$, so that $\mathbf{w} \in W$. To show that $\mathbf{w} \in \text{Sp } B$, we must establish the existence of scalars c_1 and c_2 such that $c_1(1, 0, 2) + c_2(3, 0, -1) = (a, 0, b)$. We solve $c_1 + 3c_2 = a$ with $2c_1 - c_2 = b$ and find $c_1 = (a + 3b)/7$, $c_2 = (2a - b)/7$. Thus

$$\frac{a + 3b}{7}(1, 0, 2) + \frac{2a - b}{7}(3, 0, -1) = (a, 0, b) = \mathbf{w}$$

so $\mathbf{w} \in \text{Sp } B$. Since each vector $\mathbf{w}$ of W is in $\text{Sp } B$, we have $W \subset \text{Sp } B$. Hence $W = \text{Sp } B$. That B is also linearly independent is routinely confirmed, so B is a basis for W. □

Example 8

We find a basis for the subspace W of $\mathbf{R}^3$ consisting of those vectors whose second coordinate is three times the third. We may write

$$\begin{aligned} W &= \{(a, 3b, b) | a, b \in \mathbf{R}\} \\ &= \{a(1, 0, 0) + b(0, 3, 1) | a, b \in \mathbf{R}\} \\ &= \text{Sp}\{(1, 0, 0), (0, 3, 1)\} \end{aligned}$$

Since $\{(1, 0, 0), (0, 3, 1)\}$ is linearly independent and spans W, it is a basis for W. □

It is entirely possible that for certain vector spaces no *finite* set of vectors will be large enough to span the vector space. Such vector spaces are not a major concern of this text, but we shall consider, at several later points, some examples of this phenomenon.

Two alternative, but equivalent, characterizations of a basis hinge only on a strengthened spanning condition or a strengthened independence condition. We define these conditions in turn and show that each characterizes a basis.

Definition A set of vectors $B = \{\mathbf{u}_1, \mathbf{u}_2, \ldots, \mathbf{u}_r\}$ is said to *span V minimally* provided that B spans V, but that no proper subset of B spans V.

Example 9

The set $\{(1, 0, 0), (0, 0, 1), (1, 0, 2)\}$ spans the space W of vectors in $\mathbf{R}^3$ with middle coordinate 0, but not minimally, because the proper subset $A = \{(1, 0, 0), (0, 0, 1)\}$ also spans W (see Example 5). However, no proper subset of A spans W, so A spans W minimally. □

☆ Theorem 4.7 A set of vectors $B = \{\mathbf{u}_1, \mathbf{u}_2, \ldots, \mathbf{u}_r\}$ is a basis for the vector space V if and only if B spans V minimally.

Proof. First suppose that B is a basis for V; then B spans V. We must show that the spanning is minimal. If it were not, then a proper subset, which we may say is $\{\mathbf{u}_1, \mathbf{u}_2, \ldots, \mathbf{u}_s\}$, with $s < r$, would span V. Now since $\mathbf{u}_r \in V$, we must have $\mathbf{u}_r = c_1\mathbf{u}_1 + c_2\mathbf{u}_2 + \cdots + c_s\mathbf{u}_s$, for some choice of scalars $c_1, c_2, \ldots, c_s$. This equation may be written as $c_1\mathbf{u}_1 + c_2\mathbf{u}_2 + \cdots + c_s\mathbf{u}_s - \mathbf{u}_r = \mathbf{0}$, which expresses $\mathbf{0}$ as a nontrivial combination of the vectors of B (the coefficient of $\mathbf{u}_r$ is -1), contradicting the linear independence of the basis B. Thus B must span minimally.

Now suppose B spans V minimally. We must show that B is a basis for V, so we must prove B is linearly independent. Suppose the equation $c_1\mathbf{u}_1 + c_2\mathbf{u}_2 + \cdots + c_r\mathbf{u}_r = \mathbf{0}$ has a nontrivial solution with, say, $c_1 \neq 0$. Then $\mathbf{u}_1$ can be expressed as a linear combination of the remaining vectors:

$$\mathbf{u}_1 = -\frac{c_2}{c_1}\mathbf{u}_2 - \frac{c_3}{c_1}\mathbf{u}_3 - \cdots - \frac{c_r}{c_1}\mathbf{u}_r$$

Now let $\mathbf{v}$ be any vector in V. Since B spans V, we can write $\mathbf{v} = a_1\mathbf{u}_1 + a_2\mathbf{u}_2 + \cdots + a_r\mathbf{u}_r$. Substituting for $\mathbf{u}_1$ we have

$$\begin{aligned}\mathbf{v} &= a_1\left(-\frac{c_2}{c_1}\mathbf{u}_2 - \frac{c_3}{c_1}\mathbf{u}_3 - \cdots - \frac{c_r}{c_1}\mathbf{u}_r\right) + a_2\mathbf{u}_2 + \cdots + a_r\mathbf{u}_r \\ &= \left(a_2 - \frac{a_1c_2}{c_1}\right)\mathbf{u}_2 + \left(a_3 - \frac{a_1c_3}{c_1}\right)\mathbf{u}_3 + \cdots + \left(a_r - \frac{a_1c_r}{c_1}\right)\mathbf{u}_r\end{aligned}$$

Since $\mathbf{v}$ is thus expressed as a linear combination of $\mathbf{u}_2, \mathbf{u}_3, \ldots, \mathbf{u}_r$, we conclude that $V \subset \mathrm{Sp}\{\mathbf{u}_2, \mathbf{u}_3, \ldots, \mathbf{u}_r\}$. Since V is closed under addition and scalar multiplication, $\mathrm{Sp}\{\mathbf{u}_2, \mathbf{u}_3, \ldots, \mathbf{u}_r\} \subset V$. Thus the proper subset $\{\mathbf{u}_2, \mathbf{u}_3, \ldots, \mathbf{u}_r\}$ of B spans V, contradicting our assumption that B spans V minimally. Hence B is linearly independent and is a basis for V.

The second part of the proof contains a result worth stating independently; it is recorded below.

Corollary 4.7 Let $B = \{\mathbf{u}_1, \mathbf{u}_2, \ldots, \mathbf{u}_r\}$. If, for any k between 1 and r, u_k is a linear combination of $B' = \{u_1, u_2, \ldots, u_{k-1}, u_{k+1}, \ldots, u_r\}$, then

$$\mathrm{Sp}\, B' = \mathrm{Sp}\, B$$

That is, if one vector in a set of vectors is a linear combination of the others, then the span of the set is unchanged after the removal of that vector.

The strengthened independence condition we spoke of follows.

Definition A set B of vectors is said to be *maximally independent in the vector space* V provided that B is a linearly independent set in V but B is not a proper subset of any linearly independent subset of V.

Example 10

We show that the set $B = \{(1, 3), (2, -1)\}$ is maximally independent in $\mathbf{R}^2$. First, the set is linearly independent, since neither member is a multiple of the other. Let C be any proper superset of B. Then C contains B together with at least one other vector, say (a, b). We test the set $C' = \{(1, 3), (2, -1), (a, b)\}$ for linear independence by solving $c_1(1, 3) + c_2(2, -1) + c_3(a, b) = (0, 0)$. This equation is equivalent to the homogeneous system

$$\begin{aligned} c_1 + 2c_2 + ac_3 &= 0 \\ 3c_1 - c_2 + bc_3 &= 0 \end{aligned}$$

Since the system has fewer equations than unknowns, it must have nontrivial solutions (Theorem 1.6). Thus C' is linearly dependent, and, by Theorem 4.5, so is C. Now B is maximally independent by definition. □

☆ Theorem 4.8 A set of vectors $B = \{\mathbf{u}_1, \mathbf{u}_2, \ldots, \mathbf{u}_r\}$ is a basis for the vector space V if and only if B is maximally independent in V.

Proof. Assume that B is a basis for V. Then B is linearly independent and spans V. Let $\mathbf{u}_0$ be in V but not in B. Then, since B spans V, we have $\mathbf{u}_0 = c_1\mathbf{u}_1 + c_2\mathbf{u}_2 + \cdots + c_r\mathbf{u}_r$ for certain scalars $c_1, c_2, \ldots, c_r$. This means, according to Theorem 4.4, that the proper superset $\{\mathbf{u}_0, \mathbf{u}_1, \mathbf{u}_2, \ldots, \mathbf{u}_r\}$ of B is linearly dependent; thus B is maximally independent in V.

Conversely, assume B is maximally independent in V. To show that B must also span V, let $\mathbf{v}$ be any element of V not in B. Since the proper superset $\{\mathbf{v}, \mathbf{u}_1, \mathbf{u}_2, \ldots, \mathbf{u}_r\}$ of B must be linearly dependent, we have $c_0\mathbf{v} + c_1\mathbf{u}_1 + c_2\mathbf{u}_2 + \cdots + c_r\mathbf{u}_r = \mathbf{0}$ where at least one of the scalar coefficients is not zero. In fact, c_0 is not zero, for, if it were the case that $c_0 = 0$, we would have $c_1\mathbf{u}_1 + c_2\mathbf{u}_2 + \cdots + c_r\mathbf{u}_r = \mathbf{0}$ with at least one scalar coefficient still not zero, violating the linear independence of B. Since $c_0 \neq 0$, we obtain

$$\mathbf{v} = -\frac{c_1}{c_0}\mathbf{u}_1 - \frac{c_2}{c_0}\mathbf{u}_2 - \cdots - \frac{c_r}{c_0}\mathbf{u}_r$$

This shows that $\mathbf{v}$ is in Sp B. Moreover, each $\mathbf{u}_k$ is in Sp B (since we can write $\mathbf{u}_k = 0\mathbf{u}_1 + \cdots + 1\mathbf{u}_k + \cdots + 0\mathbf{u}_r$), so that all of V is in Sp B. By closure of V, Sp B is contained in V. Hence Sp $B = V$. This, together with the automatic linear independence of B, makes B a basis for V.

We have a final alternative characterization of basis which will be especially useful in the future.

☆ Theorem 4.9 A set of vectors $B = \{\mathbf{u}_1, \mathbf{u}_2, \ldots, \mathbf{u}_r\}$ is a basis for a vector space V if and only if each vector in V is expressible in a unique way as a linear combination of the vectors in B (meaning that for each $\mathbf{v}$ in V, one and only one set of scalar coefficients in combination with the vectors in B will equal $\mathbf{v}$).

Proof. Let B be a basis for V; let $\mathbf{v}$ be in V. Since $V =$ Sp B, we have at least one way of writing $\mathbf{v} = a_1\mathbf{u}_1 + a_2\mathbf{u}_2 + \cdots + a_r\mathbf{u}_r$, with certain scalars $a_1, a_2, \ldots, a_r$. If there were another way, say $\mathbf{v} = b_1\mathbf{u}_1 + b_2\mathbf{u}_2 + \cdots + b_r\mathbf{u}_r$, we could subtract one expression for $\mathbf{v}$ from the other and get

$$\mathbf{0} = \mathbf{v} - \mathbf{v} = (a_1 - b_1)\mathbf{u}_1 + (a_2 - b_2)\mathbf{u}_2 + \cdots + (a_r - b_r)\mathbf{u}_r$$

from which the linear independence of B would give us

$$a_1 - b_1 = a_2 - b_2 = \cdots = a_r - b_r = 0$$

That is, $a_1 = b_1, a_2 = b_2, \ldots, a_r = b_r$; the two representations of $\mathbf{v}$ as a combination of vectors in B are identical.

Conversely, operating under the assumption that each vector in V has a unique representation as a linear combination of the vectors in B, we clearly have $V =$ Sp B. Furthermore, the fact that $\mathbf{0}$ can be written in only one way as a linear combination of the vectors in B means that only the trivial combination gives $\mathbf{0}$, which is to say that B is linearly independent. Thus B is a basis for V.

As an indication of the importance of this theorem, we devise the following test for a basis for $\mathbf{R}^n$.

Theorem 4.10 Let A be an $n \times n$ matrix. Then the columns of A constitute a basis for $\mathbf{R}^n$ if and only if $\det A \neq 0$. (Here, of course, we have in mind the $n \times 1$ format for elements of $\mathbf{R}^n$.)

Proof. We need to observe that, according to vector space operations in $\mathbf{R}^n$ and the way matrices multiply,

$$x_1\begin{bmatrix} A_{11} \\ A_{21} \\ \vdots \\ A_{n1} \end{bmatrix} + x_2\begin{bmatrix} A_{12} \\ A_{22} \\ \vdots \\ A_{n2} \end{bmatrix} + \cdots + x_n\begin{bmatrix} A_{1n} \\ A_{2n} \\ \vdots \\ A_{nn} \end{bmatrix} = \begin{bmatrix} A_{11} & A_{12} & \cdots & A_{1n} \\ A_{21} & A_{22} & \cdots & A_{2n} \\ & & \cdots & \\ A_{n1} & A_{n2} & \cdots & A_{nn} \end{bmatrix}\begin{bmatrix} x_1 \\ x_2 \\ \vdots \\ x_n \end{bmatrix}$$

for any scalars $x_1, x_2, \ldots, x_n$. Therefore, the columns of A constitute a basis for $\mathbf{R}^n$ if and only if, for each $\mathbf{b}$ in $\mathbf{R}^n$, the matrix equation $A\mathbf{x} = \mathbf{b}$ has a unique solution. This is by virtue of Theorem 4.9. However, in Section 1.5, in Corollary 1.9, we found this to be the case if and only if the coefficient matrix A is nonsingular. And this, in turn, is so if and only if $\det A \neq 0$.

Example 11

While it is not difficult to show directly from the definition of basis that $\{(1, 0, 0), (1, 1, 0), (1, 1, 1)\}$ is a basis for $\mathbf{R}^3$, it is even easier to use our test from Theorem 4.10. We put

$$A = \begin{bmatrix} 1 & 1 & 1 \\ 0 & 1 & 1 \\ 0 & 0 & 1 \end{bmatrix}$$

and recall that the determinant of a triangular matrix is simply the product of its main-diagonal elements. Thus $\det A = 1 \neq 0$, and our set of vectors is indeed a basis for $\mathbf{R}^3$. □

We can, with a little ingenuity, broaden the application of this result.

Example 12

We test $\{t + t^2, 1 + t, 1 + t^2\}$ to see if it is a basis for $\mathcal{P}_2$, the space of all polynomials of degree 2 or less. By Theorem 4.9, what is at issue is whether or not, for any polynomial $a + bt + ct^2$, there are unique scalars x, y, and z such that

$$x(t + t^2) + y(1 + t) + z(1 + t^2) = a + bt + ct^2$$

Collecting the coefficients of the powers of t on the left and equating them with the corresponding coefficients on the right gives the system

$$\begin{aligned} y + z &= a \\ x + y \quad &= b \\ x \quad + z &= c \end{aligned}$$

which will have for each choice of a, b, and c a unique solution exactly when

$$\det\begin{bmatrix} 0 & 1 & 1 \\ 1 & 1 & 0 \\ 1 & 0 & 1 \end{bmatrix} \neq 0$$

Since this determinant is, in fact, -2, the given set of vectors is a basis of $\mathcal{P}_2$. □

4.2 Problem Set

1. Let $\mathbf{u}_1 = (1, 2, -1)$, $\mathbf{u}_2 = (3, -1, 4)$, $\mathbf{u}_3 = (2, 1, 3)$, $\mathbf{v} = (3, -8, 11)$.
   a. Find $2\mathbf{u}_1 - \mathbf{u}_2 + 3\mathbf{u}_3$.
   b. Show that $\mathbf{v} \in \text{Sp}\{\mathbf{u}_1, \mathbf{u}_2\}$.
   c. Show that $\{\mathbf{u}_1, \mathbf{u}_2, \mathbf{u}_3\}$ is linearly independent.
   d. From b we can deduce that $\{\mathbf{u}_1, \mathbf{u}_2, \mathbf{v}\}$ is linearly dependent. Why?
   e. from c we can deduce that $\mathbf{u}_3 \notin \text{Sp}\{\mathbf{u}_1, \mathbf{u}_2\}$. Why?
   f. Show that $\{\mathbf{u}_1, \mathbf{u}_2, \mathbf{u}_3, \mathbf{v}\}$ is linearly dependent.
2. a. Show that the set $\{(1, 3, -2)\}$ is linearly independent.
   b. Show that $\{(1, 2, -1), (-2, -4, 2)\}$ is linearly dependent.

c. Argue that the set $\{\mathbf{u}\}$ is linearly independent if and only if $\mathbf{u} \neq \mathbf{0}$.

d. Argue that $\{\mathbf{u}, \mathbf{v}\}$ is linearly independent if and only if neither $\mathbf{u}$ nor $\mathbf{v}$ is a scalar multiple of the other.

3. a. Show that $\{(3, 1, 5, 2), (1, 2, 0, 1), (1, 3, 4, 1)\}$ is linearly independent.

   b. Show that $\{(3, 1, 5, 2), (5, -9, -1, 2), (1, 3, 4, 1)\}$ is linearly dependent.

4. Show that $\{(1, 0, 2), (2, 0, 3)\}$ is a basis for the subspace of $\mathbf{R}^3$ consisting of all triples with middle coordinate 0.

5. Use the definition of basis to show that $\{(1, 0, 0), (1, 1, 0), (1, 1, 1)\}$ is a basis for $\mathbf{R}^3$.

6. Determine whether or not $\{(1, 2, 3), (4, 5, 6), (7, 8, 9)\}$ is a basis for $\mathbf{R}^3$.

7. a. Recall the notation of the natural basis vectors for $\mathbf{R}^3$: $\mathbf{i} = (1, 0, 0)$, $\mathbf{j} = (0, 1, 0)$, $\mathbf{k} = (0, 0, 1)$ (Section 3.4). Use the definition of basis to show that $\{\mathbf{i}, \mathbf{j}, \mathbf{k}\}$ is a basis for $\mathbf{R}^3$.

   b. Show that $\{(1, 0), (0, 1)\}$ is a basis for $\mathbf{R}^2$. The bases for $\mathbf{R}^3$ and $\mathbf{R}^2$ given in this problem are called the "natural bases" for those spaces.

   c. Determine a natural basis for $\mathbf{R}^n$.

8. Show that $\{1 + t, 1 - t, t + t^2\}$ is a basis for $\mathcal{P}_2$, the vector space of all polynomials of degree 2 or less.

9. Use intuition to describe what might be termed a "natural basis" for $\mathcal{P}_2$; for $\mathcal{P}_3$; for $\mathcal{P}_n$, in general.

10. Use the results of Problem 9 to show that there is no upper bound on the number of vectors in a linearly independent set in the vector space of *all* real-valued, differentiable functions.

11. Give a basis for each of the following vector spaces.

    a. $\{(x, y, 2x) | x, y \in \mathbf{R}\}$

    b. $\{(x, y, x + y, x - y) | x, y \in \mathbf{R}\}$

    c. $\{(x, 2x, 3x) | x \in \mathbf{R}\}$

12. Give a basis for the space of all

    a. 3×3 matrices

    b. symmetric 3×3 matrices

    c. skew-symmetric 3×3 matrices

    d. 3×3 matrices with trace 0

    e. upper-triangular 3×3 matrices

13. Let $\mathbf{u}_1 = (1, -1, 3)$, $\mathbf{u}_2 = (-1, 0, 2)$, $\mathbf{u}_3 = (1, -2, 8)$. Which of the following are linear combinations of $\mathbf{u}_1$, $\mathbf{u}_2$, $\mathbf{u}_3$? $\mathbf{x} = (9, 2, 14)$; $\mathbf{y} = (1, -3, 13)$; $\mathbf{z} = (0, 1, 2)$; $\mathbf{w} = (-1, -2, 12)$

14. Which of these sets span $\mathbf{R}^2$? Why? Which are bases for $\mathbf{R}^2$? Why?
    a. $\{(1, 1), (0, 1)\}$
    b. $\{(1, -1), (2, -2), (0, 0)\}$
    c. $\{(0, 1), (1, 0), (1, 1)\}$

15. Which of these sets span $\mathbf{R}^3$? Why? Which are bases? Why?
    a. $\{(1, 0, 1), (1, 1, 1)\}$
    b. $\{(0, 0, 0), (1, 2, 1), (1, -1, 2), (0, 0, 1)\}$
    c. $\{(1, -1, 2), (1, 2, 3), (1, 4, -2)\}$
    d. $\{(1, 0, 0), (1, -1, 2), (0, 1, 0), (1, 1, 1)\}$

16. $\mathbf{x} = (3, 0, 2)$ is a vector in the subspace of Problem 4. Express $\mathbf{x}$ as a linear combination of the basis vectors given there.

17. $p(t) = 3 + t - 2t^2$ is a vector in $\mathcal{P}_2$. Express $p(t)$ as a linear combination of the basis vectors for $\mathcal{P}_2$ given in Problem 8.

18. Illustrate Theorem 4.4 by expressing one of the vectors in Problem 3b as a linear combination of the other two vectors.

19. Use identities or definitions from trigonometry and calculus to show that the given sets are linearly dependent in the vector space of all real-valued functions. (Assume the domain is appropriately restricted, if necessary.)
    a. $\{\sin^2 t, \cos^2 t, 5\}$
    b. $\{\sin^2 t, 1, \cos 2t\}$
    c. (sinh t, e^t, e^{-t}}

20. Let each of W_1 and W_2 be a subspace of a vector space V. Let $\{\mathbf{u}_1, \mathbf{u}_2, \ldots, \mathbf{u}_r\}$ be a basis for W_1 and let $\{\mathbf{v}_1, \mathbf{v}_2, \ldots, \mathbf{v}_s\}$ be a basis for W_2. Assume also that $W_1 \cap W_2 = \{\mathbf{0}\}$. Show that $\{\mathbf{u}_1, \mathbf{u}_2, \ldots, \mathbf{u}_r, \mathbf{v}_1, \mathbf{v}_2, \ldots, \mathbf{v}_s\}$ is a basis for $W_1 + W_2$. (See Section 4.1, Problem 15, for the definition of $W_1 + W_2$.)

21. Show that $\{(\sqrt{2}, 1), (2, \sqrt{2})\}$ constitutes a linearly independent set if only rational scalars are allowed, but a linearly dependent set if real scalars are allowed. (*Hint:* Refer to Problems 16 and 17 of Section 4.1).

22. Show that there is an upper bound on the number of vectors that can be in a linearly independent set in $\mathbf{R}^n$. (*Hint:* If $m > n$, show that a set of m vectors in $\mathbf{R}^n$ must be linearly dependent, using the fact that a homogeneous system of n equations in m unknowns has a nontrivial solution if $m > n$.)

4.3 Tests for Linear Independence

In this section we will describe ways of testing sets of vectors in $\mathbf{R}^m$ and certain sets of functions for linear independence. To begin, let $B = \{\mathbf{v}_1, \mathbf{v}_2, \ldots, \mathbf{v}_n\}$ be a set of vectors in $\mathbf{R}^m$, with $n \leq m$. Let

$$\begin{aligned} \mathbf{v}_1 &= (A_{11}, A_{12}, \ldots, A_{1m}) \\ \mathbf{v}_2 &= (A_{21}, A_{22}, \ldots, A_{2m}) \\ &\cdots \\ \mathbf{v}_n &= (A_{n1}, A_{n2}, \ldots, A_{nm}) \end{aligned}$$

Whether B is linearly dependent or independent depends on the solution set of the equation

$$c_1\mathbf{v}_1 + c_2\mathbf{v}_2 + \cdots + c_n\mathbf{v}_n = \mathbf{0}$$

which, expressed in full detail, is

$$c_1(A_{11}, A_{12}, \ldots, A_{1m}) + c_2(A_{21}, A_{22}, \ldots, A_{2m}) + \\ \cdots + c_n(A_{n1}, A_{n2}, \ldots, A_{nm}) = (0, 0, \ldots, 0)$$

Equating corresponding coordinates yields the homogeneous system

$$\begin{aligned} A_{11}c_1 + A_{21}c_2 + \cdots + A_{n1}c_n &= 0 \\ A_{12}c_1 + A_{22}c_2 + \cdots + A_{n2}c_n &= 0 \\ &\cdots \\ A_{1m}c_1 + A_{2m}c_2 + \cdots + A_{nm}c_n &= 0 \end{aligned}$$

This system has matrix representation

$$\begin{bmatrix} A_{11} & A_{21} & \cdots & A_{n1} \\ A_{12} & A_{22} & \cdots & A_{n2} \\ & & \cdots & \\ A_{1m} & A_{2m} & \cdots & A_{nm} \end{bmatrix} \begin{bmatrix} c_1 \\ c_2 \\ \vdots \\ c_n \end{bmatrix} = \begin{bmatrix} 0 \\ 0 \\ \vdots \\ 0 \end{bmatrix}$$

which is of the form $A\mathbf{c} = \mathbf{0}$. Notice that the vectors $\mathbf{v}_1, \mathbf{v}_2, \ldots, \mathbf{v}_n$ appear as the *columns* of the $m \times n$ coefficient matrix A.

To solve $A\mathbf{c} = \mathbf{0}$ for $\mathbf{c}$, we reduce A to echelon form. The result will be either

$$\left[\begin{array}{c} I \\ \hline O \end{array}\right]$$

(by which we mean an $n \times n$ identity matrix atop an $(m - n) \times n$ zero matrix), or else one of the columns of the reduced matrix, say the k-th, will have no leading one.

In the first case, the equation has only the trivial solution (why?); in the second case, there are nontrivial solutions corresponding to nonzero values assigned arbitrarily to c_k.

The following theorem summarizes this test for linear independence.

Theorem 4.11 A set B in $\mathbf{R}^m$ containing n vectors, $n \leq m$, is linearly independent if and only if the $m \times n$ matrix having the vectors in B as its columns has echelon form

$$\left[\begin{array}{c} I \\ \hline O \end{array}\right]$$

Example 1

In applying Theorem 4.11 to the set $B = \{(1, 1, 3, 0), (2, 1, -1, 4), (5, 0, 2, 1)\}$, we find that the matrix

$$\begin{bmatrix} 1 & 2 & 5 \\ 1 & 1 & 0 \\ 3 & -1 & 2 \\ 0 & 4 & 1 \end{bmatrix}$$

has echelon form

$$\begin{bmatrix} 1 & 0 & 0 \\ 0 & 1 & 0 \\ 0 & 0 & 1 \\ 0 & 0 & 0 \end{bmatrix} = \left[\begin{array}{c} I \\ \hline O \end{array}\right]$$

Hence the set B is linearly independent. □

A special case arises when $m = n$; in that event the matrix A is square.

Corollary 4.11 A set B in $\mathbf{R}^m$ of exactly m vectors is linearly independent if and only if the $m \times m$ matrix having the vectors in B as its columns has echelon form I (the $m \times m$ identity matrix), or equivalently, has a nonzero determinant (see Theorem 4.10).

Example 2

Earlier we found that the set $\{(2, 1, 3), (1, 1, -2), (8, 5, 5)\}$ was linearly dependent (Example 3, Section 4.2). We reach the same conclusion more efficiently by computing

$$\det\begin{bmatrix} 2 & 1 & 8 \\ 1 & 1 & 5 \\ 3 & -2 & 5 \end{bmatrix} = 0$$

and applying Corollary 4.11. □

We can apply Theorem 4.11 to test certain sets of functions for linear independence.

Example 3

Let us test the pair $\{f, g\}$ of functions for linear independence, where $f(x) = x^2$ and $g(x) = \sqrt[3]{x}$ in the space, say, of all functions continuous on $\mathbf{R}$. Suppose that $c_1 f + c_2 g = \Theta$. This means that $c_1 f(x) + c_2 g(x) = 0$ *for all* x *in* $\mathbf{R}$. In particular, if we choose $x = 1$ and $x = 8$ (somewhat arbitrarily), we have $c_1 + c_2 = 0$ and $64c_1 + 2c_2 = 0$. It is routine to verify that the only solution to this system is $(c_1, c_2) = (0, 0)$; thus $\{f, g\}$ is linearly independent. □

There is a simpler way to deduce linear independence of a set of functions, provided that they are all sufficiently differentiable. Let $f_1, f_2, \ldots, f_n$ be $(n - 1)$-times differentiable, and suppose

$$c_1 f_1(x) + c_2 f_2(x) + \cdots + c_n f_n(x) = 0 \tag{1}$$

Differentiating both sides of the equation repeatedly gives

$$\begin{aligned} c_1 f_1(x) + c_2 f_2(x) &+ \cdots + c_n f_n(x) &= 0 \\ c_1 f_1'(x) + c_2 f_2'(x) &+ \cdots + c_n f_n'(x) &= 0 \\ &\cdots \\ c_1 f_1^{(n-1)}(x) + c_2 f_2^{(n-1)}(x) &+ \cdots + c_n f_n^{(n-1)}(x) &= 0 \end{aligned}$$

If we regard these equations as a homogeneous system in the variables $c_1, c_2, \ldots, c_n$, then the system, and hence (1), have only the trivial solution $(c_1, c_2, \ldots, c_n) = (0, 0, \ldots, 0)$ *if*

$$\det\begin{bmatrix} f_1(x) & f_2(x) & \cdots & f_n(x) \\ f_1'(x) & f_2'(x) & \cdots & f_n'(x) \\ & \cdots & & \\ f_1^{(n-1)}(x) & f_2^{(n-1)}(x) & \cdots & f_n^{(n-1)}(x) \end{bmatrix} \neq 0 \qquad \textit{for some } x$$

The determinant above is called the *Wronskian* of $f_1, f_2, \ldots, f_n$ at x and will be denoted $\mathrm{Wr}(f_1, f_2, \ldots, f_n)(x)$.

The preceding discussion establishes the following result.

Theorem 4.12 If the Wronskian function $\mathrm{Wr}(f_1, f_2, \ldots, f_n)$ is not identically zero, then the set $\{f_1, f_2, \ldots, f_n\}$ is linearly independent.

Example 4

Let us test the set $\{x^2, \sqrt[3]{x}\}$ again for linear independence, this time using Theorem 4.12. The Wronskian here is

$$\mathrm{Wr}(x^2, \sqrt[3]{x}) = \det\begin{bmatrix} x^2 & x^{1/3} \\ 2x & \frac{1}{3}x^{-2/3} \end{bmatrix} = -\frac{5}{3}x^{4/3}$$

which is not the zero function. Hence $\{x^2, \sqrt[3]{x}\}$ is linearly independent. □

Example 5

To see that $\{x, e^x, \sin x\}$ is linearly independent, we compute the Wronskian:

$$\det\begin{bmatrix} x & e^x & \sin x \\ 1 & e^x & \cos x \\ 0 & e^x & -\sin x \end{bmatrix} = e^x(2 \sin x - x \sin x - x \cos x)$$

This function is not Θ (for example, when $x = \pi/2$ the Wronskian is $e^{\pi/2}(2 - (\pi/2))$, which is nonzero). □

Example 6

Consider the set $\{f, g\}$ where $f(x) = x^3$ and $g(x) = x^3 + |x^3|$. To find $\mathrm{Wr}(f, g)$ we first note that

$$g(x) = \begin{cases} 0 & \text{if } x < 0 \\ 2x^3 & \text{if } x \geq 0 \end{cases}$$

and that g is differentiable everywhere, including at 0. Its derivative is

$$g'(x) = \begin{cases} 0 & \text{if } x < 0 \\ 6x^2 & \text{if } x \geq 0 \end{cases}$$

Now,

$$\mathrm{Wr}(f, g)(x) = \det \begin{bmatrix} x^3 & x^3 + |x^3| \\ 3x^2 & \begin{Bmatrix} 0 & \text{if} & x < 0 \\ 6x^2 & \text{if} & x \geq 0 \end{Bmatrix} \end{bmatrix} = 0$$

for all x. Hence we may not apply Theorem 4.12 to this pair of functions because the Wronskian function is zero. Falling back on an argument like that of Example 3, we can show that $\{f, g\}$ is linearly independent. □

Example 6 illustrates that the converse of Theorem 4.12 is false; a linearly independent set of functions need not have a nonzero Wronskian. There is, however, a partial converse to Theorem 4.12, which we state here; the proof is left to the exercises.

Theorem 4.13 Suppose that $\mathrm{Wr}(f, g) = 0$ and that $g(x) \neq 0$ for all x. Then $\{f, g\}$ is linearly dependent.

4.3 Problem Set

1. Use the method from Example 1 to test these sets for independence.
   a. $\{(-2, 0, 1, 0), (4, 1, 0, 1), (14, 2, -3, 2)\}$
   b. $\{(1, 5, 0, 1, 6), (2, 1, 0, -1, 1), (4, 0, 1, -1, 1)\}$
2. Use the method from Example 2 to test these sets for independence.
   a. $\{(1, 1, 0), (1, 0, 1), (0, 1, 1)\}$
   b. $\{(1, 1, 2), (0, -1, 1), (2, 5, 1)\}$
   c. $\{(2, 3), (1, -2)\}$
   d. $\{(5, 3, 1, 1), (1, 0, 1, 0), (0, 1, 2, 0), (13, 11, 5, 3)\}$
3. Test these sets for linear independence.
   a. $\{(3, 1), (2, 2)\}$
   b. $\{(4, 3, 1), (2, 0, 6), (1, 2, 1)\}$
   c. $\{(1, 1, 2, 1, 4), (3, 0, 1, 0, 2), (-1, 2, 3, 2, 6)\}$
   d. $\{(5, 1, 2, 4), (1, 6, 0, 3), (1, -3, 1, 2)\}$

e. $\{(1, 4, 1), (3, 4, 2), (0, 8, 1)\}$
f. $\{(1, 0, 1, 0), (0, 1, 0, 0), (2, 0, 1, 0), (0, 0, 1, 1)\}$
g. $\{(4, 4, 2, 1, 6, 3, 5), (0, 1, 3, 5, 1, 2, 1), (1, 3, 5, 1, 8, 8, 0)\}$

4. Find the Wronskian of each set of functions below, and determine which sets are linearly independent.
   a. $\{\sin x, \cos x\}$
   b. $\{1, x, x^2, x^3, x^4\}$
   c. $\{x^2 - 3, 3x - 2x^2, x - 2\}$
   d. $\{e^x, e^{2x}, e^{3x}\}$
   e. $\{\sin^2 x, 2, 3\cos^2 x\}$

5. Prove Theorem 4.13. (*Hint:* Note that $\left(\frac{f}{g}\right)' = \frac{-\mathrm{Wr}(f, g)}{g^2}$)

4.4 Dimension of a Vector Space

A vector space with nonzero vectors in it must contain infinitely many vectors, since it contains all scalar multiples of any of its nonzero elements and these multiples are all different. Nevertheless, it is possible to compare sizes of vector spaces, not by counting the number of vectors in the space, but by counting the number of vectors in a basis. To see how this is done, we first need to see that the size of a basis places an upper bound on the number of vectors that a linearly independent set can contain.

Theorem 4.14 Let $B = \{\mathbf{u}_1, \mathbf{u}_2, \ldots, \mathbf{u}_n\}$ be a basis for a vector space V and let $C = \{\mathbf{v}_1, \mathbf{v}_2, \ldots, \mathbf{v}_m\}$ be a subset of V with $m > n$. Then C is linearly dependent. (That is, a linearly independent set cannot contain more vectors than a basis contains.)

Proof. Since $\mathbf{v}_k$ is in Sp B for $k = 1, 2, \ldots, m$, we have

$$\begin{aligned}
\mathbf{v}_1 &= a_{11}\mathbf{u}_1 + a_{12}\mathbf{u}_2 + \cdots + a_{1n}\mathbf{u}_n \\
\mathbf{v}_2 &= a_{21}\mathbf{u}_1 + a_{22}\mathbf{u}_2 + \cdots + a_{2n}\mathbf{u}_n \\
&\cdots \\
\mathbf{v}_m &= a_{m1}\mathbf{u}_1 + a_{m2}\mathbf{u}_2 + \cdots + a_{mn}\mathbf{u}_n
\end{aligned}$$

for certain coefficients $a_{11}, a_{12}, \ldots, a_{mn}$. To test C for linear dependence, consider the equation $b_1\mathbf{v}_1 + b_2\mathbf{v}_2 + \cdots + b_m\mathbf{v}_m = \mathbf{0}$. Substituting from above and combining coefficients of $\mathbf{u}_1, \mathbf{u}_2, \ldots, \mathbf{u}_n$, we get

$$\begin{aligned}&(b_1a_{11} + b_2a_{21} + \cdots + b_ma_{m1})\mathbf{u}_1\\ &\quad + (b_1a_{12} + b_2a_{22} + \cdots + b_ma_{m2})\mathbf{u}_2\\ &\quad + \cdots + (b_1a_{1n} + b_2a_{2n} + \cdots + b_ma_{mn})\mathbf{u}_n = \mathbf{0}\end{aligned}$$

Since B is linearly independent, all the above coefficients must be zero:

$$\begin{aligned}a_{11}b_1 + a_{21}b_2 + \cdots + a_{m1}b_m &= 0\\ a_{12}b_1 + a_{22}b_2 + \cdots + a_{m2}b_m &= 0\\ &\cdots\\ a_{1n}b_1 + a_{2n}b_2 + \cdots + a_{mn}b_n &= 0\end{aligned}$$

This is a homogeneous system of n equations in the unknowns $b_1, b_2, \ldots, b_m$. Since $m > n$, the system must have a nontrivial solution. But that means that $b_1\mathbf{v}_1 + b_2\mathbf{v}_2 + \cdots + b_m\mathbf{v}_m = \mathbf{0}$ has a nontrivial solution; that is, C is linearly dependent.

We mentioned in Section 4.2 that, from a geometrical viewpoint, it would seem impossible to have *three* vectors in a linearly independent set in $\mathbf{R}^2$. This is now confirmed, for we have also seen that the *two* vectors (1, 0) and (0, 1) constitute a basis for $\mathbf{R}^2$.

To see that the number of vectors in a basis can indicate the size of a vector space, it is necessary to know that all bases of a given vector space contain the same number of vectors. The following corollary formalizes this fact.

Corollary 4.14 If $B_1 = \{\mathbf{u}_1, \mathbf{u}_2, \ldots, \mathbf{u}_n\}$ and $B_2 = \{\mathbf{v}_1, \mathbf{v}_2, \ldots, \mathbf{v}_m\}$ are each bases for a vector space V, then $m = n$. (That is, any two bases have the same number of vectors.)

Proof. Since B_1 is a basis and B_2 is linearly independent, we obtain $m \leq n$ by Theorem 4.14. Since B_2 is a basis and B_1 is linearly independent, $n \leq m$. Thus $m = n$.

Because of this corollary, we can make the following definition unambiguously.

Definition If a vector space V has a basis, then the *dimension* of V, denoted by dim V, is the number of vectors in any basis for V. If $V = \{\mathbf{0}\}$, there is no independent set in V and hence no basis; in this case, dim $V = 0$. If dim V exists, we say that V is *finite-dimensional.*

At the other extreme, if there is no upper bound on the number of vectors that can form a linearly independent set in V, then we say that V is *infinite-dimensional.*

To find the dimension of a vector space, one must find a basis for the space and simply count the number of vectors in the basis. This is particularly easy to do in the case of $\mathbf{R}^n$.

As we have seen, $\{(1, 0), (0, 1)\}$ constitutes a basis for $\mathbf{R}^2$; thus dim $\mathbf{R}^2 = 2$. Similarly, $\{\mathbf{i}, \mathbf{j}, \mathbf{k}\}$ is a basis for $\mathbf{R}^3$, so dim $\mathbf{R}^3 = 3$. In general, in $\mathbf{R}^n$, the vectors

$$\begin{aligned} \mathbf{e}_1 &= (1, 0, 0, \ldots, 0) \\ \mathbf{e}_2 &= (0, 1, 0, \ldots, 0) \\ &\ldots \\ \mathbf{e}_n &= (0, 0, 0, \ldots, 1) \end{aligned}$$

form a basis for $\mathbf{R}^n$, so dim $\mathbf{R}^n = n$. This specific basis for $\mathbf{R}^n$ is called the *natural basis* for $\mathbf{R}^n$. (Others say "standard basis" for this.) Thus, by Corollary 4.14, any other basis for $\mathbf{R}^n$ must contain exactly n vectors.

Example 1

The set $\{(1, 1, 1), (1, 1, 0), (1, 0, 0), (1, 0, 1)\}$, even though it spans $\mathbf{R}^3$ (verify), cannot be a basis for $\mathbf{R}^3$, for it contains too many vectors. The set $\{(1, 1, 1), (1, 0, 1)\}$, even though it is linearly independent in $\mathbf{R}^3$, cannot be a basis for $\mathbf{R}^3$, for it contains too few vectors. (Consequently, our first set must be linearly dependent and the second cannot span $\mathbf{R}^3$.) □

Theorems about bases tend to come in pairs; we have seen an instance of this already in Section 4.2, where we dealt with minimal spanning sets and maximal independent sets. The reason is that a basis results exactly when *both* properties of spanning and linear independence are each optimal in some sort of way. We have two more such pairs of theorems.

Theorem 4.15 If C is a finite set of vectors spanning a vector space V, then some subset of C is a basis for V.

Proof. If C spans V *minimally*, C itself will be the desired basis. If C does not span V minimally, then some proper subset of C spans V. If this subset spans minimally, then it is the desired basis. If not, some proper subset of it spans V. This process may be continued, but only for a finite number of steps since C itself is finite. When the process stops, the end result will be a subset of C which is a minimal spanning set, and hence is a basis for V.

Example 2
The spanning set $\{(1, 1, 1), (1, 1, 0), (1, 0, 0), (0, 0, 1)\}$ does not span $\mathbf{R}^3$ minimally; however, discarding any one of the four vectors except the third one will result in a basis for $\mathbf{R}^3$. (Verify.) □

Example 3
Let $C = \{\mathbf{u}_1, \mathbf{u}_2, \mathbf{u}_3, \mathbf{u}_4\}$ where $\mathbf{u}_1 = (3, -1, 2, 1)$, $\mathbf{u}_2 = (1, 1, 2, -1)$, $\mathbf{u}_3 = (-6, 2, -4, -2)$, and $\mathbf{u}_4 = (-1, 3, 2, -3)$. Let $V = \text{Sp}\, C$. By Theorem 4.15, some subset of C is a basis for V; we will produce such a subset. First, we see that $\mathbf{u}_3 = -2\mathbf{u}_1$. By Corollary 4.7, we have $V = \text{Sp}\{\mathbf{u}_1, \mathbf{u}_2, \mathbf{u}_4\}$. Next we observe that $\mathbf{u}_1 = 2\mathbf{u}_2 - \mathbf{u}_4$, so a second application of the corollary gives $V = \text{Sp}\{\mathbf{u}_2, \mathbf{u}_4\}$. Neither $\mathbf{u}_2$ nor $\mathbf{u}_4$ is a multiple of the other, so $\{\mathbf{u}_2, \mathbf{u}_4\}$ is linearly independent. Since it spans V, we have found a basis for V. □

☆ **Theorem 4.16** If V is finite-dimensional and C is a linearly independent set of vectors in V, then C is contained in a basis for V.

Proof. Instead of discarding vectors, as in the preceding proof, the technique here is to accumulate more vectors, if necessary, to build up a basis for V. If C is *maximally* independent, no more vectors are needed; C itself is the desired basis. If C is not maximally independent, some proper superset of C is independent. If this is maximally independent, we are finished. If not, continue the process. The process must stop in a finite number of steps because, according to Theorem 4.14, we cannot reach a point where a linearly independent set contains more vectors than the dimension of V. When the process halts, we will have a maximally independent set—and hence a basis—containing C.

Example 4
The linearly independent set $\{(1, 1, 1), (1, 0, 1)\}$ is not maximally independent in $\mathbf{R}^3$. By trial and error, if necessary, a third vector can

be found in $\mathbf{R}^3$ which, together with the original two, will constitute a linearly independent set, and hence a basis. (We could even use the cross product, but only because we are given two vectors in $\mathbf{R}^3$). □

If the dimension of a vector space is known, then we can confirm that a set containing exactly the right number of vectors is a basis by simply checking only one of the defining requisites: the other requisite automatically follows, as our next pair of theorems demonstrates.

Theorem 4.17 If $\dim V = n$, then any linearly independent set of precisely n vectors is a basis for V.

Proof. If C is a linearly independent set of exactly n vectors, then, on the one hand, C is a subset of a basis for V, by Theorem 4.16. But, on the other hand, this basis contains n vectors and hence is C itself.

Theorem 4.18 If $\dim V = n$, then any spanning set of precisely n vectors is a basis for V.

We leave this short proof as an exercise.

As evidence that dimension does indeed provide a measurement of the size of a vector space, we present one final theorem.

Theorem 4.19 If V is a finite-dimensional vector space and W is a subspace of V, then W is also finite-dimensional and $\dim W \leq \dim V$. Also, $\dim W = \dim V$ only if $W = V$.

Proof. Suppose that $\dim V = n$. Trivially, if $W = \{\mathbf{0}\}$, then $\dim W = 0 \leq n = \dim V$. Now if $W \neq \{\mathbf{0}\}$, it cannot be the case that W is infinite-dimensional since any linearly independent set in W is a linearly independent set in V; so, by Theorem 4.14, an upper bound on the number of vectors in such a set is n. In particular, a basis for W is such a linearly independent set in W, so we get $\dim W \leq n = \dim V$. Finally, if $\dim W = \dim V$, then, for any basis B of W, we see that B, as a linearly independent set of n vectors in V, must also be a basis for V, according to Theorem 4.17. Thus $W = \operatorname{Sp} B = V$.

Another example of dimension as a measurement of size can be found in Problem 22.

In a certain sense, which we will explore in Section 5.2, $\mathbf{R}^n$ is essentially the only vector space of dimension n; others of dimension n are merely copies of $\mathbf{R}^n$ in different notation. (For instance, $\mathcal{P}_2$, the space of polynomials of degree ≤ 2, is essentially the same as $\mathbf{R}^3$.) However, there is a wide variety of infinite-dimensional vector spaces. Some are copies of what we called, in an earlier exercise, $\mathbf{R}^\infty$, the space of all infinite, real-valued sequences. (The space of all polynomials of whatever degree is a copy of $\mathbf{R}^\infty$.) Other infinite-dimensional vector spaces, such as the space $\mathcal{F}$ of all real-valued functions on $\mathbf{R}$, are not copies of $\mathbf{R}^\infty$; they are of an even higher order of magnitude.

4.4 Problem Set

1. Find a basis for $\mathbf{R}^2$ which contains the vector (3, 2).
2. Find a basis for $\mathbf{R}^2$ which is a subset of $\{(1, 1), (0, 4), (5, -2)\}$.
3. Find a basis for $\mathbf{R}^3$ which contains the subset
   a. $\{(4, 1, 0), (1, 2, 2)\}$
   b. $\{(4, -1, 2), (-11, 3, -2)\}$
4. Find a basis for $\mathbf{R}^3$ which is a subset of $\{(3, 2, -1), (5, -2, 1), (-3, 4, -2), (0, 2, 0), (-1, 0, -3)\}$
5. Is there a basis for $\mathbf{R}^4$ which contains the subset $\{(0, 1, 2, 3), (0, 4, 5, 6), (0, 7, 8, 9)\}$? Why or why not?
6. Find a basis for $\mathbf{R}^4$ which is a subset of $\{(1, 0, 0, 1), (2, 4, 1, 2), (3, 2, 0, 1), (1, 6, 1, 5), (4, -2, 0, 3), (0, 2, 5, 3)\}$
7. Give the dimension of each vector space below.
   a. $\{(x, y, 2x) | x, y \in \mathbf{R}\}$
   b. $\{(x, y, x + y, x - y) | x, y, \in \mathbf{R}\}$
   c. $\{(x + y, x + z, y - z) | x, y, z \in \mathbf{R}\}$
   d. all symmetric 3×3 matrices
   e. all upper-triangular 3×3 matrices
   f. all 3×3 matrices with trace 0
8. Because of Theorem 4.19, the possible dimensions of subspaces of $\mathbf{R}^2$ are 0, 1, and 2. For dimension 0, we have just the origin

$\{\mathbf{0}\}$. For dimension 2, we have the entire plane $\mathbf{R}^2$. For dimension 1, we will have a one-vector basis $\{\mathbf{u}\}$, and hence the subspace will consist of vectors satisfying $\mathbf{x} = t\mathbf{u}$, which gives a line through the origin. Analyze and give geometric descriptions of the possible types of subspaces of $\mathbf{R}^3$.

9. Find a basis for, and the dimension of, each of the subspaces given here.
   a. $\{(a, b, 0)\}$ in $\mathbf{R}^3$
   b. $\{(a, b, 2a)\}$ in $\mathbf{R}^3$
   c. $\{(0, a, a)\}$ in $\mathbf{R}^3$
   d. $\{(a, 2a, 3a)\}$ in $\mathbf{R}^3$
   e. $\{(a + b, 2a - b, a - 3b)\}$ in $\mathbf{R}^3$
   f. the intersection of b and c
   g. the intersection of c and d
   h. the intersection of a and d
   i. the intersection of a and e

10. Suppose we have two subsets $\{\mathbf{u}_1, \mathbf{u}_2, \ldots, \mathbf{u}_m\}$ and $\{\mathbf{v}_1, \mathbf{v}_2, \ldots, \mathbf{v}_n\}$ of a vector space V and that $m < n$.
    a. Is it possible that the first spans V while the second is linearly independent? Why or why not?
    b. Is it possible that the second spans V while the first is linearly independent? Why or why not?

11. Consider the space $\mathbf{R}^\infty$ of all real-valued, infinite sequences.
    a. Show that $\{(1, 0, 0, \ldots)\}$ is a linearly independent set.
    b. Show that $\{(1, 0, 0, \ldots), (0, 1, 0, \ldots)\}$ is linearly independent.
    c. Find an n-element, linearly independent subset of $\mathbf{R}^\infty$.
    d. Deduce that $\mathbf{R}^\infty$ is infinite-dimensional, using Theorem 4.14.

12. Let V be a finite-dimensional vector space; let each of W_1 and W_2 be a subspace of V. Prove that

$$\dim(W_1 + W_2) = \dim W_1 + \dim W_2 - \dim(W_1 \cap W_2)$$

(*Hint:* See Figure 4.7 and Section 4.1, Problem 15, for the definition of $W_1 + W_2$. Let $\{\mathbf{w}_1, \mathbf{w}_2, \ldots, \mathbf{w}_s\}$ be a basis for $W_1 \cap W_2$. By Theorem 4.16, we obtain a basis $\{\mathbf{u}_1, \mathbf{u}_2, \ldots, \mathbf{u}_r, \mathbf{w}_1, \mathbf{w}_2, \ldots, \mathbf{w}_s\}$ for W_1 and a basis $\{\mathbf{w}_1, \mathbf{w}_2, \ldots, \mathbf{w}_s, \mathbf{v}_1, \mathbf{v}_2, \ldots, \mathbf{v}_t\}$ for W_2. Show that $\{\mathbf{u}_1, \mathbf{u}_2, \ldots, \mathbf{u}_r, \mathbf{w}_1, \mathbf{w}_2, \ldots, \mathbf{w}_s, \mathbf{v}_1, \mathbf{v}_2, \ldots, \mathbf{v}_t\}$ is a basis for $W_1 + W_2$ by showing that it spans

$W_1 + W_2$ and is linearly independent. Then the desired equation reads $r + s + t = (r + s) + (s + t) - s$.)

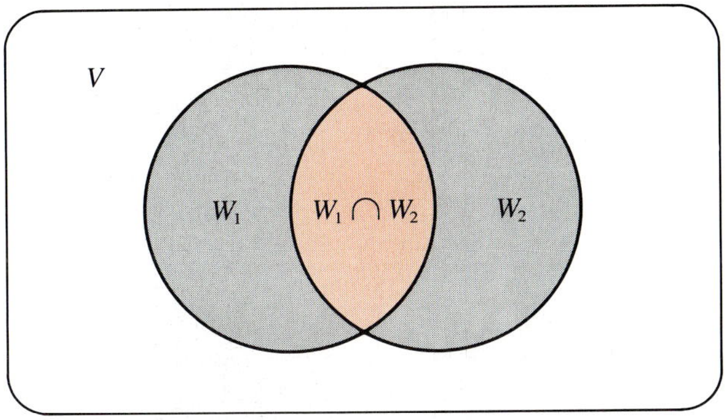

$$\dim (W_1 + W_2) = \dim W_1 + \dim W_2 - \dim (W_1 \cap W_2)$$

(Note that dimension measures the "size" of a vector space.)

Figure 4.7

13. Prove Theorem 4.18. (*Hint:* Use Theorem 4.15.)

5

LINEAR MAPS

5.1 LINEAR MAPS

Functions play a central role in all of mathematics, and linear algebra is no exception. In this chapter we shall consider functions for which vectors are both the "input" and the "output," and we shall discover that functions having two certain simple properties are the functions which are most suitable from a vector-space point of view.

Let us begin by recalling that a *function*, as encountered in calculus and pre-calculus courses, is a correspondence which assigns to each member of one set (called the *domain* of the function) a unique member of a second set (possibly the same as the first set). For instance, the formula

$$f(x) = \sqrt{x - 1}$$

is understood to define a function f which assigns to each number $x \geq 1$ that unique number $\sqrt{x-1}$. For example, $f(3) = \sqrt{2}$, so $\sqrt{2}$ is one particular *value* of this function. Recall that the *range* of f is the set consisting of all the values of the function. In our example $f(x) = \sqrt{x-1}$, we see that the domain and range are given by

$$\operatorname{dom} f = \{x \mid x \geq 1\}$$

since we cannot compute square roots of negative numbers, and

$$\operatorname{range} f = \{y \mid y \geq 0\}$$

since the radical "$\sqrt{\ }$" denotes the nonnegative square root of a number.

In calculus and pre-calculus courses, dom f and range f are very likely to be subsets of $\mathbf{R}$, as in our example. Now, however, we shall consider functions having vector spaces for their domains and ranges, and we shall be especially interested in functions which are "sensitive" to the operations present in vector spaces. This sensitivity is described in our next definition.

Definition Let each of U and V be a vector space and let f be a function having domain U and range a subset of V. Then f is said to be a *linear map from U to V* provided that

LM1. $f(\mathbf{u}_1 + \mathbf{u}_2) = f(\mathbf{u}_1) + f(\mathbf{u}_2)$ for all $\mathbf{u}_1$, $\mathbf{u}_2$ in U

LM2. $f(c\mathbf{u}) = cf(\mathbf{u})$ for all $\mathbf{u}$ in U, c in $\mathbf{R}$

See Figure 5.1.

To indicate that a function from U to V is under consideration (even if it does not satisfy conditions LM1 and LM2), we write

$$f: U \to V \quad \text{or} \quad U \xrightarrow{f} V$$

We point out here that some mathematicians say "linear function," "linear transformation," or "linear operator," instead of "linear map." Some refer to the "image of f" where we say "range of f." A few specifically call the set V in the expression $f: U \to V$ the "codomain of f." Any way it is read, V is simply any set containing the range of f; it is not uniquely determined by the function.

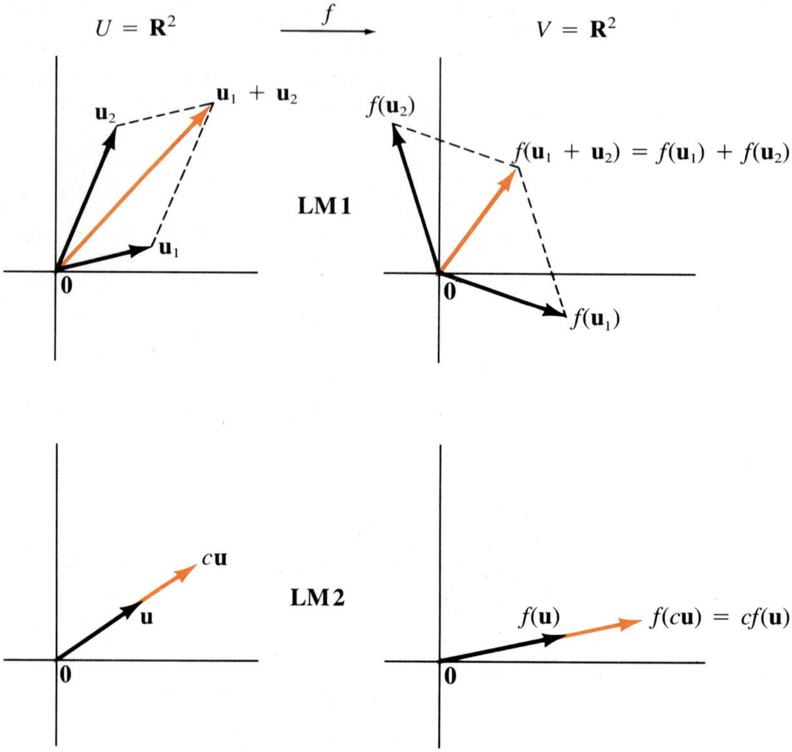

Figure 5.1

Example 1

A familiar function is given by $f(u) = \sin u$. We may recall that dom $f = \mathbf{R}$ and that range $f = \{y | -1 \leq y \leq 1\}$. Thus we write f: $\mathbf{R} \to \mathbf{R}$. Let us determine whether or not f is a linear map. We readily see that LM1 is not satisfied by observing that $\sin(u_1 + u_2) \neq \sin u_1 + \sin u_2$, for most values of u_1 and u_2. The failure of f to satisfy one of the two required conditions is enough to disqualify f as a linear map. However, it is worthwhile to test LM2 as well. We find that $\sin(cu) \neq c \sin u$, in general, and so LM2 is not satisfied. Thus f is not a linear map. □

Example 2

Let us consider an example involving $\mathbf{R}^2$. We examine the domain and the range and test LM1 and LM2 for the function given by

$$f(x, y) = (x + y, x - y) \qquad x, y \text{ in } \mathbf{R}$$

First, we note that dom $f = \mathbf{R}^2$. Moreover, the values $(x + y, x - y)$ certainly lie in $\mathbf{R}^2$, so we can say range $f \subset \mathbf{R}^2$. Thus we have $f\colon \mathbf{R}^2 \to \mathbf{R}^2$. Whether or not range f *equals* $\mathbf{R}^2$ depends on whether or not the equation

$$(x + y, x - y) = (a, b)$$

has a solution in x and y for *all* choices of a and b. Now the system

$$\begin{aligned} x + y &= a \\ x - y &= b \end{aligned}$$

most certainly does have a solution, given a and b, as we have learned in Chapter 1. This follows, for instance, from Corollary 1.9, since the coefficient matrix is nonsingular (its determinant is -2 and hence nonzero). We have yet to decide if f is a *linear map*. We must test the validity of LM1 and LM2.

For LM1, let $\mathbf{u}_1 = (x_1, y_1)$, $\mathbf{u}_2 = (x_2, y_2)$. Then

$$\begin{aligned} f(\mathbf{u}_1 + \mathbf{u}_2) &= f(x_1 + x_2, y_1 + y_2) \\ &= ((x_1 + x_2) + (y_1 + y_2), (x_1 + x_2) - (y_1 + y_2)) \end{aligned}$$

whereas

$$\begin{aligned} f(\mathbf{u}_1) + f(\mathbf{u}_2) &= f(x_1, y_1) + f(x_2, y_2) \\ &= (x_1 + y_1, x_1 - y_1) + (x_2 + y_2, x_2 - y_2) \\ &= ((x_1 + y_1) + (x_2 + y_2), (x_1 - y_1) + (x_2 - y_2)) \end{aligned}$$

Thus $f(\mathbf{u}_1 + \mathbf{u}_2)$ and $f(\mathbf{u}_1) + f(\mathbf{u}_2)$ are equal, by elementary number properties.

For LM2, let $\mathbf{u} = (x, y)$ and let c be any scalar. Then

$$f(c\mathbf{u}) = f(c(x, y)) = f(cx, cy) = (cx + cy, cx - cy)$$

while

$$cf(\mathbf{u}) = cf(x, y) = c(x + y, x - y) = (c(x + y), c(x - y))$$

Hence $f(c\mathbf{u})$ and $cf(\mathbf{u})$ are equal. Now we have confirmed both LM1 and LM2; consequently, $f(x, y) = (x + y, x - y)$ gives a linear map. □

Example 3a

Let us test the function given by $g(x) = 2x - 3$, which in elementary mathematics is called a linear function (because its graph is a straight line). In our notation, we have dom $g = \mathbf{R}$ and range $g = \mathbf{R}$; thus we have $g\colon \mathbf{R} \to \mathbf{R}$. In testing the condition LM1, we find that

$$g(x_1 + x_2) = 2(x_1 + x_2) - 3 = 2x_1 + 2x_2 - 3$$

whereas

$$g(x_1) + g(x_2) = (2x_1 - 3) + (2x_2 - 3) = 2x_1 + 2x_2 - 6$$

Therefore, $g(x_1 + x_2)$ and $g(x_1) + g(x_2)$ are not identically equal, so LM1 is not satisfied. (It is as easily seen that LM2 also does not hold, although it is not necessary to check.) Hence g is not a linear map. □

Example 3a demonstrates that linear *functions* are not always linear *maps*. However, linear functions, are in a certain way, obtainable from linear maps, as we see by modifying Example 3a slightly, to remove the constant term -3.

Example 3b

Let $f(x) = 2x$. As in Example 3a, we have dom $f = \mathbf{R}$, range $f = \mathbf{R}$ and $f\colon \mathbf{R} \to \mathbf{R}$. We test, in turn, LM1 and LM2. We find

$$f(x_1 + x_2) = 2(x_1 + x_2) = 2x_1 + 2x_2 = f(x_1) + f(x_2)$$

and

$$f(cx) = 2(cx) = c(2x) = cf(x)$$

Both LM1 and LM2 hold for f and consequently f is a linear map. □

As Examples 3a and 3b illustrate, a linear function from $\mathbf{R}$ to $\mathbf{R}$ will be a linear map when the constant term in its formula is 0. This means that the graph of such a linear function must be a straight line *through the origin* if the function is to be a linear map. Recall that straight lines in $\mathbf{R}^2$ which do not pass through the origin are *affine subspaces* of $\mathbf{R}^2$, but not subspaces of $\mathbf{R}^2$. They can be obtained from subspaces by adding a constant vector to subspaces. This suggests that we introduce the notion of *affine map*.

☆ Definition

A function $g\colon U \to V$ is called an *affine map* provided that, for some linear map $f\colon U \to V$ and some fixed element $\mathbf{b}$ of V,

$$g(\mathbf{u}) = f(\mathbf{u}) + \mathbf{b} \text{ for all } \mathbf{u} \text{ in } U$$

Note that in Examples 3a and 3b, we have a linear map given by $f(x) = 2x$ and we have $g(x) = f(x) + b$, where $b = -3$. Thus $g(x) = 2x - 3$ gives an affine map from $\mathbf{R}$ to $\mathbf{R}$. In general, to demonstrate that a function $g\colon U \to V$ is an affine map, we must produce a linear map $f: U \rightarrow V$ and an element $\mathbf{b}$ of V such that $g(\mathbf{x}) = f(\mathbf{x}) + \mathbf{b}$. An affine map may be, at the same time, a linear map, but only in case $\mathbf{b} = \mathbf{0}$.

Example 4
Let $g: \mathbf{R}^2 \rightarrow \mathbf{R}^3$ be given by

$$g(x, y) = (2x + y - 3, x - y + 1, x - 3y)$$

This is an example of an affine map which is not linear, for we can write $g(x, y) = f(x, y) + (-3, 1, 0)$, where $f(x, y) = (2x + y, x - y, x - 3y)$. It can be verified that f is a linear map, as required for g to be an affine map. □

Our next example shows how affine maps may relate certain geometric figures; we see that a parallelogram is the result of applying an affine map to the unit square. Finding such relationships is important in many applications of linear algebra.

Example 5
Let $g: \mathbf{R}^2 \rightarrow \mathbf{R}^2$ be given by

$$g(x, y) = (2x + y + 2, y + 1)$$

It is readily seen that g is an affine map, for we can write

$$g(x, y) = (2x + y, y) + (2, 1)$$

and it is easy to determine that $f(x, y) = (2x + y, y)$ gives a linear map. Now, the unit square in $\mathbf{R}^2$ is given by certain linear combinations of the natural basis vectors $(1, 0)$ and $(0, 1)$; in fact, the unit square is seen to be

$$S = \{a(1, 0) + b(0, 1) | 0 \leq a \leq 1, 0 \leq b \leq 1\}$$

We may find the effect of applying g to S if we first find the effect of applying f to S and then add the vector $(2, 1)$ to that result. Because f is a linear map, we see that

$$f(a(1, 0) + b(0, 1)) = af(1, 0) + bf(0, 1)$$

(See Theorem 5.1.) So f maps the unit square S to the set

$$f(S) = \{a(2, 0) + b(1, 1) | 0 \leq a \leq 1, 0 \leq b \leq 1\}$$

which is a parallelogram with one vertex at the origin. Upon adding the constant vector $(2, 1)$ to every vector in $f(S)$ we find that we have merely shifted everything 2 units to the right and 1 unit up. Thus g maps the unit square to the parallelogram with vertices (2, 1), (4, 1), (5, 2), (3, 2). (See Figure 5.2.) □

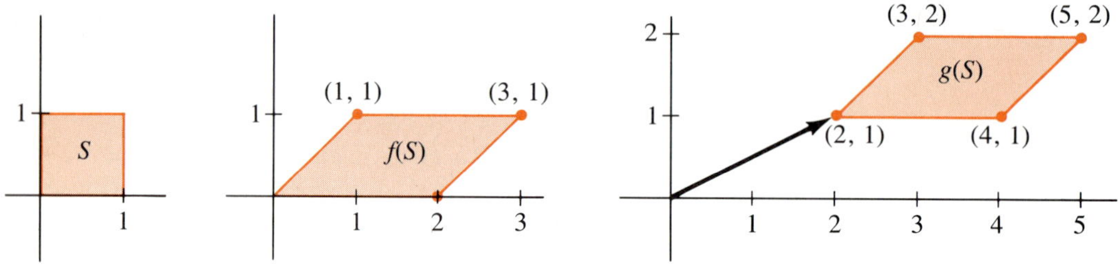

Figure 5.2

Because of conditions LM1 and LM2, linear maps possess certain special properties of computation which are not common to ordinary functions. Two such properties appear in our next two theorems.

Theorem 5.1 Suppose we have a linear map $f\colon U \to V$. Then, for vectors $\mathbf{u}_1, \mathbf{u}_2, \ldots, \mathbf{u}_r$ in U and scalars $c_1, c_2, \ldots, c_r$,

$$f(c_1\mathbf{u}_1 + c_2\mathbf{u}_2 + \cdots + c_r\mathbf{u}_r) = c_1 f(\mathbf{u}_1) + c_2 f(\mathbf{u}_2) + \cdots + c_r f(\mathbf{u}_r)$$

Proof. This is an inductive extension of LM1 and LM2, combined (see Appendix A).

Corollary 5.1a For a linear map $f\colon U \to V$,

$$f(\mathbf{u}_1 - \mathbf{u}_2) = f(\mathbf{u}_1) - f(\mathbf{u}_2)$$

Proof. Combining certain vector-space properties with Theorem 5.1, we find that

$$\begin{aligned} f(\mathbf{u}_1 - \mathbf{u}_2) &= f(\mathbf{u}_1 + (-\mathbf{u}_2)) \\ &= f(\mathbf{u}_1 + (-1)\mathbf{u}_2) \\ &= f(\mathbf{u}_1) + (-1)f(\mathbf{u}_2) \\ &= f(\mathbf{u}_1) - f(\mathbf{u}_2) \end{aligned}$$

We shall find Corollary 5.1a useful in proving Theorem 5.4.

Example 6
The "linearity" of linear maps and of vector spaces permits us to infer much information from a small amount of knowledge. For instance, suppose that, for a linear map $f\colon \mathbf{R}^3 \to \mathbf{R}^2$, we are given only the values of f for the three natural basis vectors:

$$f(1, 0, 0) = (2, -1)$$
$$f(0, 1, 0) = (3, -1)$$
$$f(0, 0, 1) = (1, 0)$$

From this knowledge we may deduce *all* of the values of f. For, if (x, y, z) is any vector in $\mathbf{R}^3$, we may write

$$\begin{aligned} f(x, y, z) &\\ &= f(x(1, 0, 0) + y(0, 1, 0) + z(0, 0, 1)) \\ &= xf(1, 0, 0) + yf(0, 1, 0) + zf(0, 0, 1) \qquad \text{(by Theorem 5.1)} \\ &= x(2, -1) + y(3, -1) + z(1, 0) \\ &= (2x + 3y + z, -x - y) \end{aligned}$$

□

By the same argument as used in Example 6, we have a more general result which we now state.

Corollary 5.1b If the values of a linear map $f\colon U \to V$ are known for vectors in a basis of U, then the values for f on all of U are uniquely determined.

A linear function f from $\mathbf{R}$ to $\mathbf{R}$ is a linear *map* only if its graph is a straight line through the origin; that is, $f(0) = 0$. This is the case for linear maps in general, as our next theorem shows.

Theorem 5.2 For a linear map $f\colon U \to V$, $f(\mathbf{0}) = \mathbf{0}$. (Here, the first $\mathbf{0}$ is the zero vector of U, and the second is that of V.)

Proof. It is possible to devise a proof using either LM1 or LM2 alone. We use LM2: $f(\mathbf{0}) = f(0\mathbf{0}) = 0f(\mathbf{0}) = \mathbf{0}$, by Theorem 4.1.

Theorem 5.2 provides a quick way of determining that certain functions are *not* linear maps. However, the converse is not true, so the condition $f(\mathbf{0}) = \mathbf{0}$ does not ensure that f is a linear map. We may illustrate these comments by returning to two earlier examples.

Example 7

We have seen that the function given by $g(x) = 2x - 3$ is not a linear map because it does not possess property LM1 (Example 3a). Note, however, that $g(0) = -3 \neq 0$, so the necessary condition of Theorem 5.2 is also not satisfied; testing the condition of Theorem 5.2 is noticeably shorter than testing LM1 (or LM2). On the other hand, consider the function given by $f(u) = \sin u$. We have found that this function is not a linear map, but it *does* satisfy the condition $f(0) = 0$. □

Example 8

Let $f\colon \mathbf{R}^2 \to \mathbf{R}^2$ be given by $f(x, y) = (x, e^y)$. We readily find that $f(\mathbf{0}) = f(0, 0) = (0, e^0) = (0, 1) \neq (0, 0) = \mathbf{0}$. Therefore, by Theorem 5.2, f is not a linear map. □

The condition of Theorem 5.2 is *necessary but not sufficient* for a function to be a linear map.

For functions in general, it is often important to know whether or not they are "one-to-one." For instance, in many calculus texts, the fact that $f(x) = \ln x$ is one-to-one is used to establish the existence of the exponential function $g(x) = e^x$. We recall the definition of "one-to-one," a condition which is no less important for linear maps.

☆ Definition

A function f is *one-to-one* (some say "reversible," "invertible," or "injective") provided that if $f(\mathbf{u}_1) = f(\mathbf{u}_2)$, then $\mathbf{u}_1 = \mathbf{u}_2$. In other words, if $\mathbf{u}_1 \neq \mathbf{u}_2$, then $f(\mathbf{u}_1) \neq f(\mathbf{u}_2)$ (that is, distinct elements have distinct images under f).

Example 9

The linear map

$$f(x, y) = (x + y, x - y) \quad x, y \text{ in } \mathbf{R}$$

is one-to-one, since equating $f(\mathbf{u}_1)$ and $f(\mathbf{u}_2)$ implies

$$(x_1 + y_1, x_1 - y_1) = (x_2 + y_2, x_2 - y_2)$$

Thus

$$x_1 + y_1 = x_2 + y_2 \quad \text{and} \quad x_1 - y_1 = x_2 - y_2$$

Adding these equations together gives $2x_1 = 2x_2$, so $x_1 = x_2$; combining this with either equation gives $y_1 = y_2$. Therefore $\mathbf{u}_1 = (x_1, y_1) = (x_2, y_2) = \mathbf{u}_2$, and the function is one-to-one. □

The method applied in Example 9 can be used on functions in general to determine if they are one-to-one. For linear maps, however, there is a simpler test which involves a certain subspace of the domain called the *kernel* of the map.

Definition Suppose we have a linear map $f\colon U \to V$. The *kernel* of f is given by

$$\ker f = \{\mathbf{u} \in U \mid f(\mathbf{u}) = \mathbf{0}\}$$

(See Figure 5.3.)

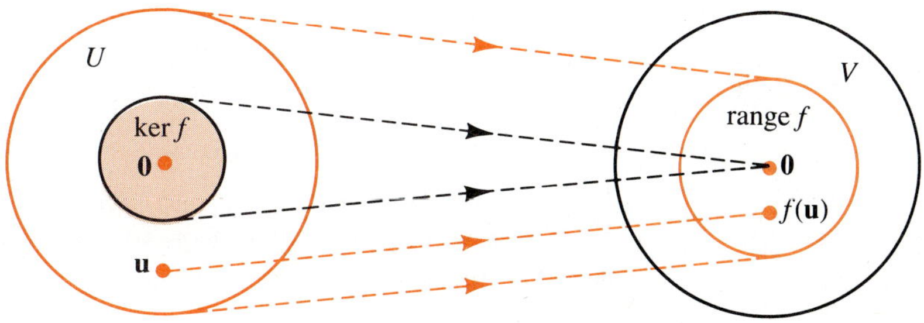

$f\colon U \to V$

Figure 5.3

Theorem 5.3 For a linear map $f\colon U \to V$, $\ker f$ is a subspace of U.

Proof. It is clear from the definition of kernel that $\ker f$ is a sub*set* of U. Furthermore, Theorem 5.2 assures that $\ker f$ contains $\mathbf{0}$. Thus we must verify the closure properties of addition and multiplication. (We are employing Theorem 4.2.) Let $\mathbf{u}_1$ and $\mathbf{u}_2$ be in $\ker f$. Then, $f(\mathbf{u}_1) = \mathbf{0}$ and $f(\mathbf{u}_2) = \mathbf{0}$. For $\mathbf{u}_1 + \mathbf{u}_2$ to belong to $\ker f$, $f(\mathbf{u}_1 + \mathbf{u}_2)$ must be $\mathbf{0}$. But $f(\mathbf{u}_1 + \mathbf{u}_2) = f(\mathbf{u}_1) + f(\mathbf{u}_2)$ (Why?), and $f(\mathbf{u}_1) + f(\mathbf{u}_2) = \mathbf{0} + \mathbf{0} = \mathbf{0}$. So, $\mathbf{u}_1 + \mathbf{u}_2 \in \ker f$. We leave the other closure property to the exercises.

Example 10
Let us examine the kernel of the linear map $f\colon \mathbf{R}^3 \to \mathbf{R}^3$ given by

$$f(x, y, z) = (x - 3y,\ y - z,\ 2x - 6z)$$

(It is routine to verify that f is a linear map.) By definition, a vector $\mathbf{x} = (x, y, z)$ is in $\ker f$ exactly in case

$$\begin{aligned} x - 3y \quad\quad &= 0 \\ y - \ z &= 0 \\ 2x \quad\quad - 6z &= 0 \end{aligned}$$

Thus the kernel is found by solving a homogeneous system of three equations in three unknowns. If we reduce the coefficient matrix to echelon form we obtain the matrix

$$\begin{bmatrix} 1 & 0 & -3 \\ 0 & 1 & -1 \\ 0 & 0 & 0 \end{bmatrix}$$

We see that the third unknown may be declared arbitrary and we have the solutions described by

$$\mathbf{x} = (3t, t, t) \qquad t \text{ arbitrary}$$

Thus $\ker f$ is given by the above parametric equation, which does describe a subspace of $\mathbf{R}^3$; in fact, we see that $\ker f$ is a straight line in $\mathbf{R}^3$ through the origin. □

We are now ready to present the test we promised.

Theorem 5.4 For a linear map $f: U \to V$, f is one-to-one if and only if

$$\ker f = \{\mathbf{0}\}$$

That is, $\ker f$ contains *only* the zero vetor of U.

Proof. To begin, suppose that f is one-to-one. If $\mathbf{u}$ is in $\ker f$, then we have $f(\mathbf{u}) = \mathbf{0}$; on the other hand, we know that $f(\mathbf{0}) = \mathbf{0}$, so $f(\mathbf{u}) = f(\mathbf{0})$. Thus, by virtue of f's being one-to-one, $\mathbf{u} = \mathbf{0}$, so $\ker f$ contains only $\mathbf{0}$.

Conversely, assume that $\ker f = \{\mathbf{0}\}$. Let $f(\mathbf{u}_1) = f(\mathbf{u}_2)$ (from which we must deduce that $\mathbf{u}_1 = \mathbf{u}_2$). We get $f(\mathbf{u}_1) - f(\mathbf{u}_2) = \mathbf{0}$, so $f(\mathbf{u}_1 - \mathbf{u}_2) = \mathbf{0}$, by Corollary 5.1a. This places $\mathbf{u}_1 - \mathbf{u}_2$ in $\ker f$, whose only member is $\mathbf{0}$; therefore $\mathbf{u}_1 - \mathbf{u}_2 = \mathbf{0}$, so $\mathbf{u}_1 = \mathbf{u}_2$. Thus f is one-to-one.

Example 11

We return to the function in Example 9, armed with our new test. In that example, if $(x, y) \in \ker f$, we have $(x + y, x - y) = (0, 0)$; thus

$$\begin{aligned} x + y &= 0 \\ x - y &= 0 \end{aligned}$$

This homogeneous system has only the trivial solution (by Theorem 1.9). Thus $(x, y) = (0, 0) = \mathbf{0}$, so the only member of $\ker f$ is $\mathbf{0}$. Thus f is one-to-one. □

Note also that the linear map of Example 10 cannot be one-to-one, since its kernel was shown there to be a straight line in $\mathbf{R}^3$ through the origin and hence the kernel contains more than merely the zero vector.

Recall that we have said that vectors in $\mathbf{R}^n$ can be displayed vertically as well as horizontally (that is, as $n \times 1$ matrices instead of as $1 \times n$ matrices or as ordered n-tuples). In subsequent sections of this chapter where we explore the close relationship between linear maps and matrices, it will be necessary to use the column format for writing vectors, so we shall begin now to use that alternate mode of expressing vectors.

Since $\ker f$ is a subspace of $\operatorname{dom} f$, and not merely a subset, it is important to be able to determine $\dim \ker f$. Thus in our next example we shall find not only $\ker f$ but also a basis for $\ker f$ and then $\dim \ker f$.

Example 12

Let f be given by

$$f\left(\begin{bmatrix} x \\ y \\ z \\ w \end{bmatrix}\right) = \begin{bmatrix} x + y - 2z \\ 2x - w \\ 2y - 4z + w \end{bmatrix}$$

Here, f: $\mathbf{R}^4 \to \mathbf{R}^3$ and it can be verified that f is a linear map. The kernel of f is, by definition, precisely the set of solutions of the homogeneous system

$$\begin{aligned} x + y - 2z &= 0 \\ 2x - w &= 0 \\ 2y - 4z + w &= 0 \end{aligned}$$

Since there are fewer equations than there are unknowns in the system, we know immediately that there is a nontrival solution, so $\ker f$ is not just $\{\mathbf{0}\}$. Therefore f is not one-to-one. We will now find the *dimension* of $\ker f$ as a subspace of $\mathbf{R}^4$. To do this, we solve the system in the usual way and obtain the echelon form for the coefficient matrix. It is

$$\begin{bmatrix} 1 & 0 & 0 & -\frac{1}{2} \\ 0 & 1 & -2 & \frac{1}{2} \\ 0 & 0 & 0 & 0 \end{bmatrix}$$

We write the solutions, according to our practice, by letting the unknowns that correspond to columns not containing leading ones be arbitrary, and we get

$$\ker f = \left\{ \begin{bmatrix} \frac{1}{2}t \\ 2s - \frac{1}{2}t \\ s \\ t \end{bmatrix} \right\} = \left\{ s\begin{bmatrix} 0 \\ 2 \\ 1 \\ 0 \end{bmatrix} + t\begin{bmatrix} \frac{1}{2} \\ -\frac{1}{2} \\ 0 \\ 1 \end{bmatrix} \right\}$$

This display shows that $\ker f$ is spanned by the two vectors

$$\begin{bmatrix} 0 \\ 2 \\ 1 \\ 0 \end{bmatrix} \quad \text{and} \quad \begin{bmatrix} \frac{1}{2} \\ -\frac{1}{2} \\ 0 \\ 1 \end{bmatrix}$$

These two vectors clearly comprise an independent set. Thus we have actually produced a *basis* for $\ker f$. Since this basis contains two vectors, $\dim \ker f = 2$. □

The range of a linear map is equally as important as the kernel, and we may subject the range to the same close scrutiny that we have given to the kernel. For instance, let us see what we can learn about the range of the linear map f in the preceding example.

By the very definition of range, the range f is the set of all vectors in $\mathbf{R}^3$ having the form

$$\begin{bmatrix} x + y - 2z \\ 2x - w \\ 2y - 4z + w \end{bmatrix}$$

Examination reveals that this set in fact consists of all vectors in $\mathbf{R}^3$ whose third coordinate is twice the first coordinate minus the second coordinate; in other words, the range of f is given by

$$\text{range } f = \left\{ \begin{bmatrix} a \\ b \\ 2a - b \end{bmatrix} \right\}$$

Therefore, vectors in range f have the form

$$a \begin{bmatrix} 1 \\ 0 \\ 2 \end{bmatrix} + b \begin{bmatrix} 0 \\ 1 \\ -1 \end{bmatrix}$$

This shows that range f is the span of

$$\begin{bmatrix} 1 \\ 0 \\ 2 \end{bmatrix} \quad \text{and} \quad \begin{bmatrix} 0 \\ 1 \\ -1 \end{bmatrix}$$

So we have the interesting result that range f is in fact a sub*space* of $\mathbf{R}^3$. Furthermore, these two spanning vectors constitute an independent set and therefore form a *basis* for range f. Consequently, dim range $f = 2$. (There are other, more systematic ways for producing a basis for the range of a linear map, which, by our next theorem, must always be a subspace, and these ways will be forthcoming; for now, however, we rely on inspection.)

Theorem 5.5 For a linear map $f: U \to V$, range f is a subspace of V.

Proof. It is clear from the definition of range that range f is a subset of V. Also, range f is clearly not empty since, by Theorem 5.2, $\mathbf{0}$ (the zero of V) is necessarily a value of f and hence in its range. All we need, according to Theorem 4.2, are the two closure properties. For addition, we let $\mathbf{v}_1$ and $\mathbf{v}_2$ be in range f and deduce that their sum $\mathbf{v}_1 + \mathbf{v}_2$ must also be in range f. By definition of range, there must exist vectors $\mathbf{u}_1$ and $\mathbf{u}_2$ in U such that $f(\mathbf{u}_1) = \mathbf{v}_1$ and $f(\mathbf{u}_2) = \mathbf{v}_2$. Now, $\mathbf{v}_1 + \mathbf{v}_2 = f(\mathbf{u}_1) + f(\mathbf{u}_2) = f(\mathbf{u}_1 + \mathbf{u}_2)$ (Why?), so $\mathbf{v}_1 + \mathbf{v}_2$ is a value of f; that is, $\mathbf{v}_1 + \mathbf{v}_2$ is in range f, so we have closure under addition for range f. We leave proving the closure under scalar multiplication as an exercise.

Example 13

We describe range f and find a basis for, and the dimension of, range f, for the linear map $f\colon \mathbf{R}^4 \to \mathbf{R}^4$ given by

$$f\left(\begin{bmatrix} x \\ y \\ z \\ w \end{bmatrix}\right) = \begin{bmatrix} x - 2y \\ 2y - z \\ x - z \\ y - w \end{bmatrix}$$

Inspection reveals that, in the formula giving the values of f, the third coordinate is the sum of the first two coordinates, while the fourth coordinate may be any number, independent of the first three coordinates. Thus range f may be described by

$$\text{range } f = \left\{\begin{bmatrix} a \\ b \\ a + b \\ c \end{bmatrix}\right\}$$

which we may display as

$$\left\{a\begin{bmatrix} 1 \\ 0 \\ 1 \\ 0 \end{bmatrix} + b\begin{bmatrix} 0 \\ 1 \\ 1 \\ 0 \end{bmatrix} + c\begin{bmatrix} 0 \\ 0 \\ 0 \\ 1 \end{bmatrix}\right\}$$

The above shows that range f is spanned by the set

$$\left\{\begin{bmatrix} 1 \\ 0 \\ 1 \\ 0 \end{bmatrix}, \begin{bmatrix} 0 \\ 1 \\ 1 \\ 0 \end{bmatrix}, \begin{bmatrix} 0 \\ 0 \\ 0 \\ 1 \end{bmatrix}\right\}$$

Moreover, the three vectors here form a linearly independent set, a fact which is readily seen since the particular placement of 1's and 0's shows that in only the trivial way can a linear combination, as displayed above, produce the zero vector. Thus we have found a basis for range f consisting of three vectors and we conclude that dim range $f = 3$. □

If we investigate ker f for the linear map f in the preceding example in the same way we have done in earlier examples, we find that ker f consists of the solutions of a homogeneous system with echelon-form coefficient matrix

$$\begin{bmatrix} 1 & 0 & 0 & -2 \\ 0 & 1 & 0 & -1 \\ 0 & 0 & 1 & -2 \\ 0 & 0 & 0 & 0 \end{bmatrix}$$

as can be confirmed in the customary way. Therefore,

$$\ker f = \left\{ \begin{bmatrix} 2t \\ t \\ 2t \\ t \end{bmatrix} \right\}$$

and a basis for $\ker f$ is

$$\left\{ \begin{bmatrix} 2 \\ 1 \\ 2 \\ 1 \end{bmatrix} \right\}$$

Consequently, $\dim \ker f = 1$.

Now, let us observe that, in Example 13, we have

$$\dim \operatorname{dom} f = 4, \qquad \dim \ker f = 1, \quad \text{and} \quad \dim \operatorname{range} f = 3$$

On the other hand, in Example 12, we find

$$\dim \operatorname{dom} f = 4, \qquad \dim \ker f = 2, \quad \text{and} \quad \dim \operatorname{range} f = 2$$

We may be tempted to conjecture that, in general, $\dim \operatorname{dom} f$ is the *sum* of $\dim \ker f$ and $\dim \operatorname{range} f$. In fact, this relationship among these three spaces does hold in every case, as our next theorem demonstrates.

Theorem 5.6 For a linear map $f: U \to V$, if U is finite-dimensional, then

$$\dim \operatorname{dom} f = \dim \ker f + \dim \operatorname{range} f$$

Proof. We begin by taking a basis $\{\mathbf{u}_1, \mathbf{u}_2, \ldots, \mathbf{u}_r\}$ for $\ker f$. This basis is an independent set in U and hence is a subset of a basis for U. We denote the additional vectors needed (if any) to expand our basis for $\ker f$ to a basis for U by $\mathbf{w}_1, \mathbf{w}_2, \ldots, \mathbf{w}_s$. Thus we have a basis $\{\mathbf{u}_1, \mathbf{u}_2, \ldots, \mathbf{u}_r, \mathbf{w}_1, \mathbf{w}_2, \ldots, \mathbf{w}_s\}$ for U. So far we have $\dim \ker f = r$ and $\dim U = r + s$. If we can show that $\dim \operatorname{range} f = s$, we will be finished; to do this, we must produce a basis for $\operatorname{range} f$ consisting of precisely s vectors. It would be quite natural to try $\{f(\mathbf{w}_1), f(\mathbf{w}_2), \ldots, f(\mathbf{w}_s)\}$. This is certainly a set of s vectors in range f, but we must verify that this set is linearly independent and that it spans range f. For independence, we set $c_1f(\mathbf{w}_1) + c_2f(\mathbf{w}_2) + \cdots + c_sf(\mathbf{w}_s) = \mathbf{0}$. This gives $f(c_1\mathbf{w}_1 + c_2\mathbf{w}_2 + \cdots + c_s\mathbf{w}_s) = \mathbf{0}$, by Theorem 5.1. Therefore $c_1\mathbf{w}_1 + c_2\mathbf{w}_2 + \cdots + c_s\mathbf{w}_s \in \ker f$; as a member of $\ker f$, it can be written as a combination of our basis vectors for $\ker f$:

$$c_1\mathbf{w}_1 + c_2\mathbf{w}_2 + \cdots + c_s\mathbf{w}_s = d_1\mathbf{u}_1 + d_2\mathbf{u}_2 + \cdots + d_r\mathbf{u}_r$$

But then,

$$c_1\mathbf{w}_1 + c_2\mathbf{w}_2 + \cdots + c_s\mathbf{w}_s + (-d_1)\mathbf{u}_1 + (-d_2)\mathbf{u}_2 + \cdots + (-d_r)\mathbf{u}_r = \mathbf{0}$$

The $\mathbf{u}$'s and the $\mathbf{w}$'s together constitute a basis for U, so independence is assured, which makes $c_1 = c_2 = \cdots = c_s = 0$. (The d's must all be zero as well, but our concern was with the c's.) Thus $\{f(\mathbf{w}_1), f(\mathbf{w}_2), \ldots, f(\mathbf{w}_s)\}$ is linearly independent.

Now let $\mathbf{v}$ be any vector in range f. For some $\mathbf{x}$ in U, $\mathbf{v} = f(\mathbf{x})$. Using our basis for U, we can write

$$\mathbf{x} = a_1\mathbf{u}_1 + a_2\mathbf{u}_2 + \cdots + a_r\mathbf{u}_r + b_1\mathbf{w}_1 + b_2\mathbf{w}_2 + \cdots + b_s\mathbf{w}_s$$

so

$$\begin{aligned}\mathbf{v} = f(\mathbf{x}) &= f(a_1\mathbf{u}_1 + \cdots + a_r\mathbf{u}_r + b_1\mathbf{w}_1 + \cdots + b_s\mathbf{w}_s) \\ &= a_1f(\mathbf{u}_1) + \cdots + a_rf(\mathbf{u}_r) + b_1f(\mathbf{w}_1) + \cdots + b_sf(\mathbf{w}_s)\end{aligned}$$

by Theorem 5.1. However, $f(\mathbf{u}_1) = f(\mathbf{u}_2) = \cdots = f(\mathbf{u}_r) = \mathbf{0}$ (Why?). Therefore,

$$\mathbf{v} = b_1f(\mathbf{w}_1) + \cdots + b_sf(\mathbf{w}_s)$$

which displays $\mathbf{v}$ as a linear combination of $f(\mathbf{w}_1), \ldots, f(\mathbf{w}_s)$, assuring that these vectors actually span range f. In short, we have produced a basis for range f consisting of exactly s vectors, so dim range $f = s$. We thus have dim dom f = dim $U = r + s$ = dim ker f + dim range f, as desired.

Observe that the proof we have given for Theorem 5.6 is slightly incomplete, since we have not dealt with the extreme cases where dim ker $f = 0$ or dim range $f = 0$. Relatively minor modifications are required for these special cases.

In certain instances, our way of finding the range of a linear map by inspection might prove inadequate. We may in such situations rely on Theorem 5.6 to find the *dimension* of the range, at least.

Example 14

Consider f: $\mathbf{R}^4 \to \mathbf{R}^3$ given by

$$f\left(\begin{bmatrix} x \\ y \\ z \\ w \end{bmatrix}\right) = \begin{bmatrix} x - 2y + w \\ y + z \\ x - z + w \end{bmatrix}$$

A simple relationship among the coordinates of the values of f seems not to be available by inspection. However, we may proceed to find $\ker f$ in the usual way, which leads to the echelon matrix

$$\begin{bmatrix} 1 & 0 & 0 & 1 \\ 0 & 1 & 0 & 0 \\ 0 & 0 & 1 & 0 \end{bmatrix}$$

We may write the solution for the corresponding homogeneous system and obtain

$$\ker f = \left\{\begin{bmatrix} -t \\ 0 \\ 0 \\ t \end{bmatrix}\right\}$$

Thus a basis for $\ker f$ is

$$\left\{\begin{bmatrix} -1 \\ 0 \\ 0 \\ 1 \end{bmatrix}\right\}$$

Consequently, $\dim \ker f = 1$. Now, $\dim \operatorname{dom} f = \dim \mathbf{R}^4 = 4$. We therefore we find $\dim \operatorname{range} f = \dim \operatorname{dom} f - \dim \ker f = 4 - 1 = 3$. In this example, $\operatorname{range} f \subset \mathbf{R}^3$, so we can conclude that range f *is* $\mathbf{R}^3$ (and, if a basis for range f is required, we may take, for instance, the natural basis for $\mathbf{R}^3$). □

It is customary to employ certain standard terminology and notation for $\dim \ker f$ and $\dim \operatorname{range} f$, as given in our next definition.

☆ **Definition**

For a linear map f: $U \to V$, the *nullity* $N(f)$ *of* f is the dimension of the kernel of f, and the *rank* $R(f)$ *of* f is the dimension of the range of f.

Therefore, the formula in Theorem 5.6 can be written

$$\dim \operatorname{dom} f = \mathrm{N}(f) + \mathrm{R}(f)$$

It may be observed, somewhat informally, that the bigger the kernel, the smaller the range, and vice versa. (Refer to Figure 5.3, page 173.)

Example 15

We use Theorem 5.6 to deduce that, for a linear map $f: U \to V$, if $\dim V < \dim U$, then f cannot be one-to-one. The range of f is a subset of V, so $\mathrm{R}(f) \leq \dim V < \dim U$; therefore, to "balance" the equation of Theorem 5.6, we must have $\mathrm{N}(f) > 0$. Hence $\ker f \neq \{\mathbf{0}\}$, so by Theorem 5.4, f must not be one-to-one. □

☆ **Definition**

For a linear map $f: U \to V$, f is said to be a linear map from U *onto* V provided that range $f = V$ (and is not merely a subset of V). The term "surjective" (*sur* is French for "on") is also used to describe this situation.

Example 16

The linear map $f: \mathbf{R}^4 \to \mathbf{R}^3$ given in Example 12 is *not* a linear map *onto* $\mathbf{R}^3$ because $\dim \mathbf{R}^3 = 3$, whereas we found that dim range $f = 2$, so range f must be a *proper* subset of $\mathbf{R}^3$. On the other hand, the linear map $f: \mathbf{R}^4 \to \mathbf{R}^3$ given in Example 14 *is* a linear map *onto* $\mathbf{R}^3$, since, as we saw there, range $f = \mathbf{R}^3$. □

In discussing whether or not the term "onto" applies to a given linear map, we must be careful to specify *what* we are saying the map is onto. Clearly, every linear map is onto *some* vector space, namely its own range. The question, then, is whether or not, for $f: U \to V$, range $f = V$. (We may not use "onto" as an adjective.)

The following theorem, although simple, provides a key link in connecting two seemingly unrelated concepts of linear algebra.

Theorem 5.7

Let A be an $m \times n$ matrix. Then $f(\mathbf{x}) = A\mathbf{x}$ gives a linear map from $\mathbf{R}^n$ to $\mathbf{R}^m$. (Here, vectors in $\mathbf{R}^n$ are $n \times 1$ matrices and vectors in $\mathbf{R}^m$ are $m \times 1$ matrices.)

Proof. To establish this fact, we need to invoke only two properties of matrix algebra:

$$\begin{aligned} f(\mathbf{x}_1 + \mathbf{x}_2) &= A(\mathbf{x}_1 + \mathbf{x}_2) \\ &= A\mathbf{x}_1 + A\mathbf{x}_2 \qquad \text{(by M8, Section 1.3)} \\ &= f(\mathbf{x}_1) + f(\mathbf{x}_2) \end{aligned}$$

and

$$\begin{aligned} f(c\mathbf{x}) &= A(c\mathbf{x}) \\ &= c(A\mathbf{x}) \qquad \text{(by M10, Section 1.3)} \\ &= cf(\mathbf{x}) \end{aligned}$$

Example 17

Let

$$A = \begin{bmatrix} 1 & 1 & -2 & 0 \\ 2 & 0 & 0 & -1 \\ 0 & 2 & -4 & 1 \end{bmatrix}$$

For vectors in $\mathbf{R}^4$ we write

$$\begin{bmatrix} x \\ y \\ z \\ w \end{bmatrix}$$

Then

$$f\left(\begin{bmatrix} x \\ y \\ z \\ w \end{bmatrix}\right) = \begin{bmatrix} 1 & 1 & -2 & 0 \\ 2 & 0 & 0 & -1 \\ 0 & 2 & -4 & 1 \end{bmatrix}\begin{bmatrix} x \\ y \\ z \\ w \end{bmatrix}$$

gives a linear map from $\mathbf{R}^4$ to $\mathbf{R}^3$. In fact, computing the matrix product involved provides precisely the same linear map as in Example 12, for, because of the way matrices multiply, we obtain

$$f\left(\begin{bmatrix} x \\ y \\ z \\ w \end{bmatrix}\right) = \begin{bmatrix} 1 & 1 & -2 & 0 \\ 2 & 0 & 0 & -1 \\ 0 & 2 & -4 & 1 \end{bmatrix}\begin{bmatrix} x \\ y \\ z \\ w \end{bmatrix} = \begin{bmatrix} x + y - 2z \\ 2x - w \\ 2y - 4z + w \end{bmatrix}$$

□

Theorem 5.7 and Example 17 show that an $m \times n$ matrix gives rise, in a certain sense, to a linear map. In the exercises, we provide certain linear maps and ask for the matrices, thus reversing the situation.

We have now introduced linear maps—those particular functions which are, by virtue of their special properties, most congenial to the structure of vector spaces. We shall find in subsequent sections that linear maps themselves can constitute vector spaces and that they correspond to matrices in a way only slightly hinted at in Theorem 5.7.

5.1 Problem Set

1. Each of the following maps is of the type $f\colon \mathbf{R}^n \to \mathbf{R}^m$. In each case specify m and n and determine whether the formula gives a *linear* map. If it is not linear, is it affine?

   a. $f(x, y, z) = (x^2, y - z)$

   b. $f(x, y) = (x, \sin y)$

   c. $f(x, y, z) = (x + y, x - y, 2x)$

   d. $f(s, t) = (1, 3, 2) + s(0, -1, 3) + t(2, 1, 0)$

   e. $f(x, y, z) = (x + y, 2z + 3)$

   f. $f(t) = (5, 2, 7) + t(4, -2, 2)$

   g. $f(x, y, z) = (x + y + z, y + 2, x + z + 3, z)$

   h. $f\left(\begin{bmatrix} x \\ y \\ z \end{bmatrix}\right) = \begin{bmatrix} x - y \\ y - z \end{bmatrix}$

   i. $f\left(\begin{bmatrix} x \\ y \\ z \end{bmatrix}\right) = \begin{bmatrix} x - y \\ y - z \\ x - z \end{bmatrix}$

   j. $f\left(\begin{bmatrix} x \\ y \\ z \\ w \end{bmatrix}\right) = \begin{bmatrix} x + y - 3w \\ 2z + w + 4 \\ 5x - 2w + 3 \end{bmatrix}$

   k. $f\left(\begin{bmatrix} x \\ y \\ z \end{bmatrix}\right) = \begin{bmatrix} 2 & -1 & 0 \\ 0 & 1 & 4 \end{bmatrix}\begin{bmatrix} x \\ y \\ z \end{bmatrix}$

l. $f(x, y, z, w) = (x - 2y, y - z, z + 2w)$

m. $f(x, y) = (|x| + |y|, x + y)$

2. Of those functions in Problem 1 that are not linear maps, which violate the necessary condition $f(\mathbf{0}) = \mathbf{0}$ of Theorem 5.2?
3. Prove Corollary 5.1b.
4. For each function in Problem 1 that is a linear map, find an appropriate matrix A such that $f(\mathbf{x}) = A\mathbf{x}$, as in Example 17. If the function is not a linear map but is affine, write $f(\mathbf{x})$ in the form $A\mathbf{x} + \mathbf{b}$.
5. For each function in Problem 1 that is a linear map, find ker f, a basis for ker f, N(f), range f, a basis for range f, and R(f). Verify the equation of Theorem 5.6. Determine whether each map is one-to-one and whether it is onto $\mathbf{R}^m$. Describe ker f and range f geometrically (for example, as a plane or a line).
6. Let A be an $m \times n$ matrix. Show that the formula $f(\mathbf{x}) = \mathbf{x}A$ gives a linear map from $\mathbf{R}^m$ to $\mathbf{R}^n$. (Here, $\mathbf{x}$ is written as a $1 \times m$ matrix.)
7. Let D be a function defined on $\mathcal{P}_3$ by the formula

$$D(p(t)) = \frac{d}{dt} p(t)$$

the ordinary first derivative. For example, $D(t - 2t^2 + 3t^3) = 1 - 4t + 9t^2$. Use facts from first-year calculus to show that D is a linear map. What is range D? Why? What is ker D? Why? Verify Theorem 5.6.

8. Let

$$I(p(t)) = \int_0^1 p(t)\, dt \quad \text{for } p(t) \text{ ε } \mathcal{P}_2$$

For instance, $I(1 - 6t^2) = -1$. Use first-year calculus facts to show that I is a linear map. What is range I? Why? What is ker I? Why? Verify Theorem 5.6.

9. Let

$$A(p(u)) = \int_0^t p(u)\, du \quad \text{for } p(u) \text{ ε } \mathcal{P}_1$$

For instance, $A(2 - 3u) = 2t - \frac{3}{2}t^2$. Follow the instructions for Problem 8.

10. Prove Theorem 5.2 using LM1. (*Hint:* Consider $f(\mathbf{0} + \mathbf{0})$.)
11. Finish the proof of Theorem 5.3 by showing that ker f is closed under scalar multiplication.

12. Finish the proof of Theorem 5.5 by showing that the range of a linear map is closed under scalar multiplication.
13. Suppose that we have a linear map $f: U \to V$. Let $B = \{\mathbf{u}_1, \mathbf{u}_2, \ldots, \mathbf{u}_n\}$ be a set of vectors in U.
    a. If B spans U, show that $\{f(\mathbf{u}_1), f(\mathbf{u}_2), \ldots, f(\mathbf{u}_n)\}$ spans range f. (*Hint:* Take $\mathbf{v} \in$ range f and show that $\mathbf{v}$ is a linear combination of $f(\mathbf{u}_1), f(\mathbf{u}_2), \ldots, f(\mathbf{u}_n)$.)
    b. If $\{f(\mathbf{u}_1), f(\mathbf{u}_2), \ldots, f(\mathbf{u}_n)\}$ is linearly independent in V, show that B is linearly independent in U.
14. a. Show that a linear map maps a subspace onto a subspace.
    b. Show that an affine map maps a subspace or an affine subspace onto an affine subspace.
15. Use the natural basis for $\mathbf{R}^4$ and the result of Problem 13a to show that

$$\left\{\begin{bmatrix}1\\2\\0\end{bmatrix}, \begin{bmatrix}1\\0\\2\end{bmatrix}, \begin{bmatrix}-2\\0\\-4\end{bmatrix}, \begin{bmatrix}0\\-1\\1\end{bmatrix}\right\}$$

spans range f for the linear map f in Example 12. Then, by virtue of Theorem 4.15, enough vectors in this spanning set can be discarded to reduce it to a basis for range f. Do this, producing such a basis. (You might not get the same basis for range f as we got in that example, but you should get the same *number* of vectors. See Section 4.4.)

16. Problem 15 gives a method for finding $\mathrm{R}(f) = \dim$ range f, for a linear map f. Apply this method to the functions in Problem 1 which are linear maps and check your results against those of Problem 5.
17. Let $f: U \to V$ satisfy LM1.
    a. If you know mathematical induction, prove, using only LM1, that $f(n\mathbf{u}) = nf(\mathbf{u})$, for any vector $\mathbf{u}$ in U and any positive integer n. (This can be found in Appendix A.) Skip to part b if you do not know this method of proof.
    b. Assume the results of part a and show that $f(m\mathbf{u}) = mf(\mathbf{u})$, for any *negative* integer m.
    c. Use the results of parts a and b to show that, for any integer $q \neq 0$,

$$f\left(\frac{1}{q}\mathbf{u}\right) = \frac{1}{q}f(\mathbf{u})$$

    d. Combine results above to prove that for any rational number r $f(r\mathbf{u}) = rf(\mathbf{u})$.

e. Does part d show that LM2 can be derived from LM1? Why or why not?

18. For the following, use the equation of Theorem 5.6, as we did in Example 9. Let $f\colon U \to V$ be a linear map. Prove:

    a. If $\dim U < \dim V$, then f cannot map *onto* V.

    b. If $\dim U = \dim V$, then f is one-to-one if and only if f maps U onto V.

19. For $f\colon U \to V$, prove that f is a linear map if and only if, for every $\mathbf{u}_1$ and $\mathbf{u}_2$ in U and every pair of scalars r and s, $f(r\mathbf{u}_1 + s\mathbf{u}_2) = rf(\mathbf{u}_1) + sf(\mathbf{u}_2)$.

20. Suppose that, for a linear map $f\colon \mathbf{R}^3 \to \mathbf{R}^2$, it is known that $f(1, 0, 0) = (5, 1)$, $f(0, 1, 0) = (1, 1)$, and $f(0, 0, 1) = (-1, 1)$.

    a. Find $f(-1, 3, 2)$.

    b. Find $f(3, 0, 1)$.

    c. Find the general formula for $f(x, y, z)$. (*Hint:* See Example 6.)

21. Suppose that, for a linear map $f\colon \mathbf{R}^3 \to \mathbf{R}^2$, it is known that $f(1, 1, 1) = (5, 1)$, $f(1, 1, 0) = (1, 1)$, and $f(1, 0, 0) = (-1, 1)$. Repeat parts a, b, and c of Problem 20. (*Hint:* Express the vectors in $\mathbf{R}^3$ in terms of the basis $\{(1, 1, 1), (1, 1, 0), (1, 0, 0)\}$.)

22. Complete the proof of Theorem 5.6 by showing that

$$\dim \operatorname{dom} f = \dim \ker f + \dim \operatorname{range} f$$

in case

a. $\dim \ker f = 0$

b. $\dim \operatorname{range} f = 0$

5.2 The Algebra of Linear Maps and Isomorphisms

In this section, we shall see that vector spaces can be encountered on several levels, for we propose to endow linear maps with an algebraic structure by introducing certain operations on them. One consequence of this will be that the set of *all* linear maps from a given vector space U to a given vector space V will itself constitute a vector space.

In this section, when we write $f: U \to V$ (or the like), we shall automatically mean that f is a linear map from a vector space U to a vector space V.

Definition

Suppose we have $f: U \to V$ and $g: U \to V$. The *sum* $f + g$ is the function given by

$$(f + g)(\mathbf{u}) = f(\mathbf{u}) + g(\mathbf{u}) \qquad \text{for all } \mathbf{u} \in U$$

(This is the usual definition of sum of two functions.)

Theorem 5.8

For $f: U \to V$ and $g: U \to V$, the sum $f + g$ is a linear map from U to V.

Proof. We must establish the validity of LM1 and LM2 for $f + g$.

$$\begin{aligned}(f + g)(\mathbf{u}_1 + \mathbf{u}_2) &= f(\mathbf{u}_1 + \mathbf{u}_2) + g(\mathbf{u}_1 + \mathbf{u}_2) \\ &\qquad \text{(definition of } f + g) \\ &= f(\mathbf{u}_1) + f(\mathbf{u}_2) + g(\mathbf{u}_1) + g(\mathbf{u}_2) \\ &\qquad \text{(LM1 for } f \text{ and } g) \\ &= f(\mathbf{u}_1) + g(\mathbf{u}_1) + f(\mathbf{u}_2) + g(\mathbf{u}_2) \\ &\qquad \text{(VS1, VS2)} \\ &= (f + g)(\mathbf{u}_1) + (f + g)(\mathbf{u}_2) \\ &\qquad \text{(definition of } f + g)\end{aligned}$$

Thus LM1 checks for $f + g$. We leave LM2 to the exercises. It is clear that $f + g$ has the same domain as do f and g, and that range$(f + g) \subset V$.

Example 1

Let $f(x, y) = (x - 2y, 3x, x + y)$ and $g(x, y) = (2x, 0, 5y)$. Then, $(f + g)(x, y) = f(x, y) + g(x, y) = (3x - 2y, 3x, x + 6y)$. All of f, g, and $f + g$ are $\mathbf{R}^2 \to \mathbf{R}^3$. □

Definition

For $f: U \to V$ and any scalar c, the *scalar multiple* cf is the function given by

$$(cf)(\mathbf{u}) = cf(\mathbf{u})$$

Theorem 5.9 For f: $U \to V$ and any scalar c, the scalar multiple cf is a linear map from U to V.

Proof. We test LM1 and LM2 again, for the function cf.

$$\begin{aligned}(cf)(\mathbf{u}_1 + \mathbf{u}_2) &= cf(\mathbf{u}_1 + \mathbf{u}_2) && \text{(definition of } cf\text{)}\\ &= c[f(\mathbf{u}_1) + f(\mathbf{u}_2)] && \text{(LM1 for } f\text{)}\\ &= cf(\mathbf{u}_1) + cf(\mathbf{u}_2) && \text{(VS5)}\\ &= (cf)(\mathbf{u}_1) + (cf)(\mathbf{u}_2) && \text{(definition of } cf\text{)}\end{aligned}$$

Thus, cf satisfies LM1. The rest of the proof is left to the exercises.

Example 2

For $f(x, y, z) = (x - y, 2y + 3z)$, the linear map $6f$ will have the formula $(6f)(x, y, z) = (6x - 6y, 12y + 18z)$, and, like f, will be $\mathbf{R}^3 \to \mathbf{R}^2$. □

Theorem 5.10 For a given vector space U and a given vector space V, the set $\mathcal{L}(U, V)$ of all linear maps from U to V, with addition and scalar multiplication as defined above, constitutes a vector space.

Proof. Theorems 5.8 and 5.9 give us closure under the operations. It remains to verify the vector-space properties VS1–VS8. We provide only a sample here. For instance, consider VS2. For f, g, h, all $U \to V$, we establish

$$(f + g) + h = f + (g + h)$$

We must show that $[(f + g) + h](\mathbf{u}) = [f + (g + h)](\mathbf{u})$, for all $\mathbf{u} \in U$. This runs as follows:

$$\begin{aligned}[(f + g) + h](\mathbf{u}) &= (f + g)(\mathbf{u}) + h(\mathbf{u})\\ &= [f(\mathbf{u}) + g(\mathbf{u})] + h(\mathbf{u})\\ &= f(\mathbf{u}) + [g(\mathbf{u}) + h(\mathbf{u})]\\ &= f(\mathbf{u}) + (g + h)(\mathbf{u})\\ &= [f + (g + h)](\mathbf{u})\end{aligned}$$

Of the other seven vector-space properties, we shall mention only VS3 before consigning them to the exercises: for a "zero vector" of $\mathcal{L}(U, V)$ we need a linear map $U \to V$ which acts as an additive identity in conjunction with other linear maps. For this we take the constant function Θ defined by $\Theta(\mathbf{u}) = \mathbf{0}$ for all $\mathbf{u} \in U$. (The $\mathbf{0}$ on the right of the equation is the zero of V.) It is routine to confirm that Θ is indeed a linear map.

Definition

Let $f: U \to V$ and $g: V \to W$ be linear maps. The *composite* $g \circ f$ is that function given by

$$(g \circ f)(\mathbf{u}) = g(f(\mathbf{u}))$$

(We have here the usual definition of composite function.)

Theorem 5.11

For $f: U \to V$ and $g: V \to W$, the composite $g \circ f$ is a linear map from U to W.

Proof. Again, we must look at LM1 and LM2.

$$\begin{aligned}(g \circ f)(\mathbf{u}_1 + \mathbf{u}_2) &= g(f(\mathbf{u}_1 + \mathbf{u}_2)) && \text{(definition of } g \circ f\text{)}\\ &= g(f(\mathbf{u}_1) + f(\mathbf{u}_2)) && \text{(LM1 for } f\text{)}\\ &= g(f(\mathbf{u}_1)) + g(f(\mathbf{u}_2)) && \text{(LM1 for } g\text{)}\\ &= (g \circ f)(\mathbf{u}_1) + (g \circ f)(\mathbf{u}_2) && \text{(definition of } g \circ f\text{)}\end{aligned}$$

The proof of LM2 is equally routine and is left as an exercise. To see that $g \circ f$ is from U to W, we suggest a glance at this:

$$U \xrightarrow{f} V \xrightarrow{g} W$$

Example 3

Let $f(x, y, z) = (2x, y - x)$ and let $g(r, s) = (r - 3s, r + s)$. We find the explicit formula for $(g \circ f)(x, y, z)$.

$$\begin{aligned}(g \circ f)(x, y, z) = g(f(x, y, z)) &= g(2x, y - x)\\ &= (2x - 3(y - x), 2x + (y - x))\\ &= (5x - 3y, x + y)\end{aligned}$$

Here, we have $f: \mathbf{R}^3 \to \mathbf{R}^2$, $g: \mathbf{R}^2 \to \mathbf{R}^2$, and $g \circ f: \mathbf{R}^3 \to \mathbf{R}^2$. □

Note that $g \circ f$ and $f \circ g$ are not, in general, the same; in fact one of these composites might not even exist, despite the existence of the other, as was the case in Example 3 (where $f \circ g$ does not exist).

Composition gives a "product" of sorts of linear maps. And, with respect to this product, there is an identity, defined as follows.

Definition Let U be a vector space. The *identity map I on U* is given by

$$I(\mathbf{u}) = \mathbf{u} \qquad \text{for all } \mathbf{u} \text{ in } U$$

(It is easily verified that this does define a linear map from U onto U.)

For complete precision, we should write something like I_U for the identity map on U; however, just I will suffice since the domain space can be inferred from the context. The next theorem shows the effect of composing a map with the identity map.

Theorem 5.12 For $f\colon U \to V$,

$$I \circ f = f \quad \text{and} \quad f \circ I = f$$

(The first I is the identity map for V and the second, for U.)

Proof. $(I \circ f)(\mathbf{u}) = I(f(\mathbf{u})) = f(\mathbf{u})$, so $I \circ f = f$. That $f \circ I = f$ is left as an exercise.

Definition Suppose we have $f\colon U \to V$. Then f is called an *isomorphism from U onto V* provided that

a. f is one-to-one.

b. f is onto V.

The identity map on a vector space U is clearly an isomorphism from U onto itself. We can find at least one other example from the previous section.

Example 4

The linear map $f(x, y) = (x + y, x - y)$ was shown in Examples 9 and 11 of Section 5.1 to be one-to-one. This map is also onto $\mathbf{R}^2$ because, by Theorem 5.6,

$$\begin{aligned} 2 &= \dim \mathbf{R}^2 \\ &= \dim \operatorname{dom} f \\ &= \dim \ker f + \dim \operatorname{range} f \\ &= 0 + \dim \operatorname{range} f \end{aligned}$$

So $\dim \operatorname{range} f = 2$, which makes range f *all* of $\mathbf{R}^2$; thus f maps onto $\mathbf{R}^2$. Hence f is an isomorphism from $\mathbf{R}^2$ onto $\mathbf{R}^2$. □

In fact, as Problem 18 of the previous section shows,

> for $f: U \to V$, if $\dim U = \dim V$, then f is an isomorphism from U onto V if *either* f is one-to-one *or* f is onto V.

Recall that any one-to-one function has an inverse function. A one-to-one linear map will have a one-to-one linear inverse, as the next theorem shows.

Theorem 5.13 Suppose we have an isomorphism f from U onto V. Then the correspondence f^{-1} given by

$$f^{-1}(\mathbf{v}) = \mathbf{u} \quad \text{if and only if} \quad f(\mathbf{u}) = \mathbf{v}$$

is a linear map from V onto U. Moreover, f^{-1} is an isomorphism from V onto U.

Proof. We first note that $\operatorname{dom} f^{-1} = V$ precisely because f is onto V; the fact that f^{-1} is a function can be confirmed. We shall test LM1 here in partial proof that f^{-1} is a linear map. We need $f^{-1}(\mathbf{v}_1 + \mathbf{v}_2) = f^{-1}(\mathbf{v}_1) + f^{-1}(\mathbf{v}_2)$. Let $\mathbf{u}_1 = f^{-1}(\mathbf{v}_1)$ and $\mathbf{u}_2 = f^{-1}(\mathbf{v}_2)$. Then

$$f(\mathbf{u}_1 + \mathbf{u}_2) = f(\mathbf{u}_1) + f(\mathbf{u}_2) = \mathbf{v}_1 + \mathbf{v}_2$$

Therefore

$$f^{-1}(\mathbf{v}_1 + \mathbf{v}_2) = \mathbf{u}_1 + \mathbf{u}_2 = f^{-1}(\mathbf{v}_1) + f^{-1}(\mathbf{v}_2)$$

Other details are left as exercises.

Example 5

For $f(x, y) = (x + y, x - y)$, already shown in Example 4 to be an isomorphism from $\mathbf{R}^2$ onto $\mathbf{R}^2$, we find a formula for f^{-1}. Let $f(x, y) = (r, s)$. Since $f^{-1}(r, s) = (x, y)$, we need a formula giving (x, y) in terms of r and s. We have $(x + y, x - y) = (r, s)$; thus

$$x + y = r$$
$$x - y = s$$

Solving for x, y in terms of r, s gives

$$x = \frac{1}{2}(r + s)$$
$$y = \frac{1}{2}(r - s)$$

Thus f^{-1} is given by

$$f^{-1}(r, s) = (x, y) = \left(\frac{1}{2}(r + s), \frac{1}{2}(r - s)\right)$$ □

The next theorem gives the relationship joining an isomorphism, its inverse, and the identity map. Its proof is left as an exercise.

Theorem 5.14 For $f: U \to V$, where f is an isomorphism from U onto V,

$$f \circ f^{-1} = I \quad \text{and} \quad f^{-1} \circ f = I$$

Definition Let U and V be vector spaces. We say that U is *isomorphic* to V provided that there is an isomorphism f from U onto V.

Example 6

Consider $\mathcal{P}_2$ and $\mathbf{R}^3$. Vectors in $\mathcal{P}_2$ are uniquely expressible in the form $a + bt + ct^2$, while those in $\mathbf{R}^3$ are uniquely expressible in the form (a, b, c). To exhibit an isomorphism from $\mathcal{P}_2$ onto $\mathbf{R}^3$, it would be natural to prescribe it as follows:

$$f(a + bt + ct^2) = (a, b, c)$$

It can be verified that this f is indeed an isomorphism from $\mathcal{P}_2$ onto $\mathbf{R}^3$. □

We conclude with some elementary properties of isomorphisms.

Theorem 5.15 For vector spaces U, V, W,

a. U is isomorphic to itself.

b. if U is isomorphic to V, then V is isomorphic to U.

c. if U is isomorphic to V and V is isomorphic to W, then U is isomorphic to W.

Proof. Part a holds because I: $U \to U$ is an isomorphism, and part b follows immediately from Theorem 5.13. Part c is left to the exercises.

Theorem 5.16 For vector spaces U and V, U is isomorphic to V if and only if $\dim U = \dim V$.

The proof is left to the exercises.

Corollary 5.16 A vector space has dimension n if and only if it is isomorphic to $\mathbf{R}^n$.

Recall that we mentioned earlier, in Section 4.4, that $\mathbf{R}^n$ is essentially the only vector space of dimension n; others of dimension n are merely copies of $\mathbf{R}^n$ in different notation. This, in effect, is exactly what Corollary 5.16 asserts.

5.2 Problem Set

1. For linear maps f and g, what, if anything, must be true of their domains and ranges
   a. in order that $f + g$ exist?
   b. in order that $g \circ f$ exist?
2. Complete the proofs of Theorems 5.8, 5.9, and 5.11 by doing the LM2 part of each; complete Theorem 5.10.

3. Let f, g: $\mathbf{R}^3 \to \mathbf{R}^2$ be given by

$$f\left(\begin{bmatrix} x \\ y \\ z \end{bmatrix}\right) = \begin{bmatrix} 2x - y \\ 3z + 2y \end{bmatrix} \quad \text{and} \quad g\left(\begin{bmatrix} x \\ y \\ z \end{bmatrix}\right) = \begin{bmatrix} x - y \\ y - z \end{bmatrix}$$

Write an explicit formula for

$$(f + g)\left(\begin{bmatrix} x \\ y \\ z \end{bmatrix}\right)$$

4. Find, by inspection, matrices A, B, and C such that, for f and g from Problem 3,

$$f(\mathbf{x}) = A\mathbf{x}, \qquad g(\mathbf{x}) = B\mathbf{x}, \quad \text{and} \quad (f + g)(\mathbf{x}) = C\mathbf{x}$$

Can you find a simple realtionship among these three matrices?

5. Let f: $\mathbf{R}^3 \to \mathbf{R}^2$ and g: $\mathbf{R}^2 \to \mathbf{R}^4$ be given by

$$f\left(\begin{bmatrix} x \\ y \\ z \end{bmatrix}\right) = \begin{bmatrix} 2x - y \\ x + z \end{bmatrix} \quad \text{and} \quad g\left(\begin{bmatrix} r \\ t \end{bmatrix}\right) = \begin{bmatrix} 3r + t \\ r - 2t \\ -r + t \\ 2r - 3t \end{bmatrix}$$

Write an explicit formula for

$$(g \circ f)\left(\begin{bmatrix} x \\ y \\ z \end{bmatrix}\right)$$

Does $f \circ g$ exist? Why or why not?

6. Find, by inspection, matrices A, B, C such that, for f and g from Problem 5,

$$g(\mathbf{x}) = A\mathbf{x} \quad \text{where } \mathbf{x} = \begin{bmatrix} r \\ t \end{bmatrix},$$

$$f(\mathbf{x}) = B\mathbf{x} \quad \text{where } \mathbf{x} = \begin{bmatrix} x \\ y \\ z \end{bmatrix},$$

$$\text{and} \quad (g \circ f)(\mathbf{x}) = C\mathbf{x} \quad \text{where } \mathbf{x} = \begin{bmatrix} x \\ y \\ z \end{bmatrix}$$

Can you find a simple relationship among the three matrices A, B, C?

7. Show that the vector spaces $\mathbf{R}^6$ and M_{23} are isomorphic by considering the function that maps

$$(a, b, c, d, e, f) \rightarrow \begin{bmatrix} a & b & c \\ d & e & f \end{bmatrix}$$

Show that the vector spaces M_{23} and M_{32} are isomorphic.

8. Verify that the identity map I on a vector space U is a linear map.
9. Prove Theorem 5.12.
10. The function $f(x, y, z) = (x - y, y - z, x + z)$ is a linear map from $\mathbf{R}^3$ to $\mathbf{R}^3$ with $\mathrm{N}(f) = 0$ and $\mathrm{R}(f) = 3$. Explain why this makes f an isomorphism from $\mathbf{R}^3$ onto $\mathbf{R}^3$. Find an explicit formula for its inverse f^{-1} (as in Example 5 of this section).
11. Verify that the map $f: \mathcal{P}_2 \rightarrow \mathbf{R}^3$ given by $f(a + bt + ct^2) = (a, b, c)$ is an isomorphism.
12. Complete the proof of Theorem 5.13 by showing that f^{-1} is a function; that it is onto U; and that LM2 holds.
13. Suppose $f: U \rightarrow V$. Prove:
    a. f is onto V if and only if f maps every spanning set for U onto a spanning set for V.
    b. f is one-to-one if and only if f maps every linearly independent set in U onto a linearly independent set in V.
    c. f is an isomorphism if and only if f maps every basis for U onto a basis for V.
14. Prove Theorem 5.14 by showing that $(f \circ f^{-1})(\mathbf{v}) = \mathbf{v}$ and $(f^{-1} \circ f)(\mathbf{u}) = \mathbf{u}$ for all $\mathbf{v} \in V$ and $\mathbf{u} \in U$. (*Hint:* Given $\mathbf{v} \in V$, let $\mathbf{u} = f^{-1}(\mathbf{v})$, and given $\mathbf{u} \in U$, let $\mathbf{v} = f(\mathbf{u})$.)
15. Finish the proof of Theorem 5.15.
16. Prove Theorem 5.16. (*Hint:* Refer to Theorem 5.6 and Problem 13c above.)
17. Show that if $f: U \rightarrow V$ and $g: V \rightarrow W$ are each one-to-one, then so is $g \circ f$. Repeat the problem with "one-to-one" replaced by "onto."
18. If $f: U \rightarrow V$, $g: V \rightarrow W$, and $g \circ f$ is
    a. one-to-one, must f be one-to-one? Must g be one-to-one?
    b. onto W, must f be onto V? Must g be onto W?
19. Show that the vector space of Problem 22 of Section 4.1 is isomorphic to $\mathbf{R}^1$.

5.3 Matrices and Linear Maps

The last section may have led to suspicions that the algebra of linear maps might be related somehow to the algebra of matrices, as developed in Section 1.3. Such suspicions would be well-founded. The relationship is so close that, in fact, *the study of linear maps is essentially the same as the study of matrices*. Our goal in this section is to confirm that assertion.

Before continuing, we remark that *all vector spaces in this section are assumed to be finite-dimensional.*

Forging the connection between linear maps and matrices begins with a link provided by Theorem 4.9, where we established that *once a basis for a vector space is chosen*, each vector in the space can be expressed in one and only one way as a linear combination of the basis vectors. Thus, if $A = \{\mathbf{u}_1, \mathbf{u}_2, \ldots, \mathbf{u}_n\}$ is a basis for a vector space V, then, for any vector $\mathbf{x}$ in V, the scalars $c_1, c_2, \ldots, c_n$ satisfying $\mathbf{x} = c_1\mathbf{u}_1 + c_2\mathbf{u}_2 + \cdots + c_n\mathbf{u}_n$ are uniquely determined by $\mathbf{x}$ and A.

Definition

If $A = \{\mathbf{u}_1, \mathbf{u}_2, \ldots, \mathbf{u}_n\}$ is a basis for a vector space V and $\mathbf{x} \in V$, then the unique scalars $c_1, c_2, \ldots, c_n$ such that

$$\mathbf{x} = c_1\mathbf{u}_1 + c_2\mathbf{u}_2 + \cdots + c_n\mathbf{u}_n$$

are called the *coordinates of* $\mathbf{x}$ *with respect to* (or, *relative to*) A. To indicate this, we write

$$\mathbf{x}_A = \begin{bmatrix} c_1 \\ c_2 \\ \vdots \\ c_n \end{bmatrix}$$

Example 1

In $\mathbf{R}^3$, let the chosen basis be

$$A = \left\{ \begin{bmatrix} 1 \\ 1 \\ 1 \end{bmatrix}, \begin{bmatrix} 1 \\ 1 \\ 0 \end{bmatrix}, \begin{bmatrix} 1 \\ 0 \\ 0 \end{bmatrix} \right\}$$

For the vector

$$\begin{bmatrix} -1 \\ 1 \\ 3 \end{bmatrix}$$

we solve

$$\begin{aligned} x + y + z &= -1 \\ x + y \quad &= 1 \\ x \quad\quad &= 3 \end{aligned}$$

and obtain

$$\begin{bmatrix} -1 \\ 1 \\ 3 \end{bmatrix} = (3)\begin{bmatrix} 1 \\ 1 \\ 1 \end{bmatrix} + (-2)\begin{bmatrix} 1 \\ 1 \\ 0 \end{bmatrix} + (-2)\begin{bmatrix} 1 \\ 0 \\ 0 \end{bmatrix}$$

Thus

$$\begin{bmatrix} -1 \\ 1 \\ 3 \end{bmatrix}_A = \begin{bmatrix} 3 \\ -2 \\ -2 \end{bmatrix}$$

On the other hand, had we chosen the *natural* basis

$$B = \left\{ \begin{bmatrix} 1 \\ 0 \\ 0 \end{bmatrix}, \begin{bmatrix} 0 \\ 1 \\ 0 \end{bmatrix}, \begin{bmatrix} 0 \\ 0 \\ 1 \end{bmatrix} \right\}$$

we would have

$$\begin{bmatrix} -1 \\ 1 \\ 3 \end{bmatrix}_B = \begin{bmatrix} -1 \\ 1 \\ 3 \end{bmatrix}$$

(which, of course, is one reason for calling B the natural basis). □

Example 2

In $\mathcal{P}_2$, let $p(t) = 4 + 5t + 5t^2$, and let the chosen basis be $A = \{1 + t, 1 - t, 1 + t^2\}$. As we have done before, we set

$$4 + 5t + 5t^2 = x(1 + t) + y(1 - t) + z(1 + t^2)$$

We obtain

$$4 + 5t + 5t^2 = (x + y + z) + (x - y)t + zt^2$$

Equating corresponding coefficients gives us three equations in the unknowns x, y, z which we solve to obtain $x = 2$, $y = -3$, $z = 5$. Thus

$$(4 + 5t + 5t^2)_A = \begin{bmatrix} 2 \\ -3 \\ 5 \end{bmatrix}$$

□

Example 3

Suppose we are given a basis

$$A = \left\{ \begin{bmatrix} -1 \\ 0 \\ 2 \\ 3 \end{bmatrix}, \begin{bmatrix} 3 \\ 1 \\ 2 \\ 0 \end{bmatrix}, \begin{bmatrix} 1 \\ 3 \\ -2 \\ 1 \end{bmatrix}, \begin{bmatrix} 1 \\ -1 \\ 2 \\ 1 \end{bmatrix} \right\}$$

for $\mathbf{R}^4$ and the *coordinates* of a vector $\mathbf{x}$ relative to the basis A:

$$\mathbf{x}_A = \begin{bmatrix} 0 \\ -2 \\ 4 \\ 3 \end{bmatrix}$$

Then $\mathbf{x}$ itself can be easily found:

$$\mathbf{x} = (0)\begin{bmatrix} -1 \\ 0 \\ 2 \\ 3 \end{bmatrix} + (-2)\begin{bmatrix} 3 \\ 1 \\ 2 \\ 0 \end{bmatrix} + (4)\begin{bmatrix} 1 \\ 3 \\ -2 \\ 1 \end{bmatrix} + (3)\begin{bmatrix} 1 \\ -1 \\ 2 \\ 1 \end{bmatrix} = \begin{bmatrix} 1 \\ 7 \\ -6 \\ 7 \end{bmatrix}$$

□

Just as a particular vector, together with a choice of basis, determines an n-tuple of coordinates which uniquely represents that vector, a particular linear map, together with a choice of bases (bas*es*, since a domain and range are involved), determines a representing *matrix*.

Definition

Let $f\colon U \to V$ be a linear map; let $A = \{\mathbf{u}_1, \mathbf{u}_2, \ldots, \mathbf{u}_n\}$ be a basis for U and $B = \{\mathbf{v}_1, \mathbf{v}_2, \ldots, \mathbf{v}_m\}$ be a basis for V. Let M_f be the matrix whose j-th column is $f(\mathbf{u}_j)_B$; that is, the j-th column of M_f consists of the coordinates of $f(\mathbf{u}_j)$ relative to the basis B. Then M_f is called the *matrix for f relative to A and B*.

Before exhibiting an example, we shall show how such a matrix interacts with the coordinates of vectors.

Theorem 5.17 If f, U, V and M_f are as in the definition above, and if A and B are bases for U and V as above (with $\dim U = n$ and $\dim V = m$), then M_f is an $m \times n$ matrix. Moreover, for any $\mathbf{x}$ in U,

$$f(\mathbf{x})_B = M_f\mathbf{x}_A$$

The right side of this equation is a matrix product; thus the theorem asserts that $f(\mathbf{x})$, or at least its coordinates, can be found by simply computing a certain matrix product.

Proof. That M_f is $m \times n$ is simple to verify: since the j-th column of M_f comes from $f(\mathbf{u}_j)$ and there are exactly n such $\mathbf{u}$'s in the basis for U, M_f must have n columns. Each column has exactly m entries since $f(\mathbf{u}_j)$ has m coordinates relative to the basis B (which consists of m vectors).

Now for the less simple part, which involves some bookkeeping. (Carefully scrutinize the example following this proof, in case the rest of the proof is obscured by subscripts.)

Let $\mathbf{x} = c_1\mathbf{u}_1 + c_2\mathbf{u}_2 + \cdots + c_n\mathbf{u}_n$. Thus

$$\mathbf{x}_A = \begin{bmatrix} c_1 \\ c_2 \\ \cdot \\ \cdot \\ \cdot \\ c_n \end{bmatrix}$$

Let $f(\mathbf{u}_j) = A_{1j}\mathbf{v}_1 + \ldots + A_{mj}\mathbf{v}_m$. Thus the j-th column of M_f is

$$\begin{bmatrix} A_{1j} \\ \cdot \\ \cdot \\ \cdot \\ A_{mj} \end{bmatrix}$$

Therefore

$$M_f = \begin{bmatrix} A_{11} & \cdots & A_{1n} \\ A_{21} & \cdots & A_{2n} \\ & \cdots & \\ A_{m1} & \cdots & A_{mn} \end{bmatrix}$$

We now find the coordinates of $f(\mathbf{x})$ relative to B as follows:

$$\begin{aligned} f(\mathbf{x}) &= f(c_1\mathbf{u}_1 + c_2\mathbf{u}_2 + \cdots + c_n\mathbf{u}_n) \\ &= c_1f(\mathbf{u}_1) + c_2f(\mathbf{u}_2) + \cdots + c_nf(\mathbf{u}_n) \\ &= c_1(A_{11}\mathbf{v}_1 + A_{21}\mathbf{v}_2 + \cdots + A_{m1}\mathbf{v}_m) \\ &\quad + c_2(A_{12}\mathbf{v}_1 + A_{22}\mathbf{v}_2 + \cdots + A_{m2}\mathbf{v}_m) \\ &\quad + \cdots \\ &\quad + c_n(A_{1n}\mathbf{v}_1 + A_{2n}\mathbf{v}_2 + \cdots + A_{mn}\mathbf{v}_m) \\ &= (A_{11}c_1 + A_{12}c_2 + \cdots + A_{1n}c_n)\mathbf{v}_1 \\ &\quad + (A_{21}c_1 + A_{22}c_2 + \cdots + A_{2n}c_n)\mathbf{v}_2 \\ &\quad + \cdots \\ &\quad + (A_{m1}c_1 + A_{m2}c_2 + \cdots + A_{mn}c_n)\mathbf{v}_m \end{aligned}$$

Hence

$$f(\mathbf{x})_B = \begin{bmatrix} A_{11}c_1 + \cdots + A_{1n}c_n \\ A_{21}c_1 + \cdots + A_{2n}c_n \\ \cdots \\ A_{m1}c_1 + \cdots + A_{mn}c_n \end{bmatrix}$$

We compare this with the matrix product $M_f\mathbf{x}_A$:

$$M_f\mathbf{x}_A = \begin{bmatrix} A_{11} & \cdots & A_{1n} \\ A_{21} & \cdots & A_{2n} \\ & \cdots & \\ A_{m1} & \cdots & A_{mn} \end{bmatrix} \begin{bmatrix} c_1 \\ c_2 \\ \vdots \\ c_n \end{bmatrix}$$

These are the same, so we have established that $f(\mathbf{x})_B = M_f\mathbf{x}_A$.

Example 4

We take $f\colon \mathbf{R}^3 \to \mathbf{R}^2$ given by

$$f\left(\begin{bmatrix} x \\ y \\ z \end{bmatrix}\right) = \begin{bmatrix} 2x - 3y \\ x + 2z \end{bmatrix}$$

For $\mathbf{R}^3$, we choose the basis

$$A = \left\{\begin{bmatrix} 1 \\ 1 \\ 1 \end{bmatrix}, \begin{bmatrix} 1 \\ 1 \\ 0 \end{bmatrix}, \begin{bmatrix} 1 \\ 0 \\ 0 \end{bmatrix}\right\}$$

and for $\mathbf{R}^2$ we use

$$B = \left\{ \begin{bmatrix} 1 \\ 1 \end{bmatrix}, \begin{bmatrix} 1 \\ -1 \end{bmatrix} \right\}$$

As prescribed in Theorem 5.17, we find M_f by finding the coordinates of each $f(\mathbf{u}_j)$ relative to the basis B for $\mathbf{R}^2$:

$$f(\mathbf{u}_1) = f\left(\begin{bmatrix} 1 \\ 1 \\ 1 \end{bmatrix} \right) = \begin{bmatrix} -1 \\ 3 \end{bmatrix} = (1)\begin{bmatrix} 1 \\ 1 \end{bmatrix} + (-2)\begin{bmatrix} 1 \\ -1 \end{bmatrix}$$

$$f(\mathbf{u}_2) = f\left(\begin{bmatrix} 1 \\ 1 \\ 0 \end{bmatrix} \right) = \begin{bmatrix} -1 \\ 1 \end{bmatrix} = (0)\begin{bmatrix} 1 \\ 1 \end{bmatrix} + (-1)\begin{bmatrix} 1 \\ -1 \end{bmatrix}$$

$$f(\mathbf{u}_3) = f\left(\begin{bmatrix} 1 \\ 0 \\ 0 \end{bmatrix} \right) = \begin{bmatrix} 2 \\ 1 \end{bmatrix} = (\tfrac{3}{2})\begin{bmatrix} 1 \\ 1 \end{bmatrix} + (\tfrac{1}{2})\begin{bmatrix} 1 \\ -1 \end{bmatrix}$$

Thus

$$f(\mathbf{u}_1)_B = \begin{bmatrix} 1 \\ -2 \end{bmatrix} \qquad f(\mathbf{u}_2)_B = \begin{bmatrix} 0 \\ -1 \end{bmatrix} \qquad \text{and} \quad f(\mathbf{u}_3)_B = \begin{bmatrix} \frac{3}{2} \\ \frac{1}{2} \end{bmatrix}$$

These are the columns of M_f, so

$$M_f = \begin{bmatrix} 1 & 0 & \frac{3}{2} \\ -2 & -1 & \frac{1}{2} \end{bmatrix}$$

To test our result, we pick *any* vector $\mathbf{x}$ in $\mathbf{R}^3$; let us select

$$\mathbf{x} = \begin{bmatrix} -1 \\ 1 \\ 3 \end{bmatrix}$$

since we already used this vector in Example 1, where we had the same basis for $\mathbf{R}^3$, and hence we already have in hand $\mathbf{x}_A$:

$$\mathbf{x}_A = \begin{bmatrix} 3 \\ -2 \\ -2 \end{bmatrix}$$

We find

$$f(\mathbf{x})_B = M_f\mathbf{x}_A = \begin{bmatrix} 1 & 0 & \frac{3}{2} \\ -2 & -1 & \frac{1}{2} \end{bmatrix} \begin{bmatrix} 3 \\ -2 \\ -2 \end{bmatrix} = \begin{bmatrix} 0 \\ -5 \end{bmatrix}$$

Therefore, according to Theorem 5.17, 0 and -5 should be the coordinates of $f(\mathbf{x})$ relative to the basis B. We can check this by direct computation:

$$f(\mathbf{x}) = f\left(\begin{bmatrix} -1 \\ 1 \\ 3 \end{bmatrix}\right) = \begin{bmatrix} -5 \\ 5 \end{bmatrix}$$

from the formula for f. But

$$(0)\begin{bmatrix} 1 \\ 1 \end{bmatrix} + (-5)\begin{bmatrix} 1 \\ -1 \end{bmatrix}$$

does equal

$$\begin{bmatrix} -5 \\ 5 \end{bmatrix}$$

Thus 0 and -5 are indeed the coordinates of $f(\mathbf{x})$ relative to B. □

Example 5

Let $f\colon \mathbf{R}^2 \to \mathbf{R}^3$ be given by

$$f\left(\begin{bmatrix} x \\ y \end{bmatrix}\right) = \begin{bmatrix} x + 2y \\ y - x \\ x + y \end{bmatrix}$$

and choose the basis

$$A = \left\{\begin{bmatrix} 1 \\ 2 \end{bmatrix}, \begin{bmatrix} 2 \\ 1 \end{bmatrix}\right\}$$

for $\mathbf{R}^2$ and the basis

$$B = \left\{\begin{bmatrix} 1 \\ 1 \\ 0 \end{bmatrix}, \begin{bmatrix} 0 \\ 1 \\ 1 \end{bmatrix}, \begin{bmatrix} 1 \\ 0 \\ 1 \end{bmatrix}\right\}$$

for $\mathbf{R}^3$. To find the matrix for f relative to A and B, we need the coordinates relative to B of

$$f\left(\begin{bmatrix} 1 \\ 2 \end{bmatrix}\right) \quad \text{and} \quad f\left(\begin{bmatrix} 2 \\ 1 \end{bmatrix}\right)$$

We compute

$$f\left(\begin{bmatrix}1\\2\end{bmatrix}\right) = \begin{bmatrix}5\\1\\3\end{bmatrix} = (\tfrac{3}{2})\begin{bmatrix}1\\1\\0\end{bmatrix} + (-\tfrac{1}{2})\begin{bmatrix}0\\1\\1\end{bmatrix} + (\tfrac{7}{2})\begin{bmatrix}1\\0\\1\end{bmatrix}$$

$$f\left(\begin{bmatrix}2\\1\end{bmatrix}\right) = \begin{bmatrix}4\\-1\\3\end{bmatrix} = (0)\begin{bmatrix}1\\1\\0\end{bmatrix} + (-1)\begin{bmatrix}0\\1\\1\end{bmatrix} + (4)\begin{bmatrix}1\\0\\1\end{bmatrix}$$

Thus

$$f\left(\begin{bmatrix}1\\2\end{bmatrix}\right)_B = \begin{bmatrix}\frac{3}{2}\\-\frac{1}{2}\\\frac{7}{2}\end{bmatrix} \quad\text{and}\quad f\left(\begin{bmatrix}2\\1\end{bmatrix}\right)_B = \begin{bmatrix}0\\-1\\4\end{bmatrix}$$

We use these for the columns of M_f and obtain

$$M_f = \begin{bmatrix}\frac{3}{2} & 0\\-\frac{1}{2} & -1\\\frac{7}{2} & 4\end{bmatrix}$$

which is more conveniently written as

$$M_f = (\tfrac{1}{2})\begin{bmatrix}3 & 0\\-1 & -2\\7 & 8\end{bmatrix}$$

□

Example 6

Let $D\colon \mathcal{P}_2 \to \mathcal{P}_1$ be the *derivative* map; select the basis $A = \{t^2, t + 1, 2\}$ for $\mathcal{P}_2$, and for $\mathcal{P}_1$ choose $B = \{t - 1, 3\}$. Now,

$$D(t^2) = 2t = (2)(t - 1) + (\tfrac{2}{3})(3)$$

$$D(t + 1) = 1 = (0)(t - 1) + (\tfrac{1}{3})(3)$$

$$D(2) = 0 = (0)(t - 1) + (0)(3)$$

Thus the matrix for D relative to A and B is

$$M_D = \begin{bmatrix}2 & 0 & 0\\\frac{2}{3} & \frac{1}{3} & 0\end{bmatrix}$$

□

Computing M_f is significantly simpler if natural bases are used, as in the next example.

Example 7
We examine the same linear map we saw in Example 4, but we use natural bases for $\mathbf{R}^3$ and $\mathbf{R}^2$. Since

$$f\left(\begin{bmatrix} x \\ y \\ z \end{bmatrix}\right) = \begin{bmatrix} 2x - 3y \\ x + 2z \end{bmatrix}$$

$$f\left(\begin{bmatrix} 1 \\ 0 \\ 0 \end{bmatrix}\right) = \begin{bmatrix} 2 \\ 1 \end{bmatrix} \quad f\left(\begin{bmatrix} 0 \\ 1 \\ 0 \end{bmatrix}\right) = \begin{bmatrix} -3 \\ 0 \end{bmatrix} \quad \text{and} \quad f\left(\begin{bmatrix} 0 \\ 0 \\ 1 \end{bmatrix}\right) = \begin{bmatrix} 0 \\ 2 \end{bmatrix}$$

As we saw in Example 1, these vectors actually display their coordinates relative to the *natural* basis. Therefore

$$M_f = \begin{bmatrix} 2 & -3 & 0 \\ 1 & 0 & 2 \end{bmatrix}$$

□

Observe that the matrix obtained in Example 7 in fact consists precisely of the coefficients in the formula for f. Thus, relative to the *natural* bases, M_f can be written down immediately upon inspection of the formula for f. Let us summarize some of what we have developed.

If A is an $m \times n$ matrix, then the formula

$$f(\mathbf{x}) = A\mathbf{x} \quad \mathbf{x} \in \mathbf{R}^n$$

defines a linear map f from $\mathbf{R}^n$ to $\mathbf{R}^m$ (Theorem 5.7).

Conversely, if f is a linear map from $\mathbf{R}^n$ to $\mathbf{R}^m$, then there is an $m \times n$ matrix A such that

$$f(\mathbf{x}) = A\mathbf{x} \quad \mathbf{x} \in \mathbf{R}^n$$

This matrix A is the matrix for f relative to the natural bases for $\mathbf{R}^n$ and $\mathbf{R}^m$.

To illustrate this last point, note that in Example 7,

$$f\left(\begin{bmatrix} x \\ y \\ z \end{bmatrix}\right) = \begin{bmatrix} 2x - 3y \\ x + 2z \end{bmatrix} = \begin{bmatrix} 2 & -3 & 0 \\ 1 & 0 & 2 \end{bmatrix} \begin{bmatrix} x \\ y \\ z \end{bmatrix}$$

If we have a linear map f from U to U, then U is the domain and also contains the range ($U = V$ in the discussion above). So we can use the same basis in both roles, as the next example illustrates.

Example 8

Let f: $\mathbf{R}^3 \to \mathbf{R}^3$ be given by

$$f\left(\begin{bmatrix} x \\ y \\ z \end{bmatrix}\right) = \begin{bmatrix} 2x - z \\ x + y \\ 3y - 2z \end{bmatrix}$$

Relative to the natural basis for $\mathbf{R}^3$, the matrix for f is

$$\begin{bmatrix} 2 & 0 & -1 \\ 1 & 1 & 0 \\ 0 & 3 & -2 \end{bmatrix}$$

However, relative to the non-natural basis

$$\left\{\begin{bmatrix} 1 \\ 1 \\ 1 \end{bmatrix}, \begin{bmatrix} 1 \\ 1 \\ 0 \end{bmatrix}, \begin{bmatrix} 1 \\ 0 \\ 0 \end{bmatrix}\right\}$$

a different matrix emerges, namely

$$\begin{bmatrix} 1 & 3 & 0 \\ 1 & -1 & 1 \\ -1 & 0 & 1 \end{bmatrix}$$

as can be verified. □

It is natural to wonder if there is any simple matrix relationship between these two matrices that represent the same linear map. We shall address that question immediately after recording the next lemma, whose proof is an exercise.

Lemma 5.18 If each of A and B is an $m \times n$ matrix, and if $A\mathbf{x} = B\mathbf{x}$ for every vector $\mathbf{x}$ in $\mathbf{R}^n$, then $A = B$.

Theorem 5.18 Two $n \times n$ matrices M and N represent the same linear map from $\mathbf{R}^n$ to $\mathbf{R}^n$ (relative to possibly different bases) if and only if, for some nonsingular $n \times n$ matrix C, we have $C^{-1}MC = N$.

Proof. Let us first prove this under the following circumstances:

Let f: $\mathbf{R}^n \to \mathbf{R}^n$;
let M be the matrix for f relative to the *natural* basis for $\mathbf{R}^n$;
let N be the matrix for f relative to an *arbitrary* basis, say
$B = \{\mathbf{u}_1, \mathbf{u}_2, \ldots, \mathbf{u}_n\}$ for $\mathbf{R}^n$.

The desired matrix C is that matrix whose j-th column is the vector $\mathbf{u}_j$. Thus

$$C = \begin{bmatrix} C_{11} & \cdots & C_{1n} \\ C_{21} & \cdots & C_{2n} \\ & \cdots & \\ C_{n1} & \cdots & C_{nn} \end{bmatrix} \quad \text{where } \mathbf{u}_j = \begin{bmatrix} C_{1j} \\ C_{2j} \\ \cdot \\ \cdot \\ \cdot \\ C_{nj} \end{bmatrix}$$

The matrix C is nonsingular since its columns constitute an independent set in $\mathbf{R}^n$ (see Section 4.3). C will be called a "change-of-basis matrix" for the following reason: C changes coordinates relative to B to natural coordinates. To see this, let

$$\mathbf{x} = \begin{bmatrix} x_1 \\ \cdot \\ \cdot \\ \cdot \\ x_n \end{bmatrix}$$

The coordinates $x_1, x_2, \ldots, x_n$ are thus the natural coordinates of $\mathbf{x}$. Suppose that, in terms of the non-natural basis B, we have

$$\mathbf{x} = a_1\mathbf{u}_1 + \cdots + a_n\mathbf{u}_n$$

and hence

$$\mathbf{x}_B = \begin{bmatrix} a_1 \\ \cdot \\ \cdot \\ \cdot \\ a_n \end{bmatrix}$$

We get

$$\begin{bmatrix} x_1 \\ \cdot \\ \cdot \\ \cdot \\ x_n \end{bmatrix} = \mathbf{x} = a_1\begin{bmatrix} C_{11} \\ \cdot \\ \cdot \\ \cdot \\ C_{n1} \end{bmatrix} + \cdots + a_n\begin{bmatrix} C_{1n} \\ \cdot \\ \cdot \\ \cdot \\ C_{nn} \end{bmatrix}$$

or,

$$\begin{bmatrix} x_1 \\ \cdot \\ \cdot \\ \cdot \\ x_n \end{bmatrix} = \begin{bmatrix} a_1C_{11} + \cdots + a_nC_{1n} \\ \cdots \\ a_1C_{n1} + \cdots + a_nC_{nn} \end{bmatrix} = \begin{bmatrix} C_{11} & \cdots & C_{1n} \\ & \cdots & \\ C_{n1} & \cdots & C_{nn} \end{bmatrix}\begin{bmatrix} a_1 \\ \cdot \\ \cdot \\ \cdot \\ a_n \end{bmatrix}$$

Hence $\mathbf{x} = C\mathbf{x}_B$. So C changes B-coordinates to natural coordinates.

It is immediate that $\mathbf{x}_B = C^{-1}\mathbf{x}$, so C^{-1} changes natural coordinates to non-natural coordinates. Now we link everything together to obtain our result. For each $\mathbf{x} \in \mathbf{R}^n$,

$$\begin{aligned} C\mathbf{x}_B &= \mathbf{x} \\ MC\mathbf{x}_B &= M\mathbf{x} = f(\mathbf{x}) \\ C^{-1}MC\mathbf{x}_B &= C^{-1}f(\mathbf{x}) = f(\mathbf{x})_B \end{aligned}$$

Therefore, $C^{-1}MC$ plays the role assigned to N, so (by Lemma 5.18) $C^{-1}MC = N$.

Conversely, if we have $C^{-1}MC = N$, we can write $M = CNC^{-1}$; we define a map g: $\mathbf{R}^n \to \mathbf{R}^n$ by the rule $g(\mathbf{x}) = M\mathbf{x}$. The columns of C are independent, since C is nonsingular, so they constitute a basis B for $\mathbf{R}^n$. As before, $C^{-1}MC\mathbf{x}_B$ must be $g(\mathbf{x})_B$; thus M and N are matrices for the same linear map g, with M relative to the basis B and N to the natural basis.

To establish the validity of Theorem 5.18 in cases where neither of the bases involved is the natural one, we can pass from one basis to the other by way of the natural basis. The relationship between the matrices is then of the form

$$C_1^{-1}C_2MC_2^{-1}C_1 = N$$

that is,

$$(C_2^{-1}C_1)^{-1}M(C_2^{-1}C_1) = N$$

Example 9

We illustrate Theorem 5.18 with the map and matrices from Example 8. There,

$$M = \begin{bmatrix} 2 & 0 & -1 \\ 1 & 1 & 0 \\ 0 & 3 & -2 \end{bmatrix} \qquad N = \begin{bmatrix} 1 & 3 & 0 \\ 1 & -1 & 1 \\ -1 & 0 & 1 \end{bmatrix}$$

Since the non-natural basis in this case is

$$\left\{ \begin{bmatrix} 1 \\ 1 \\ 1 \end{bmatrix}, \begin{bmatrix} 1 \\ 1 \\ 0 \end{bmatrix}, \begin{bmatrix} 1 \\ 0 \\ 0 \end{bmatrix} \right\}$$

the change-of-basis matrix is

$$C = \begin{bmatrix} 1 & 1 & 1 \\ 1 & 1 & 0 \\ 1 & 0 & 0 \end{bmatrix}$$

in which case

$$C^{-1} = \begin{bmatrix} 0 & 0 & 1 \\ 0 & 1 & -1 \\ 1 & -1 & 0 \end{bmatrix}$$

Routine computation will verify that $C^{-1}MC = N$. □

Example 10

We find the matrix for

$$f\left(\begin{bmatrix} x \\ y \end{bmatrix}\right) = \begin{bmatrix} 3x - 2y \\ x + 3y \end{bmatrix}$$

relative to the basis

$$B = \left\{\begin{bmatrix} 1 \\ -1 \end{bmatrix}, \begin{bmatrix} 1 \\ 2 \end{bmatrix}\right\}$$

The change-of-basis matrix is

$$C = \begin{bmatrix} 1 & 1 \\ -1 & 2 \end{bmatrix}$$

so the matrix for f relative to B is

$$C^{-1}\begin{bmatrix} 3 & -2 \\ 1 & 3 \end{bmatrix}C = \frac{1}{3}\begin{bmatrix} 2 & -1 \\ 1 & 1 \end{bmatrix}\begin{bmatrix} 3 & -2 \\ 1 & 3 \end{bmatrix}\begin{bmatrix} 1 & 1 \\ -1 & 2 \end{bmatrix}$$

$$= \frac{1}{3}\begin{bmatrix} 12 & -9 \\ 3 & 6 \end{bmatrix} = \begin{bmatrix} 4 & -3 \\ 1 & 2 \end{bmatrix}$$

To check this, we note that

$$f\left(\begin{bmatrix} 1 \\ -1 \end{bmatrix}\right) = \begin{bmatrix} 5 \\ -2 \end{bmatrix} = (4)\begin{bmatrix} 1 \\ -1 \end{bmatrix} + (1)\begin{bmatrix} 1 \\ 2 \end{bmatrix}$$

and

$$f\left(\begin{bmatrix} 1 \\ 2 \end{bmatrix}\right) = \begin{bmatrix} -1 \\ 7 \end{bmatrix} = (-3)\begin{bmatrix} 1 \\ -1 \end{bmatrix} + (2)\begin{bmatrix} 1 \\ 2 \end{bmatrix}$$ □

We have chosen not to prove Theorem 5.18 in its most general form. However, we shall state it here in the hope that the student will see that, in fact, little modification would be required in the proof we offered to establish the more general case.

Theorem 5.19 *Generalized version of Theorem 5.18.* Two $m \times n$ matrices M and N represent the same linear map (relative to possibly different selections of bases) from an n-dimensional vector space U to an m-dimensional vector space V if and only if there exist a nonsingular $m \times m$ matrix C and a nonsingular $n \times n$ matrix D such that

$$C^{-1}MD = N$$

C is the change-of-basis matrix in V and D is the change-of-basis matrix in U.

Definition Let each of M and N be an $n \times n$ matrix. M is said to be *similar to* N provided that there is a nonsingular $n \times n$ matrix C such that $C^{-1}MC = N$.

In Example 10 we found that

$$C^{-1}\begin{bmatrix} 3 & -2 \\ 1 & 3 \end{bmatrix}C = \begin{bmatrix} 4 & -3 \\ 1 & 2 \end{bmatrix}$$

Therefore,

$$\begin{bmatrix} 3 & -2 \\ 1 & 3 \end{bmatrix}$$

is similar to

$$\begin{bmatrix} 4 & -3 \\ 1 & 2 \end{bmatrix}$$

for example.

Note that Theorem 5.18 now asserts that M and N represent the same linear map if and only if M is similar to N. As our next theorem demonstrates, the relationship of similarity possesses the same three properties we have seen several times before (for instance, in the case of row equivalence and isomorphisms). Such a relationship is called an *equivalence relation*. The proof is left as an exercise.

Theorem 5.20 For $n \times n$ matrices P, Q, and R,

a. P is similar to itself.

b. If P is similar to Q, then Q is similar to P.

c. If P is similar to Q and Q is similar to R, then P is similar to R.

We promised at the outset of this section to confirm that the study of matrices is the same as the study of linear maps, at least in some sense. We are prepared to do so now.

Theorem 5.21 In each of the vector spaces U, V, and W, assume that a basis has been chosen and that the matrices involved are relative to the selected bases.

a. For linear maps f and g from U to V,

$$M_{(f+g)} = M_f + M_g$$

b. For a linear map f from U to V and any scalar c,

$$M_{(cf)} = cM_f$$

c. For a linear map f: $U \to V$ and a linear map g: $V \to W$,

$$M_{g \circ f} = M_g M_f$$

d. For a linear map f: $U \to V$, f is an isomorphism from U onto V if and only if M_f is nonsingular; in this case,

$$M_{(f^{-1})} = M_f^{-1}$$

Proof. We confine our attention to circumstances in which the vector spaces are of the form $\mathbf{R}^n$ and the selected bases are the natural bases (although matters are only marginally more complicated otherwise). For part a, therefore, assume that we have f: $\mathbf{R}^n \to \mathbf{R}^m$ and g: $\mathbf{R}^n \to \mathbf{R}^m$. Since we are using natural bases, we know that $f(\mathbf{x}) = M_f\mathbf{x}$ and $g(\mathbf{x}) = M_g\mathbf{x}$ for all $\mathbf{x}$ in $\mathbf{R}^n$. Therefore

$$\begin{aligned}(f + g)(\mathbf{x}) &= f(\mathbf{x}) + g(\mathbf{x}) = M_f\mathbf{x} + M_g\mathbf{x} \\ &= (M_f + M_g)\mathbf{x} \quad \text{for all } \mathbf{x} \text{ in } \mathbf{R}^n\end{aligned}$$

On the other hand, $(f + g)(\mathbf{x}) = M_{(f+g)}\mathbf{x}$ for all $\mathbf{x}$ in $\mathbf{R}^n$. Therefore, by Lemma 5.18, $M_{(f+g)} = M_f + M_g$. The simplicity of this argument is due to the simple connection between matrices and linear maps.

We prove one more part to make sure the mode of proof is clear. We choose part d and let f: $\mathbf{R}^n \to \mathbf{R}^n$ (we must have $U = V = \mathbf{R}^n$ in this part). Now, if f is invertible, then its inverse map f^{-1} exists. Therefore, $f^{-1}(f(\mathbf{x})) = \mathbf{x}$ for all $\mathbf{x}$ in $\mathbf{R}^n$. The linear map f^{-1} has a matrix $M_{(f^{-1})}$, so we have

$$M_{(f^{-1})}(M_f\mathbf{x}) = \mathbf{x} \quad \text{for all } \mathbf{x} \text{ in } \mathbf{R}^n$$

This is equivalent to

$$(M_{(f^{-1})}M_f)\mathbf{x} = \mathbf{x} \quad \text{for all } \mathbf{x} \text{ in } \mathbf{R}^n$$

so $M_{(f^{-1})}M_f = I$, the identity matrix. Consequently,

$$M_{(f^{-1})} = M_f^{-1}$$

Conversely, if M_f is nonsingular, then the inverse matrix M_f^{-1} exists and we can introduce the linear map g given by

$$g(\mathbf{x}) = M_f^{-1}\mathbf{x} \quad \text{for all } \mathbf{x} \text{ in } \mathbf{R}^n$$

This gives

$$g(f(\mathbf{x})) = M_f^{-1}(M_f\mathbf{x}) = (M_f^{-1}M_f)\mathbf{x} = I\mathbf{x} = \mathbf{x} \quad \text{for all } \mathbf{x}$$

Therefore, $g = f^{-1}$, and it follows that f is an isomorphism.

We leave the other parts of the proof to the exercises.

We illustrate Theorem 5.21, part d, with an example.

Example 11

We consider f: $\mathbf{R}^2 \to \mathbf{R}^2$ given by $f(x, y) = (x + y, x - y)$. We found in Example 5 of Section 5.2 that f, which is an isomorphism from $\mathbf{R}^2$ onto $\mathbf{R}^2$, has for its inverse map the linear map given by

$$f^{-1}(r, s) = \left(\frac{1}{2}(r + s), \frac{1}{2}(r - s)\right)$$

Now, using the natural basis for $\mathbf{R}^2$, we get

$$M_f = \begin{bmatrix} 1 & 1 \\ 1 & -1 \end{bmatrix} \quad \text{and} \quad M_{(f^{-1})} = \begin{bmatrix} \frac{1}{2} & \frac{1}{2} \\ \frac{1}{2} & -\frac{1}{2} \end{bmatrix} = \frac{1}{2}\begin{bmatrix} 1 & 1 \\ 1 & -1 \end{bmatrix}$$

That $M_{(f^{-1})}$ is indeed the inverse of M_f (as Theorem 5.21 asserts) is easy to verify by direct matrix multiplication, or, if one prefers, by finding the inverse of M_f by either row operations or the adjoint method. □

Based on the developments in this section, it is evident that there is a significant correspondence between matrices and linear maps, once bases for the vector spaces involved have been fixed.

> There is a one-to-one correspondence between matrices and linear maps (as expressed in the formula $f(\mathbf{x}) = A\mathbf{x}$) such that, by virtue of Theorem 5.21, sums correspond to sums, scalar multiples to scalar multiples, products to composites, and inverses to inverses. Thus the algebraic properties of matrices are mirrored in linear maps and vice versa.

Therefore, the properties M1–11 in Section 1.3 must apply to linear maps as well as to matrices. For instance, it is not necessary, because of the correspondence, to prove an associative law for composition of maps:

$$(f \circ g) \circ h = f \circ (g \circ h)$$

Its validity is assured by the associative law for matrix multiplication. On the other hand, had we developed the algebra of linear maps first, certain laws for matrices could have been obtained with considerably less effort by invoking the corresponding law for linear maps. The associative law, in fact, is a good example. The only computation required to establish that law for linear maps is this:

$$\begin{aligned} ((f \circ g) \circ h)(\mathbf{x}) &= (f \circ g)(h(\mathbf{x})) = f(g(h(\mathbf{x}))) \\ &= f((g \circ h)(\mathbf{x})) = (f \circ (g \circ h))(\mathbf{x}) \end{aligned}$$

(assuming appropriate domains and ranges). This is in sharp contrast to the effort required in Section 1.3 to deduce associativity for matrix multiplication (Theorem 1.3, page 17).

On page 32 we asserted (without proof) that, for square matrices of the same size, if $AB = I$, then $BA = I$. The proof is difficult, but the proof of the corresponding theorem for linear maps is less formidable. Let us look at the linear-map version: if $f \circ g = I$, then $g \circ f = I$, for f, g: $U \to U$, and I, the identity map on U. We'll sketch the proof here and leave some details to the exercises. (Remember that in this section, all vector spaces are finite-dimensional.) For all $\mathbf{u}$ in U, we have $\mathbf{u} = f(g(\mathbf{u}))$, since $f \circ g = I$. Thus f is a linear map onto U and hence must be one-to-one as well. On the other hand, g must be one-to-one (verify) and consequently is a linear map onto U. Then, for each $\mathbf{u}$ in U, there is a vector $\mathbf{v}$ in U such that $\mathbf{u} = g(\mathbf{v})$. From $f \circ g = I$, we get $f(\mathbf{u}) = \mathbf{v}$. Hence $(g \circ f)(\mathbf{u}) = g(f(\mathbf{u})) = g(\mathbf{v}) = \mathbf{u}$ so $g \circ f = I$ on U.

5.3 Problem Set

1. Find $\mathbf{x}_A$ for

a. $\mathbf{x} = \begin{bmatrix} 1 \\ 2 \end{bmatrix}$; $A = \left\{ \begin{bmatrix} 2 \\ 3 \end{bmatrix}, \begin{bmatrix} 1 \\ 1 \end{bmatrix} \right\}$

b. $\mathbf{x} = \begin{bmatrix} 1 \\ 0 \\ 1 \end{bmatrix}$; $A = \left\{ \begin{bmatrix} 1 \\ -1 \\ 1 \end{bmatrix}, \begin{bmatrix} 0 \\ 1 \\ 1 \end{bmatrix}, \begin{bmatrix} 1 \\ 0 \\ 0 \end{bmatrix} \right\}$

c. $\mathbf{x} = \begin{bmatrix} -1 \\ 1 \\ 0 \\ 1 \end{bmatrix}$; $A = \left\{ \begin{bmatrix} 1 \\ 1 \\ 1 \\ 1 \end{bmatrix}, \begin{bmatrix} 1 \\ 1 \\ 1 \\ 0 \end{bmatrix}, \begin{bmatrix} 1 \\ 1 \\ 0 \\ 0 \end{bmatrix}, \begin{bmatrix} 1 \\ 0 \\ 0 \\ 0 \end{bmatrix} \right\}$

2. Find the matrix for f relative to A and B.

a. $f\left(\begin{bmatrix} x \\ y \end{bmatrix}\right) = \begin{bmatrix} x - y \\ x + y \end{bmatrix}$, $A = \left\{ \begin{bmatrix} 1 \\ 0 \end{bmatrix}, \begin{bmatrix} 0 \\ 1 \end{bmatrix} \right\}$, $B = \left\{ \begin{bmatrix} 2 \\ 3 \end{bmatrix}, \begin{bmatrix} 1 \\ 1 \end{bmatrix} \right\}$

b. $f\left(\begin{bmatrix} x \\ y \\ z \end{bmatrix}\right) = \begin{bmatrix} x + 2y + z \\ y - z \end{bmatrix}$, $A = \left\{ \begin{bmatrix} 1 \\ -1 \\ 1 \end{bmatrix}, \begin{bmatrix} 0 \\ 1 \\ 1 \end{bmatrix}, \begin{bmatrix} 1 \\ 0 \\ 0 \end{bmatrix} \right\}$,

$B = \left\{ \begin{bmatrix} 1 \\ 1 \end{bmatrix}, \begin{bmatrix} 1 \\ 0 \end{bmatrix} \right\}$

c. $f\left(\begin{bmatrix} x \\ y \\ z \end{bmatrix}\right) = \begin{bmatrix} x + y \\ x - z \\ y + z \\ y - x \end{bmatrix}$, $A = \left\{ \begin{bmatrix} 1 \\ 1 \\ 1 \end{bmatrix}, \begin{bmatrix} 1 \\ 1 \\ 0 \end{bmatrix}, \begin{bmatrix} 1 \\ 0 \\ 0 \end{bmatrix} \right\}$,

$B = \left\{ \begin{bmatrix} 1 \\ 1 \\ 1 \\ 1 \end{bmatrix}, \begin{bmatrix} 1 \\ 0 \\ 1 \\ 1 \end{bmatrix}, \begin{bmatrix} 1 \\ 1 \\ 0 \\ 1 \end{bmatrix}, \begin{bmatrix} 1 \\ 1 \\ 1 \\ 0 \end{bmatrix} \right\}$

3. For each part of Problem 2, find $f(\mathbf{x})$ directly from the formula for f where $\mathbf{x}$ is given below. Then find $\mathbf{x}_A$ and use the matrix for the

map to obtain $f(\mathbf{x})_B$. Check your result by finding $f(\mathbf{x})_B$ directly from $f(\mathbf{x})$.

a. $\mathbf{x} = \begin{bmatrix} -2 \\ 1 \end{bmatrix}$ b. $\mathbf{x} = \begin{bmatrix} 1 \\ 1 \\ 1 \end{bmatrix}$ c. $\mathbf{x} = \begin{bmatrix} 1 \\ 2 \\ -1 \end{bmatrix}$

4. In each of the following, write the matrix for f relative to the indicated basis, using the basis for both the domain and the range-containing space.

a. $f\left(\begin{bmatrix} x \\ y \end{bmatrix}\right) = \begin{bmatrix} 2x + y \\ 3x - y \end{bmatrix}$; the natural basis

b. f as in part a; basis: $\left\{\begin{bmatrix} 1 \\ 1 \end{bmatrix}, \begin{bmatrix} 1 \\ -1 \end{bmatrix}\right\}$

c. $f\left(\begin{bmatrix} x \\ y \\ z \end{bmatrix}\right) = \begin{bmatrix} x - y \\ y - z \\ x + z \end{bmatrix}$; the natural basis

d. f as in part c; basis: $\left\{\begin{bmatrix} 1 \\ 0 \\ 1 \end{bmatrix}, \begin{bmatrix} 1 \\ 1 \\ 0 \end{bmatrix}, \begin{bmatrix} 0 \\ 1 \\ 1 \end{bmatrix}\right\}$

5. For Problem 4, parts a and b, and for Problem 4, parts c and d, find the change-of-basis matrix C; verify that $C^{-1}MC = N$, where M and N are as in Theorem 5.18.

6. Prove Lemma 5.18: If each of A and B is an $n \times n$ matrix, and if $A\mathbf{x} = B\mathbf{x}$ for every vector $\mathbf{x}$ in $\mathbf{R}^n$, then $A = B$. (*Hint:* Let $\mathbf{x}$ be, successively, each natural basis vector of $\mathbf{R}^n$, and see what that yields.)

7. Prove Theorem 5.20.

8. Prove parts b and c of Theorem 5.21.

9. Let f: $\mathbf{R}^m \to \mathbf{R}^p$, g, h: $\mathbf{R}^n \to \mathbf{R}^m$ be linear maps. Prove this distributive law: $f \circ (g + h) = f \circ g + f \circ h$.

10. For the linear map f in Problem 4a, find a formula for f^{-1}. Use the formula for f^{-1} to write the matrix for f^{-1} relative to the natural basis for $\mathbf{R}^2$. Check that this is the inverse of the matrix found in 4a. Repeat this process using the non-natural basis given in 4b.

11. Follow the instructions for Problem 10, applying them to Problems 4c and 4d.

12. Show that two similar matrices have the same trace.
13. Let A and B be matrices such that A is similar to B.
    a. Show that if A is nonsingular, then B is nonsingular and A^{-1} is similar to B^{-1}.
    b. Show that A^t is similar to B^t.
14. Show that the only matrix similar to an identity matrix is that identity matrix itself.
15. Complete the details of the proof, at the close of this section, that $f \circ g = I$ implies $g \circ f = I$.

5.4 Nullspace, Columnspace, and Rank of a Matrix

We explore further, in this section, the correlation between matrices and linear maps. Associated with any matrix are three special subspaces.

Definition Let A be an $m \times n$ matrix.

a. The *nullspace* of A is the set $\{\mathbf{x} \in \mathbf{R}^n | A\mathbf{x} = \mathbf{0}\}$.

b. The *columnspace* of A is the subspace of $\mathbf{R}^m$ spanned by the columns of A.

c. The *rowspace* of A is the subspace of $\mathbf{R}^n$ spanned by the rows of A.

Two of the subspaces just defined have already been examined in the context of linear maps.

Theorem 5.22 Let A be an $m \times n$ matrix and let $f: \mathbf{R}^n \to \mathbf{R}^m$ be the linear map given by $f(\mathbf{x}) = A\mathbf{x}$.

a. The nullspace of A is $\ker f$. (Hence the nullspace of A is a subspace of $\mathbf{R}^n$.)

b. The columnspace of A is range f.

Proof. Part a is left to the exercises, but we shall prove part b. Let

$$A = \begin{bmatrix} A_{11} & A_{12} & \cdots & A_{1n} \\ A_{21} & A_{22} & \cdots & A_{2n} \\ & & \cdots & \\ A_{m1} & A_{m2} & \cdots & A_{mn} \end{bmatrix} \qquad \mathbf{x} = \begin{bmatrix} x_1 \\ x_2 \\ \vdots \\ x_n \end{bmatrix}$$

Then it is not difficult to see that

$$f(\mathbf{x}) = A\mathbf{x} = x_1\begin{bmatrix} A_{11} \\ A_{21} \\ \vdots \\ A_{m1} \end{bmatrix} + x_2\begin{bmatrix} A_{12} \\ A_{22} \\ \vdots \\ A_{m2} \end{bmatrix} + \cdots + x_n\begin{bmatrix} A_{1n} \\ A_{2n} \\ \vdots \\ A_{mn} \end{bmatrix}$$

Thus a vector in $\mathbf{R}^n$ is in range f if and only if it is a linear combination of the columns of A.

Definition

For an $m \times n$ matrix A,

a. the *nullity* of A, denoted $\mathrm{N}(A)$, is the dimension of the nullspace of A.

b. the *column rank* of A is the dimension of the columnspace of A.

c. the *row rank* of A is the dimension of the rowspace of A.

Now a theorem from Section 5.2 about linear maps immediately translates to the matrix setting.

Theorem 5.23

For an $m \times n$ matrix A,

$$\mathrm{N}(A) + \text{column rank } A = n$$

Proof. This merely reiterates Theorem 5.6, for if we consider the linear map f: $\mathbf{R}^n \to \mathbf{R}^m$ given by $f(\mathbf{x}) = A\mathbf{x}$, then the assertion of that theorem that dim dom f = dim ker f + dim range f becomes, by virtue of Theorem 5.22, $n = \mathrm{N}(A)$ + column rank A.

Example 1

We will find the nullity, row rank, and column rank of the echelon matrix

$$A = \begin{bmatrix} 1 & 2 & 0 & 3 & 0 \\ 0 & 0 & 1 & -1 & 0 \\ 0 & 0 & 0 & 0 & 1 \\ 0 & 0 & 0 & 0 & 0 \end{bmatrix}$$

The nullspace of A must be the set of solutions of the homogeneous system $A\mathbf{x} = \mathbf{0}$, of which the matrix A is the coefficient matrix. By our well-established procedure for declaring certain unknowns to be arbitrary, we have nullspace $A = \{(-2r - 3s, r, s, s, 0)\} = \{r(-2, 1, 0, 0, 0) + s(-3, 0, 1, 1, 0)\}$. Thus a basis for nullspace A is $\{(-2, 1, 0, 0, 0), (-3, 0, 1, 1, 0)\}$. Hence, $\mathrm{N}(A) = 2$. Inspection reveals that the first three rows constitute an independent set which spans rowspace A; therefore a basis for rowspace A is $\{(1, 2, 0, 3, 0), (0, 0, 1, -1, 0), (0, 0, 0, 0, 1)\}$, so the row rank of A is 3. Finally, inspection also reveals that the first, the third, and the fifth columns of A constitute a basis for the columnspace of A, so the column rank of A is also 3. □

It is no mere coincidence that, in the preceding example, the row rank and the column rank turned out to be equal. This is the case not only for echelon matrices, as the example illustrates, but also for arbitrary matrices. Later we will establish this fact but we record the echelon case now.

Lemma 5.24a For an echelon matrix A, row rank A = column rank A.

Proof. Let the echelon matrix A be $m \times n$. By the method of writing nullspace A, the dimension of nullspace A is equal to the number of columns of A not containing leading 1's. Thus, by Theorem 5.23,

$$\begin{aligned} \text{column rank } A &= n - \mathrm{N}(A) \\ &= (\text{number of columns}) \\ &\quad - (\text{number of columns } \textit{without} \text{ leading 1's}) \\ &= \text{number of columns } \textit{with} \text{ leading 1's} \\ &= \text{number of leading 1's} \end{aligned}$$

Now the nonzero rows of A form a basis for rowspace A since they clearly span the rowspace and constitute a linearly independent set.

Hence

$$\text{row rank } A = \text{number of nonzero rows} = \text{number of leading 1's}$$

Consequently, row rank A = column rank A.

Note that we do not say that row*space* A = column*space* A; this is clearly incorrect since these subspaces, in general, do not even lie in the same vector space. In Example 1, for instance, the former is a subspace of $\mathbf{R}^4$ while the latter lies in $\mathbf{R}^5$.

To establish equality of row rank and column rank in general, we need two more lemmas.

Lemma 5.24b An elementary row operation does not change the row rank of a matrix.

Proof. Since the row rank of a matrix is the dimension of its rowspace, it will be sufficient to show that each of the three types of elementary row operations does not change the rowspace. Each type of operation must be considered in turn, and we leave this to the exercises.

Lemma 5.24c An elementary row operation does not change the column rank of a matrix.

Proof. This proposition profits from translation to the linear-map setting, as we discussed at the end of the last section. We let f: $\mathbf{R}^n \to \mathbf{R}^m$ be the linear map given by $f(\mathbf{x}) = A\mathbf{x}$, where A is $m \times n$. As we have seen, the columnspace of A is range f, so column rank A = dim range f. Now an elementary row operation on A corresponds to an $m \times m$ elementary matrix E which in turn corresponds to an isomorphism g: $\mathbf{R}^m \to \mathbf{R}^m$, since E is nonsingular. Thus column rank EA = dim range $g \circ f$. But, range $g \circ f$ is the range of g when g is applied to range f. In fact, because of g, range f and range $g \circ f$ are isomorphic and hence have the same dimension. Thus column rank EA = dim range $g \circ f$ = dim range f = column rank A, as desired.

Theorem 5.24 For any matrix A, row rank A = column rank A.

Proof. We have seen that, for an *echelon* matrix C, row rank C = column rank C. But, if C is the echelon form for A, then the preceding lemmas guarantee that row rank A = row rank C = column rank C = column rank A.

Corollary 5.24 *How to find a basis for the columnspace of a matrix.* Let A be a matrix and let C be the echelon form for A. If columns numbered $j_1, \ldots, j_r$ constitute a basis for the columnspace of C, then columns numbered $j_1, \ldots, j_r$ constitute a basis for the columnspace of A. In particular, those columns of A corresponding to the columns of C containing leading 1's form a basis for the columnspace of A.

Proof. As in the proof of Lemma 5.24c, applying an elementary row operation does not change the dimension of a columnspace. What is more, since an isomorphism maps a basis onto a basis, each row operation transforms a basis for the columnspace to a basis for the columnspace of the resulting matrix. Since A can be converted to C (or C can be converted to A) by a finite sequence of such operations, a basis for the columnspace of A corresponds to a basis for the columnspace of C.

Definition The *rank* of a matrix A, denoted $\mathrm{R}(A)$, is the common value of row rank A and column rank A.

Theorem 5.25 For an $m \times n$ matrix A,

$$n = \mathrm{N}(A) + \mathrm{R}(A)$$

Proof. This merely recapitulates Theorem 5.23.

Example 2

We illustrate all of the above theorems and lemmas with the matrix

$$A = \begin{bmatrix} 2 & -4 & -3 & -7 \\ 3 & -6 & 1 & 6 \end{bmatrix}$$

For this matrix, the echelon form is

$$C = \begin{bmatrix} 1 & -2 & 0 & 1 \\ 0 & 0 & 1 & 3 \end{bmatrix}$$

We find nullspace $A = \{(2s - t, s, -3t, t)\} = \{s(2, 1, 0, 0) + t(-1, 0, -3, 1)\}$. Hence a basis for the nullspace is $\{(2, 1, 0, 0), (-1, 0, -3, 1)\}$, so $\mathrm{N}(A) = 2$. Since the first and third columns contain the leading 1's in C, the first and third of A will comprise a basis for the columnspace of A: $\{(2, 3), (-3, 1)\}$. Thus $\mathrm{R}(A) = 2$, but this fact could have been deduced from Theorem 5.25 if we did not want to exhibit an actual basis for the columnspace. □

If f is given by $f(\mathbf{x}) = A\mathbf{x}$, then the work above automatically has produced a basis for $\ker f$ (since $\ker f =$ nullspace A), and $\mathrm{N}(f)$ (since $\mathrm{N}(f) = \mathrm{N}(A)$), a basis for range f (since range $f =$ columnspace A), and $\mathrm{R}(f)$ (since $\mathrm{R}(f) = \mathrm{R}(A)$).

Determinants can provide an alternative means of determining the rank of a matrix.

Definition A *submatrix* of a matrix is a matrix obtained by deleting from the original matrix selected rows and/or columns. A *subdeterminant* is the determinant of a square submatrix. (A special type of submatrix is a *minor*, which was introduced on page 54.)

Theorem 5.26 For a matrix A, if s is the largest integer such that there is a nonzero $s \times s$ subdeterminant of A, then $s = \mathrm{R}(A)$.

Proof. Let s and A be as in the statement of the theorem. The s rows of A which contain a submatrix of largest size having nonzero determinant must constitute a linearly independent set, for linear dependence of these rows of A would imply linear dependence if some coordinates were deleted to form the rows of the submatrix. But since the subdeterminant is nonzero, this cannot be. Therefore any basis for rowspace A must contain at least s vectors, so

$$\mathrm{R}(A) = \dim \text{rowspace } A \geq s$$

Let $r = \mathrm{R}(A)$. Then some set of r rows of A is linearly independent. The submatrix consisting of these r rows must therefore have rank r.

But r will also be the column rank of this submatrix, so some set of r columns of this submatrix must be linearly independent. We have produced an $r \times r$ submatrix of A whose columns form a linearly independent set, so its determinant is nonzero. That is, if $r = \mathrm{R}(A)$, then there is an $r \times r$ submatrix having nonzero determinant. Since the largest possible size for such a submatrix is $s \times s$, $\mathrm{R}(A) = r \leq s$. With the earlier inequality, this implies $\mathrm{R}(A) = s$.

Example 3

We will compute the rank of

$$A = \begin{bmatrix} 1 & 1 & 0 & -1 \\ 0 & 2 & 1 & 3 \\ 1 & 1 & 0 & -1 \\ 1 & 1 & 0 & -1 \end{bmatrix}$$

It is a simple matter to spot a 2×2 submatrix having nonzero determinant. Take, for instance,

$$\begin{bmatrix} 1 & 1 \\ 0 & 2 \end{bmatrix}$$

Thus $\mathrm{R}(A)$ is at least 2. To conclude that $\mathrm{R}(A)$ is *exactly* 2, however, we must show that *every* submatrix larger than 2×2 has zero determinant. We deliberately made our example simple enough to make that task easier than would typically be the case. Note, however, that there are 16 different 3×3 submatrices in a 4×4 matrix. □

Example 4

Let us find the rank of

$$A = \begin{bmatrix} 2 & 1 & 0 & 3 \\ 1 & 3 & 1 & 5 \\ 1 & -7 & -3 & -9 \end{bmatrix}$$

Clearly, a nonzero 2×2 subdeterminant can be found—there is one in the upper left corner, for instance. Direct computation shows that all of the following 3×3 subdeterminants are 0:

$$\begin{vmatrix} 2 & 1 & 0 \\ 1 & 3 & 1 \\ 1 & -7 & -3 \end{vmatrix} \quad \begin{vmatrix} 1 & 0 & 3 \\ 3 & 1 & 5 \\ -7 & -3 & -9 \end{vmatrix} \quad \begin{vmatrix} 2 & 0 & 3 \\ 1 & 1 & 5 \\ 1 & -3 & -9 \end{vmatrix} \quad \begin{vmatrix} 2 & 1 & 3 \\ 1 & 3 & 5 \\ 1 & -7 & -9 \end{vmatrix}$$

Therefore $\mathrm{R}(A) = 2$ and, by Theorem 5.25, $\mathrm{N}(A) = 2$.

If more information is desired, such as bases for the rowspace, columnspace, and nullspace, A must be reduced to echelon form. Actually, we can stop one step short of that, with

$$\begin{bmatrix} 1 & 3 & 1 & 1 \\ 0 & 1 & \frac{2}{5} & \frac{7}{5} \\ 0 & 0 & 0 & 0 \end{bmatrix}$$

since at this stage we can already locate the leading 1's. The two nonzero rows form a basis for the rowspace: $\{(1, 3, 1, 1), (0, 1, \frac{2}{5}, \frac{7}{5})\}$. A basis for the columnspace consists of the first two columns (although other choices are possible): $\{(1, 0, 0), (3, 1, 0)\}$. We let the student find a basis for the nullspace. □

Throughout our development of the theory of linear algebra we have encountered a common idea in a number of different formulations. It will be helpful to pause to review some of these variations of that idea.

Let A be an $n \times n$ matrix and let $f: \mathbf{R}^n \to \mathbf{R}^n$ be the linear map given by $f(\mathbf{x}) = A\mathbf{x}$. Then, the following are equivalent; that is, each implies any of the others. (Of course, the negations are all equivalent to each other, too.)

1. A is nonsingular.
2. A has an inverse matrix A^{-1}.
3. $\det A \neq 0$.
4. The homogeneous system $A\mathbf{x} = \mathbf{0}$ has only the trivial solution.
5. The system $A\mathbf{x} = \mathbf{b}$ has a unique solution.
6. The columns of A constitute a basis for $\mathbf{R}^n$.
7. The rows of A constitute a basis for $\mathbf{R}^n$.
8. The echelon form for A is the identity matrix I.
9. A is a product of elementary matrices.
10. The nullspace of A is $\{\mathbf{0}\}$.
11. $\mathrm{N}(A) = 0$.
12. The columnspace of A is $\mathbf{R}^n$.
13. $\mathrm{R}(A) = n$.

14. The rowspace of A is $\mathbf{R}^n$.
15. The rows (or columns) of A constitute a linearly independent set in $\mathbf{R}^n$.
16. The rows (or columns) of A span $\mathbf{R}^n$.
17. f is one-to-one.
18. f is onto $\mathbf{R}^n$.
19. f is an isomorphism from $\mathbf{R}^n$ onto $\mathbf{R}^n$.
20. $\ker f = \{\mathbf{0}\}$.
21. $\text{range } f = \mathbf{R}^n$.
22. f has an inverse linear map f^{-1}.
23. $N(f) = 0$.
24. $R(f) = n$.

5.4 Problem Set

1. For each of the following, find the nullspace, a basis for the nullspace, and N(A); a basis for the rowspace, a basis for the columnspace, and R(A). Corroborate Theorems 5.25 and 5.26.

   a. $A = \begin{bmatrix} 1 & 2 & 3 & 4 \\ 1 & 9 & 10 & 10 \\ -2 & 3 & 1 & -2 \\ -3 & 8 & 5 & 0 \end{bmatrix}$

   b. $A = \begin{bmatrix} 1 & -1 & 2 & 3 & 3 \\ 0 & 1 & 0 & 0 & 1 \\ 2 & 1 & 4 & -1 & 2 \end{bmatrix}$

   c. $A = \begin{bmatrix} 1 & 2 & 1 & 2 \\ 2 & 0 & 1 & 1 \\ -1 & 1 & 2 & -2 \end{bmatrix}$

2. For each part of Problem 1, if f is given by $f(\mathbf{x}) = A\mathbf{x}$, what are a basis for ker f and a basis for range f? Why?

3. Find a basis for, and the dimension of, the subspace of $\mathbf{R}^3$ spanned by

$$\left\{\begin{bmatrix}1\\2\\3\end{bmatrix}, \begin{bmatrix}1\\1\\-1\end{bmatrix}, \begin{bmatrix}2\\1\\0\end{bmatrix}, \begin{bmatrix}2\\0\\-4\end{bmatrix}, \begin{bmatrix}3\\1\\-4\end{bmatrix}\right\}$$

(*Hint:* Use the given vectors as columns of a matrix and consider the columnspace.)

4. Follow the instructions of Problem 3 for the subspace of $\mathbf{R}^4$ spanned by

$$\left\{\begin{bmatrix}1\\2\\0\\-1\end{bmatrix}, \begin{bmatrix}3\\1\\-1\\0\end{bmatrix}, \begin{bmatrix}-7\\1\\3\\-2\end{bmatrix}, \begin{bmatrix}1\\-3\\1\\2\end{bmatrix}\right\}$$

5. Prove Theorem 5.22, part a.

6. Show that the first, third, and fifth columns of the echelon matrix in Example 1 constitute a basis for the columnspace.

7. Show that the first three rows of the echelon matrix in Example 1 constitute a basis for the rowspace.

8. a. Show that if $\{\mathbf{u}, \mathbf{v}, \mathbf{w}\}$ spans a vector space U, then $\{\mathbf{u}, \mathbf{v}, 3\mathbf{w}\}$ also spans U.

   b. Show that if $\{\mathbf{u}, \mathbf{v}, \mathbf{w}\}$ spans a vector space U, then $\{\mathbf{u} - 2\mathbf{v}, \mathbf{v}, \mathbf{w}\}$ also spans U.

   c. Show that if $\{\mathbf{u}, \mathbf{v}, \mathbf{w}\}$ spans a vector space U, then $\{\mathbf{v}, \mathbf{u}, \mathbf{w}\}$ also spans U.

   d. Let A be a $3 \times n$ matrix and let the rows of A be the vectors $\mathbf{u}$, $\mathbf{v}$, and $\mathbf{w}$ in $\mathbf{R}^n$. Show that parts a, b, and c above demonstrate that the rowspace (and hence the row rank) of A is unchanged by three particular elementary row operations (as asserted by Lemma 5.24b).

9. For a 3×5 matrix A,

   a. What is the greatest possible value of R(A)? of N(A)?

   b. What is the smallest possible value of R(A)? of N(A)?

10. Answer the questions in Problem 9 for a 5×3 matrix A.

11. Show that for any matrix A, $R(A^t) = R(A)$.
12. For matrices A and B, compatible for multiplication as required,
    a. show that $R(AB) \leq R(A)$. (*Hint:* Convert to the linear map version and show that dim range $g \circ f \leq$ dim range g.)
    b. show that $R(AB) \leq R(B)$. (*Hint:* Use Problems 11 and 12a.)
13. By Lemma 5.24b an elementary row operation does not change the rank of a matrix. Thus if E is an elementary matrix of appropriate dimensions, then $R(EB) = R(B)$. With A and B compatible for multiplication as required,
    a. if A is nonsingular, show that $R(AB) = R(B)$.
    b. if B is nonsingular, show that $R(AB) = R(A)$.
14. Show that two similar matrices have the same rank and the same nullity.

6

Inner Products, Eigenvalues, and Symmetric Matrices

6.1 Inner Products

In Section 3.2 we introduced the dot product of vectors in $\mathbf{R}^n$ and described some of its properties and uses. There are other important "products," in other vector spaces, which share certain of these properties. They, along with the dot product, are called *inner products*, or *scalar products*.

Definition

An *inner product on a vector space* V is a rule for multiplying pairs of vectors $\mathbf{u}, \mathbf{v} \in V$ to produce scalars, denoted $\langle \mathbf{u}, \mathbf{v} \rangle$, subject to the following rules, where $\mathbf{u}, \mathbf{v}, \mathbf{w} \in V$ and $c \in \mathbf{R}$ are arbitrary:

SP1. $\langle \mathbf{u}, \mathbf{v} \rangle = \langle \mathbf{v}, \mathbf{u} \rangle$

SP2. $\langle \mathbf{u} + \mathbf{w}, \mathbf{v} \rangle = \langle \mathbf{u}, \mathbf{v} \rangle + \langle \mathbf{w}, \mathbf{v} \rangle$

SP3. $\langle c\mathbf{u}, \mathbf{v} \rangle = c\langle \mathbf{u}, \mathbf{v} \rangle$

SP4. $\langle \mathbf{u}, \mathbf{u} \rangle > 0 \quad \text{for } \mathbf{u} \neq \mathbf{0}$

A vector space on which a specific inner product is defined is called an *inner product space*.

Example 1

Let $V = \mathbf{R}^n$. For $\mathbf{u}, \mathbf{v} \in V$ define $\langle \mathbf{u}, \mathbf{v} \rangle = \mathbf{u} \cdot \mathbf{v}$ (the dot product of Chapter 3). That rules SP1–4 are satisfied is proved in Theorem 3.1. □

Example 2

Let $V = \mathscr{C}$ (the vector space of continuous functions $[0, 1] \to \mathbf{R}$). For $f, g \in V$ define

$$\langle f, g \rangle = \int_0^1 f(x)g(x)\, dx$$

Elementary properties of integrals may be used to confirm SP1–4. □

The notions of "length," "angle," "projection," and "orthogonal" can be described entirely in terms of an inner product.

Definition

Let V be an inner product space and let $\mathbf{u}, \mathbf{v} \in V$. The *length* of $\mathbf{u}$ is the number $|\mathbf{u}| = \langle \mathbf{u}, \mathbf{u} \rangle^{1/2}$, and the *angle between* $\mathbf{u}$ *and* $\mathbf{v}$ is

$$\arccos \frac{\langle \mathbf{u}, \mathbf{v} \rangle}{|\mathbf{u}|\,|\mathbf{v}|}$$

We say that $\mathbf{u}$ and $\mathbf{v}$ are *orthogonal* if $\langle \mathbf{u}, \mathbf{v} \rangle = 0$. An *orthogonal set* of vectors is one whose elements are pairwise orthogonal. An *orthonormal set* is an orthogonal set of unit vectors.

Example 3

We readily see that, under the dot product in $\mathbf{R}^3$, the set $\{(1, -2, 1), (2, 1, 0), (-1, 2, 5)\}$ is an orthogonal set, for the dot product of any two vectors in this set is zero. If we multiply each of the vectors by the reciprocal of its length, we can convert each vector to a unit vector without altering their pairwise orthogonality. Thus

$$\left\{ \frac{1}{\sqrt{6}}(1, -2, 1), \frac{1}{\sqrt{5}}(2, 1, 0), \frac{1}{\sqrt{30}}(-1, 2, 5) \right\}$$

is an orthonormal set in $\mathbf{R}^3$. □

Example 4

In the inner product space of Example 2, the functions given by $f(x) = x$ and $g(x) = 2 - 3x$ are orthogonal because

$$\int_0^1 x(2 - 3x)\, dx = 0$$

We may compute the lengths of the vectors f and g in the inner product space and we find

$$|f| = \left(\int_0^1 x^2\, dx\right)^{1/2} = \frac{1}{\sqrt{3}}$$

and

$$|g| = \left(\int_0^1 (2 - 3x)^2\, dx\right)^{1/2} = 1$$

Thus

$$\{\sqrt{3}x, 2 - 3x\}$$

is an orthonormal set. □

Theorem 6.1 If $\{\mathbf{u}_1, \mathbf{u}_2, \ldots, \mathbf{u}_n\}$ is an orthogonal set of nonzero vectors, then it is linearly independent.

Proof. Suppose $c_1\mathbf{u}_1 + c_2\mathbf{u}_2 + \cdots + c_n\mathbf{u}_n = \mathbf{0}$. The inner product of $\mathbf{u}_k$ $(1 \leq k \leq n)$ with each side of this equation gives, on the left,

$$\begin{aligned}
&\langle c_1\mathbf{u}_1 + c_2\mathbf{u}_2 + \cdots + c_k\mathbf{u}_k + \cdots + c_n\mathbf{u}_n, \mathbf{u}_k\rangle \\
&\quad = \langle c_1\mathbf{u}_1, \mathbf{u}_k\rangle + \langle c_2\mathbf{u}_2, \mathbf{u}_k\rangle + \cdots + \langle c_k\mathbf{u}_k, \mathbf{u}_k\rangle \\
&\qquad + \cdots + \langle c_n\mathbf{u}_n, \mathbf{u}_k\rangle && \text{(by SP2)} \\
&\quad = c_1\langle \mathbf{u}_1, \mathbf{u}_k\rangle + c_2\langle \mathbf{u}_2, \mathbf{u}_k\rangle + \cdots + c_k\langle \mathbf{u}_k, \mathbf{u}_k\rangle \\
&\qquad + \cdots + c_n\langle \mathbf{u}_n, \mathbf{u}_k\rangle && \text{(by SP3)} \\
&\quad = 0 + 0 + \cdots + c_k\langle \mathbf{u}_k, \mathbf{u}_k\rangle + \cdots + 0 && \text{(by hypothesis)} \\
&\quad = c_k\langle \mathbf{u}_k, \mathbf{u}_k\rangle
\end{aligned}$$

and on the right, $\langle \mathbf{0}, \mathbf{u}_k\rangle$, which is zero, as an exercise will show. So we have $c_k\langle \mathbf{u}_k, \mathbf{u}_k\rangle = 0$. Now $\langle \mathbf{u}_k, \mathbf{u}_k\rangle > 0$ by SP4. Thus $c_k = 0$ for each k, so we conclude that $\{\mathbf{u}_1, \mathbf{u}_2, \ldots, \mathbf{u}_n\}$ is linearly independent, as desired.

The proofs of the Schwarz Inequality (Section 3.2) and its corollaries can easily be modified to apply to general inner product spaces, since only properties SP1–4 were invoked in their proofs. Thus we may assume hereafter the general Schwarz Inequality $|\langle \mathbf{u}, \mathbf{v}\rangle| \leq |\mathbf{u}||\mathbf{v}|$, which assures us that our generalized concept of angle is well-defined.

Example 5

Let us test the Schwarz Inequality relative to the dot product in $\mathbf{R}^4$ with $\mathbf{u} = (1, 0, -1, 2)$ and $\mathbf{v} = (-3, 1, 2, 1)$. We find that

$$|\mathbf{u} \cdot \mathbf{v}| = |-3| = 3$$

whereas

$$|\mathbf{u}| = \sqrt{6} \quad \text{and} \quad |\mathbf{v}| = \sqrt{15}$$

The assertion that $|\mathbf{u} \cdot \mathbf{v}| \leq |\mathbf{u}||\mathbf{v}|$ is confirmed by the fact that $3 \leq \sqrt{6}\sqrt{15}$. The computations here can also be used to see that the angle between $\mathbf{u}$ and $\mathbf{v}$ is

$$\arccos \frac{-3}{\sqrt{6}\sqrt{15}} \approx 108.4^\circ$$

□

Example 6

The Schwarz Inequality applied to the inner product of Example 2, with $f(x) = x^2$ and $g(x) = x^3$, leads us to the inequality

$$\left|\int_0^1 x^2x^3\,dx\right| \leq \left(\int_0^1 x^2x^2\,dx\right)^{1/2}\left(\int_0^1 x^3x^3\,dx\right)^{1/2}$$

which we may confirm independently. Furthermore, we find that the "angle" between the functions $f(x) = x^2$ and $g(x) = x^3$ is

$$\arccos \frac{\sqrt{35}}{6} \approx 9.6^\circ$$

□

It is often desirable to have an orthogonal basis for a given inner product space. (Note that this is the case with the natural basis for $\mathbf{R}^n$.) Fortunately, for finite-dimensional inner product spaces, such a basis always exists; in the proof of the next theorem we will construct one, using a well-known *orthogonalization process*, which replaces a given basis with an orthogonal one.

Theorem 6.2 *The Gram–Schmidt Process.* Every finite-dimensional inner product space V has an orthogonal basis.

Proof. Let $\{\mathbf{v}_1, \mathbf{v}_2, \ldots, \mathbf{v}_n\}$ be a basis for V. We will obtain an orthogonal basis for V as follows. First, let $\mathbf{u}_1 = \mathbf{v}_1$. Then define

$$\begin{aligned}\mathbf{u}_2 &= \mathbf{v}_2 - \text{proj}_{\mathbf{u}_1}\mathbf{v}_2 \\ &= \mathbf{v}_2 - \frac{\langle \mathbf{v}_2, \mathbf{u}_1\rangle}{\langle \mathbf{u}_1, \mathbf{u}_1\rangle}\mathbf{u}_1\end{aligned}$$

(see Figure 6.1). Now

$$\begin{aligned}\langle \mathbf{u}_1, \mathbf{u}_2\rangle &= \left\langle \mathbf{v}_1, \mathbf{v}_2 - \frac{\langle \mathbf{v}_2, \mathbf{v}_1\rangle}{\langle \mathbf{v}_1, \mathbf{v}_1\rangle}\mathbf{v}_1\right\rangle && (\text{since } \mathbf{u}_1 = \mathbf{v}_1) \\ &= \langle \mathbf{v}_1, \mathbf{v}_2\rangle + \left\langle \mathbf{v}_1, \frac{-\langle \mathbf{v}_2, \mathbf{v}_1\rangle}{\langle \mathbf{v}_1, \mathbf{v}_1\rangle}\mathbf{v}_1\right\rangle && (\text{by SP1 and 2}) \\ &= \langle \mathbf{v}_1, \mathbf{v}_2\rangle - \frac{\langle \mathbf{v}_2, \mathbf{v}_1\rangle}{\langle \mathbf{v}_1, \mathbf{v}_1\rangle}\langle \mathbf{v}_1, \mathbf{v}_1\rangle && (\text{by SP1 and 3}) \\ &= \langle \mathbf{v}_1, \mathbf{v}_2\rangle - \langle \mathbf{v}_2, \mathbf{v}_1\rangle \\ &= 0 && (\text{by SP1})\end{aligned}$$

So $\{\mathbf{u}_1, \mathbf{u}_2\}$ is orthogonal.

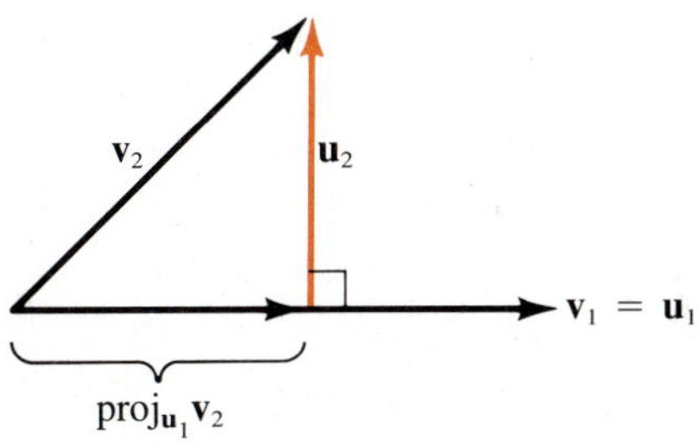

Figure 6.1

We must have $\mathbf{u}_1 \neq \mathbf{0}$, because $\mathbf{u}_1 = \mathbf{v}_1$, and $\mathbf{v}_1$, a member of a linearly independent set, cannot be zero. In addition, $\mathbf{u}_2 \neq \mathbf{0}$ because $\mathbf{u}_2 = \mathbf{0}$ implies

$$\mathbf{v}_2 = \frac{\langle \mathbf{v}_2, \mathbf{v}_1\rangle}{\langle \mathbf{v}_1, \mathbf{v}_1\rangle}\mathbf{v}_1$$

which contradicts the linear independence of $\{\mathbf{v}_1, \mathbf{v}_2\}$.

Thus $\{\mathbf{u}_1, \mathbf{u}_2\}$ is an orthogonal set of *nonzero* vectors.

Now let

$$\mathbf{u}_3 = \mathbf{v}_3 - \text{proj}_{\mathbf{u}_1}\mathbf{v}_3 - \text{proj}_{\mathbf{u}_2}\mathbf{v}_3$$

Using an argument like that above, we can show that $\{\mathbf{u}_1, \mathbf{u}_2, \mathbf{u}_3\}$ is also an orthogonal set of nonzero vectors. The process is continued inductively, ending with

$$\mathbf{u}_n = \mathbf{v}_n - \text{proj}_{\mathbf{u}_1}\mathbf{v}_n - \text{proj}_{\mathbf{u}_2}\mathbf{v}_n - \cdots - \text{proj}_{\mathbf{u}_{n-1}}\mathbf{v}_n$$

The resulting set $\{\mathbf{u}_1, \mathbf{u}_2, \ldots, \mathbf{u}_n\}$, an orthogonal set of nonzero vectors, will be linearly independent (by Theorem 6.1) and thus will be the desired basis of V (by Theorem 4.15).

If an orthonormal basis is desired, simply take

$$\mathbf{w}_k = \frac{1}{|\mathbf{u}_k|}\mathbf{u}_k$$

Then $\{\mathbf{w}_1, \mathbf{w}_2, \ldots, \mathbf{w}_n\}$ will suffice.

Example 7

Let $V = \text{Sp}\{(1, 1, 0, 1), (1, -1, 1, 1), (0, 1, 1, 1)\}$. We apply the Gram–Schmidt process to obtain an orthogonal basis for V and then we convert to an orthonormal basis. Take

$$\begin{aligned}\mathbf{u}_2 &= (1, -1, 1, 1) - \frac{(1, -1, 1, 1)\cdot(1, 1, 0, 1)}{(1, 1, 0, 1)\cdot(1, 1, 0, 1)}(1, 1, 0, 1)\\ &= (1, -1, 1, 1) - \frac{1}{3}(1, 1, 0, 1) = \left(\frac{2}{3}, -\frac{4}{3}, 1, \frac{2}{3}\right)\end{aligned}$$

and

$$\begin{aligned}\mathbf{u}_3 &= (0, 1, 1, 1) - \frac{(0, 1, 1, 1)\cdot(1, 1, 0, 1)}{(1, 1, 0, 1)\cdot(1, 1, 0, 1)}(1, 1, 0, 1)\\ &\quad - \frac{(0, 1, 1, 1)\cdot\left(\frac{2}{3}, -\frac{4}{3}, 1, \frac{2}{3}\right)}{\left(\frac{2}{3}, -\frac{4}{3}, 1, \frac{2}{3}\right)\cdot\left(\frac{2}{3}, -\frac{4}{3}, 1, \frac{2}{3}\right)}\left(\frac{2}{3}, -\frac{4}{3}, 1, \frac{2}{3}\right)\\ &= (0, 1, 1, 1) - \frac{2}{3}(1, 1, 0, 1) - \frac{1}{11}\left(\frac{2}{3}, -\frac{4}{3}, 1, \frac{2}{3}\right)\\ &= \left(\frac{-8}{11}, \frac{5}{11}, \frac{10}{11}, \frac{3}{11}\right)\end{aligned}$$

Then

$$\left\{(1, 1, 0, 1), \left(\frac{2}{3}, -\frac{4}{3}, 1, \frac{2}{3}\right), \left(\frac{-8}{11}, \frac{5}{11}, \frac{10}{11}, \frac{3}{11}\right)\right\}$$

is an orthogonal basis for V, and

$$\left\{\frac{1}{\sqrt{3}}(1, 1, 0, 1), \frac{1}{\sqrt{33}}(2, -4, 3, 2), \frac{1}{3\sqrt{22}}(-8, 5, 10, 3)\right\}$$

is the corresponding orthonormal basis for V. □

Example 8

We find an orthonormal basis for $\mathrm{Sp}\{x^2, x^3\}$, where the inner product is given by $\langle f, g\rangle = \int_0^1 f(x)g(x)\,dx$. With the notation of Theorem 6.2, we have

$$\mathbf{v}_1 = x^2, \qquad \mathbf{v}_2 = x^3, \qquad \mathbf{u}_1 = x^2, \quad \text{and} \quad \mathbf{u}_2 = \mathbf{v}_2 - \frac{\langle \mathbf{v}_2, \mathbf{u}_1\rangle}{\langle \mathbf{u}_1, \mathbf{u}_1\rangle}\mathbf{u}_1$$

Now,

$$\langle \mathbf{v}_2, \mathbf{u}_1\rangle = \int_0^1 x^3x^2\,dx = \frac{1}{6} \quad \text{and} \quad \langle \mathbf{u}_1, \mathbf{u}_1\rangle = \int_0^1 x^2x^2\,dx = \frac{1}{5}$$

so $\mathbf{u}_2 = x^3 - \frac{5}{6}x^2$. An orthogonal basis is then $\{x^2, x^3 - \frac{5}{6}x^2\}$. To make it orthonormal, we need

$$|\mathbf{u}_1| = \left(\int_0^1 x^4\,dx\right)^{1/2} = \frac{1}{\sqrt{5}} \quad \text{and}$$

$$|\mathbf{u}_2| = \left(\int_0^1 \left(x^3 - \frac{5}{6}x^2\right)^2 dx\right)^{1/2} = \frac{1}{\sqrt{252}}$$

Thus we find that $\{\sqrt{5}x^2, \sqrt{252}(x^3 - \frac{5}{6}x^2)\}$ is the desired orthonormal basis. □

6.1 Problem Set

1. Relative to the dot product on $\mathbf{R}^3$, determine which of the following sets of vectors are orthogonal.
   a. $\{(1, 0, 1), (0, 1, 0), (-1, 0, 1)\}$
   b. $\{(1, 1, 0), (1, 0, 1), (0, 1, 0)\}$
   c. $\{(2, 0, -2), (3, 2, 3), (1, -3, 1)\}$
   d. $\{(4, 1, 3), (2, 0, -3), (1, 4, 1)\}$

2. Relative to the dot product on $\mathbf{R}^4$, find the angle between each of the following pairs of vectors.
   a. (1, 2, 0, 1) and (2, 0, 1, 3)
   b. (3, −1, 0, 0) and (2, −3, 5, 7)
   c. (2, 6, 0, −4) and (−1, −3, 0, 2)
   d. (1, 1, 3, 4) and (−4, 1, 0, 2)
3. Relative to the inner product given by $\langle f, g\rangle = \int_0^1 f(x)g(x)\,dx$, find the angles between the following pairs of functions.
   a. x and x^2
   b. $\sin \pi x$ and C (a nonzero constant)
   c. x and x^n $(n \geq 0)$
   d. x and e^x
4. Show that $\{1, \sin x, \sin 2x, \ldots, \sin nx, \ldots; \cos x, \cos 2x, \ldots, \cos mx, \ldots\}$ is orthogonal relative to the inner product

$$\langle f, g\rangle = \int_{-\pi}^{\pi} f(x)g(x)\,dx$$

   for continuous functions $[-\pi, \pi] \to \mathbf{R}$.
5. With the indicated inner product, find an orthogonal basis for each of the following subspaces.
   a. Sp{(2, 1, 3), (−1, 3, 2)}; the dot product of $\mathbf{R}^3$
   b. Sp{(1, 0, 1, 2), (−1, 3, 1, 1), (0, 1, 0, 2)}; the dot product of $\mathbf{R}^4$
   c. Sp$\{x, x^2\}$; the inner product of Problem 3
6. Relative to the indicated inner products, find orthonormal bases for the spaces given.
   a. Sp{(2, 0, 2, 1), (−1, 3, 3, 1)}; the dot product of $\mathbf{R}^4$
   b. Sp$\{2x, x^3, x^5\}$; the inner product of Problem 3
7. In an arbitrary inner product space V, prove that $\langle \mathbf{u}, \mathbf{0}\rangle = 0$
   a. using SP3.
   b. using SP2.
8. Prove that the vectors $\mathbf{u}_1$, $\mathbf{u}_2$, $\mathbf{u}_3$ as obtained in Theorem 6.2 are nonzero and pairwise orthogonal.
9. Let $B = \{\mathbf{u}_1, \mathbf{u}_2, \ldots, \mathbf{u}_n\}$ be an orthonormal basis for an inner product space V. For each $\mathbf{v} \in V$ we may write $\mathbf{v} = a_1\mathbf{u}_1 + a_2\mathbf{u}_2 + \cdots + a_n\mathbf{u}_n$. Show that the coefficients $a_1, a_2, \ldots, a_n$

(the coordinates of $\mathbf{v}$ relative to B) are given by $a_k = \langle \mathbf{u}_k, \mathbf{v} \rangle$, $k = 1, 2, \ldots, n$.

10. a. Show that

$$\left\{ \frac{1}{\sqrt{2}}(1, 0, -1), \frac{1}{\sqrt{3}}(1, -1, 1), \frac{1}{\sqrt{6}}(1, 2, 1) \right\}$$

is an orthonormal basis for $\mathbf{R}^3$, using the usual dot product in $\mathbf{R}^3$.

b. Find the coordinates of each of the following vectors relative to the basis in part a.

$$\mathbf{x}_1 = (1, 0, -2) \qquad \mathbf{x}_2 = (1, 1, 1)$$
$$\mathbf{x}_3 = (4, 2, 0) \qquad \mathbf{x}_4 = (1, 0, -1)$$

11. Let $\{\mathbf{u}_1, \mathbf{u}_2, \ldots, \mathbf{u}_n\}$ be a set of unit vectors in a vector space V. Prove that $\{\mathbf{u}_1, \mathbf{u}_2, \ldots, \mathbf{u}_n\}$ is an orthonormal basis for V if and only if, for every vector $\mathbf{v}$ in V,

$$|\mathbf{v}|^2 = \langle \mathbf{u}_1, \mathbf{v} \rangle^2 + \langle \mathbf{u}_2, \mathbf{v} \rangle^2 + \cdots + \langle \mathbf{u}_n, \mathbf{v} \rangle^2$$

(*Hint:* Use the result of Problem 9.)

12. Let A be a symmetric 2×2 matrix with $\det A > 0$ and $A_{11} > 0$. (Such a matrix is called *positive definite*.) For $\mathbf{x}, \mathbf{y} \in \mathbf{R}^2$ define $\langle \mathbf{x}, \mathbf{y} \rangle = \mathbf{x}^t A \mathbf{y}$. ($\mathbf{x}$ and $\mathbf{y}$ are written in column format as 2×1 matrices.) Show that $\langle \mathbf{x}, \mathbf{y} \rangle$ satisfies SP1–4 and hence gives an inner product on $\mathbf{R}^2$.

13. If $\langle \mathbf{x}, \mathbf{y} \rangle$ is any inner product on $\mathbf{R}^2$, prove that there is a symmetric 2×2 matrix A with $\det A > 0$ and $A_{11} > 0$ such that $\langle \mathbf{x}, \mathbf{y} \rangle = \mathbf{x}^t A \mathbf{y}$. (*Hint:* Let $\mathbf{i} = (1, 0)$ and $\mathbf{j} = (0, 1)$. Write $\mathbf{x} = x_1\mathbf{i} + x_2\mathbf{j}$ and $\mathbf{y} = y_1\mathbf{i} + y_2\mathbf{j}$. Show that

$$A = \begin{bmatrix} \langle \mathbf{i}, \mathbf{i} \rangle & \langle \mathbf{i}, \mathbf{j} \rangle \\ \langle \mathbf{i}, \mathbf{j} \rangle & \langle \mathbf{j}, \mathbf{j} \rangle \end{bmatrix}$$

meets the requirements.)

14. Recall that a sequence $a = (a_1, a_2, \ldots)$ is *square-summable* if $\sum_{k=1}^{\infty} a_k^2$ is finite. Prove that the product defined by $\langle a, b \rangle = \sum_{k=1}^{\infty} a_k b_k$ is an inner product on the vector space of all square-summable sequences. (*Hint:* Review Problem 20, Section 4.1.)

15. Recall that the *trace* of an $n \times n$ matrix A is the number $\operatorname{tr} A = \sum_{k=1}^{n} A_{kk}$. Define $\langle A, B \rangle = \operatorname{tr} AB$. Show that $\langle A, B \rangle$ is an inner product on M_{nn}. (*Hint:* Use the results of Problem 8, Section 2.2.)

6.2 EIGENVALUES AND EIGENVECTORS

For a given linear map $f: V \to V$, it is useful on occasion to have a basis $\{\mathbf{u}_1, \mathbf{u}_2, \ldots, \mathbf{u}_n\}$ for V such that $f(\mathbf{u}_k)$ is a scalar multiple of $\mathbf{u}_k$ for each k. It is especially useful, as we shall soon see, to have such a basis which is also an orthonormal basis, as described in the previous section. *In this section and the next, we shall employ only the standard dot product in* $\mathbf{R}^n$*; thus "orthonormal" and "orthogonal" are with respect to that dot product.*

Definition Let $f: V \to V$ be a linear map. We say that $\mathbf{u} \in V$ is an *eigenvector* of f provided that $\mathbf{u} \neq \mathbf{0}$ and $f(\mathbf{u}) = \lambda\mathbf{u}$ for some scalar λ. Such a scalar λ is called *eigenvalue* of f.

Calculus provides an important example of an eigenvalue for a linear map.

Example 1
Let D be the derivative map on the space of all differentiable functions; that is $D(f(x)) = f'(x)$. Then any function of the form $f(x) = e^{\lambda x}$ is an eigenvector of D, with eigenvalue λ, because $D(e^{\lambda x}) = \lambda e^{\lambda x}$. We shall learn in a later section that the *only* eigenvectors of D are those functions of the form $f(x) = ke^{\lambda x}$, for arbitrary scalars k and λ. □

We shall be primarily concerned with the situation where $f: \mathbf{R}^n \to \mathbf{R}^n$. Since for such a linear map f we can write $f(\mathbf{x}) = A\mathbf{x}$, where A is a certain $n \times n$ matrix, we may say that $\mathbf{u} \neq \mathbf{0}$ is an eigenvector of A and λ is an eigenvalue of A provided that $A\mathbf{u} = \lambda\mathbf{u}$. We shall work in this matrix setting; to find the eigenvectors of A we must first find its eigenvalues, aided by the next theorem.

Theorem 6.3 For an $n \times n$ matrix A, the eigenvalues are precisely the solutions of the equation $\det(A - \lambda I) = 0$.

Proof. By definition, λ is an eigenvalue of A if and only if $A\mathbf{u} = \lambda\mathbf{u}$ for some $\mathbf{u} \in \mathbf{R}^n$, $\mathbf{u} \neq \mathbf{0}$. Now $A\mathbf{u} = \lambda\mathbf{u}$ may be written as $A\mathbf{u} -$

$\lambda\mathbf{u} = \mathbf{0}$, or as $(A - \lambda I)\mathbf{u} = \mathbf{0}$, where I is the $n \times n$ identity matrix. Since $\mathbf{u} \neq \mathbf{0}$, this means that the nullspace of the matrix $A - \lambda I$ is nontrivial or, equivalently, that $\det(A - \lambda I) = 0$.

Definition

The equation $\det(A - \lambda I) = 0$ is called the *characteristic equation* of A, and $\det(A - \lambda I)$ is called the *characteristic polynomial* of A.

Example 2

We find the eigenvalues and eigenvectors of

$$A = \begin{bmatrix} 1 & 2 \\ 5 & 4 \end{bmatrix}$$

The eigenvalues are the solutions for the characteristic equation $\det(A - \lambda I) = 0$, which becomes

$$\det\left(\begin{bmatrix} 1 & 2 \\ 5 & 4 \end{bmatrix} - \begin{bmatrix} \lambda & 0 \\ 0 & \lambda \end{bmatrix}\right) = 0 \quad \text{or} \quad \det\begin{bmatrix} 1 - \lambda & 2 \\ 5 & 4 - \lambda \end{bmatrix} = 0$$

or $(1 - \lambda)(4 - \lambda) - 10 = 0$, or $(\lambda - 6)(\lambda + 1) = 0$. Thus A has eigenvalues 6 and -1. To find the eigenvectors corresponding to $\lambda = 6$, we solve the matrix equation

$$(A - 6I)\mathbf{u} = \mathbf{0} \quad \text{or} \quad \begin{bmatrix} -5 & 2 \\ 5 & -2 \end{bmatrix}\mathbf{u} = \begin{bmatrix} 0 \\ 0 \end{bmatrix}$$

by our usual method for finding nullspaces, obtaining $\mathbf{u} = t(2, 5)$. That is, any nonzero multiple of $(2, 5)$ is an eigenvector of A corresponding to $\lambda = 6$. We may check by calculating

$$A\mathbf{u} = \begin{bmatrix} 1 & 2 \\ 5 & 4 \end{bmatrix}\begin{bmatrix} 2 \\ 5 \end{bmatrix} = \begin{bmatrix} 12 \\ 30 \end{bmatrix} = 6\begin{bmatrix} 2 \\ 5 \end{bmatrix} = \lambda\mathbf{u}$$

In a similar manner, we find the eigenvectors corresponding to $\lambda = -1$ are of the form $t(1, -1)$. □

Definition

Let A be a square matrix and let λ be an eigenvalue of A. The *eigenspace* of A for the eigenvalue λ is the set of all eigenvectors corresponding to λ, together with the zero vector.

Note that the eigenspace of A for the eigenvalue λ is simply the nullspace of $A - \lambda I$ and hence is automatically a subspace of $\mathbf{R}^n$ (if A is $n \times n$). For instance, in Example 2 we found that the eigenvectors of

$$A = \begin{bmatrix} 1 & 2 \\ 5 & 4 \end{bmatrix}$$

corresponding to the eigenvalue $\lambda = -1$ are just the vectors in $\mathbf{R}^2$ of the form $t(1, -1)$, with $t \neq 0$. If t is permitted to assume the value 0, then we have the eigenspace $\{t(1, -1) | t \in \mathbf{R}\}$, which we recognize is a subspace of $\mathbf{R}^2$, and the nullspace of

$$A - (-1)I = \begin{bmatrix} 2 & 2 \\ 5 & 5 \end{bmatrix}$$

Since the characteristic polynomial of an $n \times n$ matrix can be seen to be a polynomial of degree n, there can be no more than n eigenvalues. However, there may be fewer than n eigenvalues, as our next three examples demonstrate.

Example 3

It can be verified that the characteristic equation of

$$A = \begin{bmatrix} 2 & 1 \\ -1 & 4 \end{bmatrix}$$

is $\lambda^2 - 6\lambda + 9 = 0$, which produces only one eigenvalue, $\lambda = 3$; for the eigenspace, we get $\{t(1, 1)\}$. □

Example 4

For

$$A = \begin{bmatrix} 3 & -1 \\ 1 & 2 \end{bmatrix}$$

$\lambda^2 + 5\lambda + 7 = 0$ yields no real eigenvalues. □

Example 5

$$A = \begin{bmatrix} 1 & -1 & -1 \\ 0 & 2 & 1 \\ 0 & 0 & 1 \end{bmatrix}$$

has eigenvalues 1 and 2. Corresponding to $\lambda = 1$ is the eigenspace $\{r(1, 0, 0) + s(0, 1, -1)\}$; for $\lambda = 2$ we get the eigenspace $\{t(1, -1, 0)\}$, which can be verified. □

Now we consider how we might obtain a basis consisting of eigenvectors of a given matrix.

Theorem 6.4 Let $\mathbf{u}_1, \mathbf{u}_2, \ldots, \mathbf{u}_m$ be eigenvectors of a matrix A corresponding to *distinct* eigenvalues $\lambda_1, \lambda_2, \ldots, \lambda_m$ respectively. Then $\{\mathbf{u}_1, \mathbf{u}_2, \ldots, \mathbf{u}_m\}$ is a linearly independent set.

Proof. Suppose that $\{\mathbf{u}_1, \mathbf{u}_2, \ldots, \mathbf{u}_m\}$ is not a linearly independent set. Then there is a positive integer $k < m$ such that $\{\mathbf{u}_1, \mathbf{u}_2, \ldots, \mathbf{u}_k\}$ is linearly independent but $\{\mathbf{u}_1, \mathbf{u}_2, \ldots, \mathbf{u}_k, \mathbf{u}_{k+1}\}$ is linearly dependent. (Why?) It must then follow that

$$\mathbf{u}_{k+1} = c_1\mathbf{u}_1 + c_2\mathbf{u}_2 + \cdots + c_k\mathbf{u}_k \tag{1}$$

for some coefficients $c_1, c_2, \ldots, c_k$. (Why?) Multiplying both sides of (1) by A gives

$$A\mathbf{u}_{k+1} = c_1A\mathbf{u}_1 + c_2A\mathbf{u}_2 + \cdots + c_kA\mathbf{u}_k$$

and since $\mathbf{u}_1, \mathbf{u}_2, \ldots, \mathbf{u}_{k+1}$ are eigenvectors of A, we have

$$\lambda_{k+1}\mathbf{u}_{k+1} = c_1\lambda_1\mathbf{u}_1 + c_2\lambda_2\mathbf{u}_2 + \cdots + c_k\lambda_k\mathbf{u}_k$$

Now multiply each side of (1) by λ_{k+1}:

$$\lambda_{k+1}\mathbf{u}_{k+1} = c_1\lambda_{k+1}\mathbf{u}_1 + c_2\lambda_{k+1}\mathbf{u}_2 + \cdots + c_k\lambda_{k+1}\mathbf{u}_k$$

Subtracting this equation from the one above it gives

$$\mathbf{0} = c_1(\lambda_1 - \lambda_{k+1})\mathbf{u}_1 + \cdots + c_k(\lambda_k - \lambda_{k+1})\mathbf{u}_k$$

By linear independence of $\{\mathbf{u}_1, \mathbf{u}_2, \ldots, \mathbf{u}_k\}$, all coefficients above are 0; that is,

$$c_1(\lambda_1 - \lambda_{k+1}) = c_2(\lambda_2 - \lambda_{k+1}) = \cdots = c_k(\lambda_k - \lambda_{k+1}) = 0$$

Since the eigenvalues are distinct by hypothesis, we infer that $c_1 = c_2 = \cdots = c_k = 0$. But by (1), that would give $\mathbf{u}_{k+1} = \mathbf{0}$, which is impossible because eigenvectors are nonzero. Thus the assumption that $\{\mathbf{u}_1, \mathbf{u}_2, \ldots, \mathbf{u}_m\}$ is linearly dependent leads to a contradiction; hence $\{\mathbf{u}_1, \mathbf{u}_2, \ldots, \mathbf{u}_m\}$ must be linearly independent, as asserted.

If we combine Theorem 6.4 with Theorem 4.17, we have the following important result.

Corollary 6.4 If the $n \times n$ matrix A has n distinct eigenvalues $\lambda_1, \lambda_2, \ldots, \lambda_n$ and $\mathbf{u}_1, \mathbf{u}_2, \ldots, \mathbf{u}_n$ are corresponding eigenvectors, respectively, then $\{\mathbf{u}_1, \mathbf{u}_2, \ldots, \mathbf{u}_n\}$ is a basis for $\mathbf{R}^n$.

To illustrate Corollary 6.4, we cite the eigenvectors (2, 5) and (1, −1) for the matrix

$$A = \begin{bmatrix} 1 & 2 \\ 5 & 4 \end{bmatrix}$$

As we saw in Example 2, these eigenvectors correspond, respectively, to the distinct eigenvectors $\lambda_1 = 6$ and $\lambda_2 = -1$. Clearly, the set $\{(2, 5), (1, -1)\}$ constitutes a basis for $\mathbf{R}^2$. On the other hand, we may not obtain a basis for $\mathbf{R}^2$ consisting of eigenvectors

$$A = \begin{bmatrix} 2 & 1 \\ -1 & 4 \end{bmatrix}$$

since, as Example 3 shows, all the eigenvectors of this matrix lie in the one-dimensional eigenspace $\{t(1, 1)\}$.

We have deduced that if A has n distinct (real) eigenvalues, a basis for $\mathbf{R}^n$ can be found consisting of eigenvectors of a given $n \times n$ matrix A. However, as Examples 3, 4, and 5 show, an $n \times n$ matrix need not have n distinct eigenvalues. In particular, if a matrix has any nonreal (that is, imaginary) roots for its characteristic polynomial, then a full set of real roots is not available, distinct or not. (In the final sections of this book, we shall briefly consider some of the possibilities that arise when there are imaginary roots.)

If the roots of the characteristic polynomial are not all distinct but are all real, we can, if the matrix in question is of a certain type, nevertheless obtain a basis—even an orthonormal basis—of eigenvectors. The type of matrix which assures this occurrence is *symmetric*, and we now consider eigenvectors and eigenvalues of such matrices.

Theorem 6.5 If A is symmetric and $\mathbf{u}_1$ and $\mathbf{u}_2$ are eigenvectors corresponding to different eigenvalues λ_1 and λ_2, then $\mathbf{u}_1$ and $\mathbf{u}_2$ are orthogonal.

Proof. Let us consider $\lambda_1(\mathbf{u}_1 \cdot \mathbf{u}_2)$. We are assuming that $A\mathbf{u}_1 = \lambda_1\mathbf{u}_1$ and $A\mathbf{u}_2 = \lambda_2\mathbf{u}_2$, which means that we are writing vectors as columns. In that case, $\mathbf{u}_1 \cdot \mathbf{u}_2 = \mathbf{u}_1^t\mathbf{u}_2$, in terms of matrix multiplication. We therefore get

$$\begin{aligned}
\lambda_1(\mathbf{u}_1 \cdot \mathbf{u}_2) &= (\lambda_1\mathbf{u}_1) \cdot (\mathbf{u}_2) \\
&= (A\mathbf{u}_1)^t\mathbf{u}_2 \\
&= (\mathbf{u}_1^tA^t)\mathbf{u}_2 \qquad \text{(Theorem 1.10)} \\
&= \mathbf{u}_1^t(A^t\mathbf{u}_2) \\
&= \mathbf{u}_1^t(A\mathbf{u}_2) \qquad \text{(symmetry of } A\text{)} \\
&= \mathbf{u}_1^t(\lambda_2\mathbf{u}_2) \\
&= \mathbf{u}_1 \cdot (\lambda_2\mathbf{u}_2) \\
&= \lambda_2(\mathbf{u}_1 \cdot \mathbf{u}_2)
\end{aligned}$$

Now that we know that $\lambda_1(\mathbf{u}_1 \cdot \mathbf{u}_2) = \lambda_2(\mathbf{u}_1 \cdot \mathbf{u}_2)$, it is simple to conclude that $\mathbf{u}_1 \cdot \mathbf{u}_2 = 0$; we shall ask for a reason in the exercises.

To continue our analysis of the symmetric situation, we offer the following theorem, which will be proved in stages as we progress through the text.

Theorem 6.6 The characteristic polynomial of a symmetric matrix can be factored as the product of first-degree factors (that is, all of its roots are real). If an eigenvalue λ is a root of multiplicity m, then the dimension of the eigenspace for λ is m.

This theorem says that the characteristic polynomial for a symmetric matrix must have the form

$$\pm(\lambda - a_1)^{m_1}(\lambda - a_2)^{m_2} \cdots (\lambda - a_r)^{m_r}$$

and that the dimension of the eigenspace for $\lambda = a_k$ is m_k.

Example 6

Let

$$A = \begin{bmatrix} 6 & 0 & 4 \\ 0 & -2 & 0 \\ 4 & 0 & 0 \end{bmatrix}$$

To find the eigenvalues of A and to illustrate Theorems 6.5 and 6.6, we must first solve $\det(A - \lambda I) = 0$, which, in this case, is

$$\det\begin{bmatrix} 6-\lambda & 0 & 4 \\ 0 & -2-\lambda & 0 \\ 4 & 0 & -\lambda \end{bmatrix} = 0$$

Expanding by cofactors across the second row gives

$$\begin{aligned}(-2-\lambda)(-\lambda(6-\lambda)-16) &= 0\\ (-2-\lambda)(\lambda^2-6\lambda-16) &= 0\\ (-2-\lambda)(\lambda+2)(\lambda-8) &= 0\\ -(\lambda+2)^2(\lambda-8) &= 0\end{aligned}$$

Note that there are two eigenvalues, both real: $\lambda = -2$, of multiplicity 2, and $\lambda = 8$, of multiplicity 1. The theorem predicts that the eigenspace for $\lambda = -2$ is of dimension 2 and that for $\lambda = 8$ is of dimension 1. We compute these. For $\lambda = -2$, we find the nullspace of

$$\begin{bmatrix} 8 & 0 & 4 \\ 0 & 0 & 0 \\ 4 & 0 & 2 \end{bmatrix}$$

which has echelon form

$$\begin{bmatrix} 1 & 0 & \frac{1}{2} \\ 0 & 0 & 0 \\ 0 & 0 & 0 \end{bmatrix}$$

Following our usual procedure, we have the eigenspace $\{t(-\frac{1}{2}, 0, 1) + r(0, 1, 0)\}$, with basis $\{(-\frac{1}{2}, 0, 1), (0, 1, 0)\}$. Note the dimension is 2.

For $\lambda = 8$, $A - \lambda I$ reduces to

$$\begin{bmatrix} 1 & 0 & -2 \\ 0 & 1 & 0 \\ 0 & 0 & 0 \end{bmatrix}$$

leading to the eigenspace $\{(2t, 0, t)\}$, having basis $\{(2, 0, 1)\}$ and therefore dimension 1.

Let us test Theorem 6.5 as well by testing the orthogonality of an eigenvector for $\lambda = 8$ with an eigenvector for $\lambda = -2$; it suffices to test only basis vectors in the eigenspaces. We find that $(2, 0, 1) \cdot (-\frac{1}{2}, 0, 1) = 0$ and that $(2, 0, 1) \cdot (0, 1, 0) = 0$. (Note that the two basis vectors $(-\frac{1}{2}, 0, 1)$ and $(0, 1, 0)$ for the eigenspace of $\lambda = -2$ are also orthogonal; this is *not* assured by the theorem and, in general, is not the case.) □

Example 7

This example illustrates a useful technique in getting the characteristic equation into a solvable form. Let

$$A = \begin{bmatrix} 2 & 2 & 1 \\ 2 & 5 & 2 \\ 1 & 2 & 2 \end{bmatrix}$$

Then

$$A - \lambda I = \begin{bmatrix} 2-\lambda & 2 & 1 \\ 2 & 5-\lambda & 2 \\ 1 & 2 & 2-\lambda \end{bmatrix}$$

Note that, in order to find the roots of $\det(A - \lambda I)$, we need to have it in *factored* form; simply getting the determinant by cofactors along some arbitrarily chosen row or column and multiplying out will give a third-degree polynomial which might well be difficult to factor. Instead, we look for an advantageous type III row or column operation (which, we have learned, will not change the determinant) to perform *before* getting the determinant. A little review of the possibilities reveals that subtracting the third row from the first will cause the first row to contain the entries $1 - \lambda$, 0, and $-1 + \lambda$; a cofactor expansion across that row will automatically give a factor of $1 - \lambda$ and another factor of only *second* degree which will succumb to the quadratic formula, at worst. We proceed:

$$\begin{aligned}
\det\begin{bmatrix} 2-\lambda & 2 & 1 \\ 2 & 5-\lambda & 2 \\ 1 & 2 & 2-\lambda \end{bmatrix} &= \det\begin{bmatrix} 1-\lambda & 0 & -1+\lambda \\ 2 & 5-\lambda & 2 \\ 1 & 2 & 2-\lambda \end{bmatrix} \\
&= (1-\lambda)((5-\lambda)(2-\lambda)-4) \\
&\quad + (-1+\lambda)(4-5+\lambda) \\
&= (1-\lambda)(\lambda^2 - 7\lambda + 6) + (1-\lambda)^2 \\
&= (1-\lambda)^2(6-\lambda) + (1-\lambda)^2 \\
&= (1-\lambda)^2(7-\lambda)
\end{aligned}$$

Our two eigenvalues are $\lambda = 1$, of multiplicity 2, and $\lambda = 7$, of multiplicity 1. We continue as in Example 6:

For $\lambda = 1$, the echelon form of $A - \lambda I$ is

$$\begin{bmatrix} 1 & 2 & 1 \\ 0 & 0 & 0 \\ 0 & 0 & 0 \end{bmatrix}$$

which leads to the eigenspace $\{r(-2, 1, 0) + s(-1, 0, 1)\}$, a basis for which is $\{(-2, 1, 0), (-1, 0, 1)\}$; its dimension is the expected 2.

For $\lambda = 7$, the echelon form of $A - \lambda I$ is

$$\begin{bmatrix} 1 & 0 & -1 \\ 0 & 1 & -2 \\ 0 & 0 & 0 \end{bmatrix}$$

giving us the eigenspace $\{t(1, 2, 1)\}$ with basis $\{(1, 2, 1)\}$ and dimension 1. If we check for orthogonality this time, we find each basis vector for the first eigenspace is indeed orthogonal with that of the second. Note that, this time, the two basis vectors for the first eigenspace are *not* orthogonal, however. (As we said in the comment in Example 6, they will not be orthogonal, in general.) □

For a given symmetric $n \times n$ matrix A, it is possible to find a basis for $\mathbf{R}^n$ that consists of eigenvectors of A and, even more, to have that basis be orthonormal. All we need to do within each eigenspace is apply the Gram–Schmidt method.

Example 8

We continue from where we left off in Example 7, and apply the Gram–Schmidt method within each eigenspace. Applying the process to the basis $\{(-2, 1, 0), (-1, 0, 1)\}$ for the eigenspace for $\lambda = 1$ yields the orthonormal basis

$$\left\{\frac{1}{\sqrt{5}}(-2, 1, 0), \frac{1}{\sqrt{30}}(-1, -2, 5)\right\}$$

for that subspace. Since the other eigenspace is only one-dimensional, we need only change its single basis vector $(1, 2, 1)$ to a unit vector in the same direction, namely

$$\frac{1}{\sqrt{6}}(1, 2, 1)$$

Combining the results gives us our orthonormal basis for $\mathbf{R}^3$ consisting of eigenvectors for A:

$$\left\{\frac{1}{\sqrt{5}}(-2, 1, 0), \frac{1}{\sqrt{30}}(-1, -2, 5), \frac{1}{\sqrt{6}}(1, 2, 1)\right\}$$ □

We outline the procedure followed in Example 8, since this procedure is used for all symmetric matrices. First, we find all the eigenvalues of a given symmetric matrix; they must be real, by Theorem

6.6. We next find the corresponding eigenspaces; their dimensions are predicted by the multiplicities of the eigenvalues. Then we examine, for each eigenspace, a basis, and by the Gram–Schmidt method, construct an orthonormal basis in each eigenspace. The union of these orthonormal bases must also be orthonormal by Theorem 6.5, so this union (consisting of n vectors, by Theorem 6.6) will form an orthornormal—and hence linearly independent—set. Therefore, by Theorem 4.17, the set will be an orthonormal basis for $\mathbf{R}^n$ consisting of eigenvectors of A. This discussion proves the following theorem.

Theorem 6.7 For an $n \times n$ symmetric matrix A, there exists an orthonormal basis for $\mathbf{R}^n$ consisting of eigenvectors of A.

We have now seen two situations in which we can be certain of obtaining a basis for $\mathbf{R}^n$ consisting of eigenvectors of a given $n \times n$ matrix A:

a. The eigenvalues of A are all real and distinct.

b. The matrix A is symmetric.

In other situations it may or may not be possible to find such a basis, depending on the nature of the matrix. However, when such a basis exists, the matrix A can be transformed into a notably simple form. This involves converting to the eigenvector basis and writing the matrix similar to A, as discussed in Section 5.3.

Theorem 6.8 An $n \times n$ matrix A is similar to an $n \times n$ diagonal matrix D (meaning that we can write $C^{-1}AC = D$ with $D_{ij} = 0$ for $i \neq j$) if and only if there exists a basis for $\mathbf{R}^n$ consisting of eigenvectors of A. What is more, in the relation $C^{-1}AC = D$, the eigenvalues of A appear on the diagonal of D, each according to its multiplicity, and the columns of C are eigenvectors constituting a basis for $\mathbf{R}^n$, appearing in the same order as their corresponding eigenvalues appear on the diagonal of D.

Proof. First, suppose we have a basis $\{\mathbf{u}_1, \mathbf{u}_2, \ldots, \mathbf{u}_n\}$ for $\mathbf{R}^n$ consisting of eigenvectors of A, with $A\mathbf{u}_1 = \lambda_1\mathbf{u}_1$, and so on. We let $f\colon \mathbf{R}^n \to \mathbf{R}^n$ be given by $f(\mathbf{x}) = A\mathbf{x}$; this makes A the matrix for f relative to the *natural* basis for $\mathbf{R}^n$. What is the matrix for f relative to the non-natural *eigenvector* basis? We learned in Section 5.3 how to compute it; since $f(\mathbf{u}_1) = A\mathbf{u}_1 = \lambda\mathbf{u}_1$, and so on, the matrix will be

$$D = \begin{bmatrix} \lambda_1 & 0 & 0 & \cdots & 0 \\ 0 & \lambda_2 & 0 & \cdots & 0 \\ 0 & 0 & \lambda_3 & \cdots & 0 \\ & & \cdots & & \\ 0 & 0 & 0 & \cdots & \lambda_n \end{bmatrix}$$

Moreover, by Section 5.3, the *change-of-basis matrix* C will have for its columns the eigenvectors $\mathbf{u}_1, \mathbf{u}_2, \ldots, \mathbf{u}_n$, and we have $C^{-1}AC = D$, as desired. Conversely, suppose we are given the relationship $C^{-1}AC = D$, where D is a diagonal matrix and C is nonsingular. We could invoke Theorem 5.18 to conclude that D has the eigenvalues of A running down its diagonal and that C has corresponding eigenvectors for its columns; that would involve a certain linear map for which A and D are both matrices, but with respect to different bases. Instead of this, let us try a purely matrix approach, which in fact is simpler. From $C^{-1}AC = D$, we can write

$$AC = CD$$

Let the first entry on the diagonal of D be λ_1. Now what is the first *column* of AC? By the way matrices multiply, it is the matrix product of A and the first column of C. On the other side, the first column of CD is C times the first column of D; but this latter is just λ_1 times the first column of C. Equating, we have: A times the first column of C equals λ_1 times the first column of C. This makes the first column of C an *eigenvector of A corresponding to the eigenvalue* λ_1. We analyze the other columns in the same way, and we have our result.

Example 9

Let us *diagonalize* the matrix

$$A = \begin{bmatrix} 2 & 2 & 1 \\ 2 & 5 & 2 \\ 1 & 2 & 2 \end{bmatrix}$$

That is, we intend to produce a matrix C such that

$$C^{-1}AC = D$$

for a certain diagonal matrix D.

In Example 8, we found an orthonormal basis for $\mathbf{R}^3$ consisting of eigenvectors of A, namely

$$\left\{\frac{1}{\sqrt{5}}(-2, 1, 0), \frac{1}{\sqrt{30}}(-1, -2, 5), \frac{1}{\sqrt{6}}(1, 2, 1)\right\}$$

To illustrate Theorem 6.8, we need merely a basis of eigenvectors; the eigenvectors need not be unit vectors, so we may use

$$\{(-2, 1, 0), (-1, -2, 5), (1, 2, 1)\}$$

as our eigenvector basis for $\mathbf{R}^3$.

According to Theorem 6.8, we may write

$$C = \begin{bmatrix} -2 & -1 & 1 \\ 1 & -2 & 2 \\ 0 & 5 & 1 \end{bmatrix}$$

Note that the first two columns are eigenvectors corresponding to $\lambda = 1$ and that the third column corresponds to $\lambda = 7$. Therefore

$$C^{-1}AC = \begin{bmatrix} 1 & 0 & 0 \\ 0 & 1 & 0 \\ 0 & 0 & 7 \end{bmatrix} = D$$

With a certain amount of labor, we could confirm our result by calculating

$$C^{-1} = \frac{1}{30}\begin{bmatrix} -12 & 6 & 0 \\ -1 & -2 & 5 \\ 5 & 10 & 5 \end{bmatrix}$$

and then computing the matrix product $C^{-1}AC$ directly.

While it may seem that deleting the scalar factors which make the basis vectors be unit vectors—as we did above—simplifies computation, we shall learn later that if C^{-1} is to be calculated, it is in fact wiser to use the unit vectors as they are. □

Once again, note that not every matrix can be diagonalized in this way. We present here a complete characterization of 2×2 matrices with all real eigenvalues to reinforce this fact.

Theorem 6.9 If A is a 2×2 real matrix with only real eigenvalues, then exactly one of the following is true.

a. A is similar to

$$\begin{bmatrix} \lambda_1 & 0 \\ 0 & \lambda_2 \end{bmatrix} \quad \text{with } \lambda_1 \neq \lambda_2$$

b. A is similar to

$$\begin{bmatrix} \lambda & 0 \\ 0 & \lambda \end{bmatrix}$$

c. A is similar to

$$\begin{bmatrix} \lambda & 1 \\ 0 & \lambda \end{bmatrix}$$

and *not* to any diagonal matrix.*

Proof. The three cases arise accordingly as A has two distinct eigenvalues or a single eigenvalue of multiplicity 2. If A has two distinct eigenvalues λ_1 and λ_2, then we have already seen that A is similar to

$$\begin{bmatrix} \lambda_1 & 0 \\ 0 & \lambda_2 \end{bmatrix}$$

If A has a single eigenvalue λ of multiplicity 2, then the single eigenspace can be of dimension 2 or dimension 1. In the former case, the eigenspace is all of $\mathbf{R}^2$, and we again have a basis for $\mathbf{R}^2$ consisting of eigenvectors, which yields situation b. In fact, in case the eigenspace for λ is of dimension 2, the matrix A must actually be *equal* to

$$\begin{bmatrix} \lambda & 0 \\ 0 & \lambda \end{bmatrix}$$

(Verify.) Finally, if the eigenspace for λ is of dimension 1, we can choose an eigenvector $\mathbf{u}_1$ and form a basis $\{\mathbf{u}_1, \mathbf{u}_2\}$, where $\mathbf{u}_2$ is not an eigenvector. Let f: $\mathbf{R}^2 \to \mathbf{R}^2$ be given by $f(\mathbf{x}) = A\mathbf{x}$, so that A is the matrix for f relative to the natural basis. Since $f(\mathbf{u}_1) = \lambda\mathbf{u}_1$ and $f(\mathbf{u}_2) = c_1\mathbf{u}_1 + c_2\mathbf{u}_2$, for certain scalars c_1 and c_2, relative to the basis $\{\mathbf{u}_1, \mathbf{u}_2\}$ the matrix for f is

$$\begin{bmatrix} \lambda & c_1 \\ 0 & c_2 \end{bmatrix}$$

So A is similar to

$$\begin{bmatrix} \lambda & c_1 \\ 0 & c_2 \end{bmatrix}$$

However, similar matrices have the same eigenvalues (see exercises), so $c_2 = \lambda$. Note that $c_1 \neq 0$ (Why?). We form a new basis $\{\mathbf{v}_1, \mathbf{v}_2\}$,

*For a given matrix A with only real eigenvalues, that uniquely determined triangular matrix to which A is similar, as shown in Theorem 6.9, is called the *Jordan canonical form* for A. In more advanced texts, a Jordan canonical form is developed for $n \times n$ matrices in general.

where $\mathbf{v}_1 = c_1\mathbf{u}_1$, $\mathbf{v}_2 = \mathbf{u}_2$. Relative to this basis, we find that the matrix for f is

$$\begin{bmatrix} \lambda & 1 \\ 0 & \lambda \end{bmatrix}$$

(verify), so we have our result. Proof that this matrix is not similar to any diagonal matrix is left as an exercise.

Example 10

We can easily exhibit the similar matrices called for in Theorem 6.9 for the matrices in Examples 2 and 3. From Example 2, we may write

$$C^{-1}\begin{bmatrix} 1 & 2 \\ 5 & 4 \end{bmatrix} C = \begin{bmatrix} 6 & 0 \\ 0 & -1 \end{bmatrix}$$

where

$$C = \begin{bmatrix} 2 & 1 \\ 5 & -1 \end{bmatrix}$$

For the columns of C we have used the eigenvector (2, 5) corresponding to $\lambda = 6$ and the eigenvector $(1, -1)$ corresponding to $\lambda = -1$. From Example 3, where the matrix in question has only one eigenvalue $\lambda = 3$, we obtain

$$C^{-1}\begin{bmatrix} 2 & 1 \\ -1 & 4 \end{bmatrix} C = \begin{bmatrix} 3 & 1 \\ 0 & 3 \end{bmatrix}$$

where

$$C = \begin{bmatrix} -1 & 1 \\ -1 & 0 \end{bmatrix}$$

To find C in this case, we begin with the eigenvector $\mathbf{u}_1 = (1, 1)$, let $\mathbf{u}_2 = (1, 0)$, and follow the method used in the proof of Theorem 6.9. □

We shall present more examples and continue the study of diagonalizability in the next section, where we consider an application to analytic geometry. We close this section with an application to forecasting.

Example 11

We see in this example how diagonalization can be used to simplify calculations. Assume that no person drinks both diet cola and ordinary cola. Suppose also that the total number of cola drinkers is stable. Suppose now that long-term statistics show that each year 20% of the regular cola drinkers switch to diet cola while 10% of the diet cola drinkers change to regular. We shall find out what proportion of the cola drinkers, in the long run, will be diet cola drinkers.

For some given year, let y_0 be the number of diet cola drinkers and z_0 be the number of regular cola drinkers. Then, for the next year, we will have

$$y_1 = \text{number of diet cola drinkers} = .9y_0 + .2z_0$$
$$z_1 = \text{number of regular cola drinkers} = .1y_0 + .8z_0$$

In matrix format, we have

$$\begin{bmatrix} y_1 \\ z_1 \end{bmatrix} = \begin{bmatrix} .9 & .2 \\ .1 & .8 \end{bmatrix}\begin{bmatrix} y_0 \\ z_0 \end{bmatrix}$$

Let

$$A = \begin{bmatrix} .9 & .2 \\ .1 & .8 \end{bmatrix}$$

This is called the *transition matrix*. If we maintain the notation we have used already, then in the k-th year we will have

$$\begin{bmatrix} y_k \\ z_k \end{bmatrix} = \begin{bmatrix} .9 & .2 \\ .1 & .8 \end{bmatrix}\begin{bmatrix} y_{k-1} \\ z_{k-1} \end{bmatrix} = A\begin{bmatrix} y_{k-1} \\ z_{k-1} \end{bmatrix}$$

It is easy to see that this gives

$$\begin{bmatrix} y_k \\ z_k \end{bmatrix} = A^k\begin{bmatrix} y_0 \\ z_0 \end{bmatrix}$$

We want to know what happens as $k \to \infty$. Computing higher and higher powers of A would be unpleasant; however, if we could write $A = CDC^{-1}$, for some *diagonal* matrix D and some nonsingular matrix C, then the task would become considerably easier. According to Corollary 6.4 and Theorem 6.8, we can do this if we find that A has two different eigenvalues. It can be verified that this is the case; the eigenvalues of A are 1 and .7. Now, the eigenspace for $\lambda = 1$ is $\{(2r, r)\}$ and for $\lambda = .7$ it is $\{(t, -t)\}$. For C we can use (2, 1) and (1, −1) as columns:

$$C = \begin{bmatrix} 2 & 1 \\ 1 & -1 \end{bmatrix}$$

For D we have

$$D = \begin{bmatrix} 1 & 0 \\ 0 & .7 \end{bmatrix}$$

According to Theorem 6.8, $C^{-1}AC = D$, which is the same as $A = CDC^{-1}$. Hence $A^k = (CDC^{-1})^k = CD^kC^{-1}$; moreover

$$D^k = \begin{bmatrix} 1^k & 0 \\ 0 & (.7)^k \end{bmatrix}$$

so, as $k \to \infty$,

$$D^k \to \begin{bmatrix} 1 & 0 \\ 0 & 0 \end{bmatrix}$$

Thus the long range trend will be given by

$$C\begin{bmatrix} 1 & 0 \\ 0 & 0 \end{bmatrix} C^{-1}\begin{bmatrix} y_0 \\ z_0 \end{bmatrix}$$

Computing C^{-1} and then

$$C\begin{bmatrix} 1 & 0 \\ 0 & 0 \end{bmatrix} C^{-1}$$

we get

$$\frac{1}{3}\begin{bmatrix} 2 & 2 \\ 1 & 1 \end{bmatrix}$$

Upon multiplying that and

$$\begin{bmatrix} y_0 \\ z_0 \end{bmatrix}$$

we obtain

$$\begin{bmatrix} \frac{2}{3}y_0 + \frac{2}{3}z_0 \\ \frac{1}{3}y_0 + \frac{1}{3}z_0 \end{bmatrix}$$

which is

$$\begin{bmatrix} \frac{2}{3}(y_0 + z_0) \\ \frac{1}{3}(y_0 + z_0) \end{bmatrix}$$

Recall that $y_0 + z_0$ is the total cola-drinking population, assumed to be constant. The first coordinate above gives the long-range number of diet cola drinkers and the second, the number of regular cola drinkers.

We conclude that, in the long run, two-thirds of the cola drinkers will be drinking diet cola. It is perhaps surprising that this conclusion applies, regardless of how many diet cola drinkers there were at the outset. □

6.2 Problem Set

1. Find the eigenvalues and the corresponding eigenspaces for the following matrices.

   a. $\begin{bmatrix} 1 & -2 \\ 1 & -1 \end{bmatrix}$ b. $\begin{bmatrix} 3 & -1 \\ 2 & 0 \end{bmatrix}$

   c. $\begin{bmatrix} 2 & -6 \\ -2 & 1 \end{bmatrix}$ d. $\begin{bmatrix} 2 & 7 \\ -1 & -6 \end{bmatrix}$

   e. $\begin{bmatrix} 2 & 4 \\ -4 & -6 \end{bmatrix}$

2. Repeat Problem 1 for these matrices.

   a. $\begin{bmatrix} 1 & -1 & 0 \\ -1 & 2 & -1 \\ 0 & -1 & 1 \end{bmatrix}$ b. $\begin{bmatrix} 1 & -2 & 16 \\ 6 & -12 & 6 \\ 0 & 0 & 11 \end{bmatrix}$

3. Find an orthonormal basis for $\mathbf{R}^3$ consisting of eigenvectors of the matrix in Problem 2a.
4. Prove that the eigenspace of an $n \times n$ matrix A, corresponding to a given eigenvalue λ, constitutes a subspace of $\mathbf{R}^n$.
5. Prove that if λ is an eigenvalue of A, then λ^2 is an eigenvalue of A^2.
6. Prove that if λ is an eigenvalue of A and A is nonsingular, then $\lambda \neq 0$, and $1/\lambda$ is an eigenvalue of A^{-1}.
7. Justify adding the following statement to the list of equivalences at the end of Section 5.4:

   25. $\lambda = 0$ is not an eigenvalue of A.

8. Complete the proof of Theorem 6.5 by deducing that $\mathbf{u}_1 \cdot \mathbf{u}_2 = 0$.
9. Let A be the matrix in Problem 2a; let f: $\mathbf{R}^3 \to \mathbf{R}^3$ be given by $f(\mathbf{x}) = A\mathbf{x}$. By the methods of Section 5.3, write the matrix for f relative to the orthonormal basis found in Problem 3.

10. Find the eigenvalues and eigenspaces:

a. $\begin{bmatrix} -4 & -2 & 1 \\ -2 & -1 & 2 \\ -3 & -2 & 0 \end{bmatrix}$ (*Hint:* In the matrix $A - \lambda I$, subtract row 3 from row 1 before obtaining the determinant.)

b. $\begin{bmatrix} 2 & 3 & 3 & 4 \\ -1 & 0 & -3 & -2 \\ -1 & -3 & 0 & -2 \\ 2 & 0 & 0 & 1 \end{bmatrix}$ (*Hint:* The characteristic polynomial should be $(\lambda - 3)^2(\lambda + 3)\lambda$.)

11. For each of the given matrices, find an orthonormal basis for $\mathbf{R}^3$ consisting of eigenvectors.

a. $\begin{bmatrix} 0 & 2 & 2 \\ 2 & 0 & 2 \\ 2 & 2 & 0 \end{bmatrix}$ b. $\begin{bmatrix} 0 & 3 & 0 \\ 3 & 0 & 4 \\ 0 & 4 & 0 \end{bmatrix}$ c. $\begin{bmatrix} 1 & 1 & 0 \\ 1 & 1 & 0 \\ 0 & 0 & 1 \end{bmatrix}$

12. For each matrix A in Problem 11, find a nonsingular matrix C such that $C^{-1}AC$ is diagonal. Check your answer.

13. For each eigenvalue λ in Example 7, verify that the echelon forms for $A - \lambda I$ are as given.

14. Verify, by direct computation, that the orthonormal basis vectors obtained in Example 8 are orthonormal and that they are eigenvectors for the corresponding eigenvalues.

15. In Example 9, calculate C^{-1} and compute $C^{-1}AC$ to see if the result is D.

16. Show that each entry on the main diagonal of a triangular matrix is an eigenvalue of that matrix. (*Hint:* What form does $A - \lambda I$ take? What is $\det(A - \lambda I)$?)

17. (The converse to Problem 16.) Show that a number different from any entry on the main diagonal of a triangular matrix cannot be an eigenvalue. (*Hint:* If μ is not equal to any main-diagonal entry, determine that the rows of $A - \mu I$ must constitute an independent set. What does this imply about $\det(A - \mu I)$?)

18. Prove: If the columns of a matrix each have entries adding up to 1, then 1 must be an eigenvalue of that matrix. (*Hint:* Show that the hypothesis implies that the rows of the matrix $A - I$ constitute a dependent set.)

19. Prove: If A and B are similar matrices, say $B = C^{-1}AC$, then A and B have the same eigenvalues. (*Hint:* What is $\det(Q^{-1}PQ)$? What is $\det C^{-1}(A - \lambda I)C$?)

20. If each of A and B is an $n \times n$ matrix, if each is similar to a diagonal matrix, and if both have the same eigenvalues with the same multiplicities, show that A and B are similar to each other. (*Hint:* What must the diagonal matrices involved be like?)

21. Can

$$\begin{bmatrix} 1 & 0 & 0 \\ 0 & 2 & 0 \\ 0 & 0 & 3 \end{bmatrix} \quad \text{and} \quad \begin{bmatrix} 1 & -1 & 2 \\ 0 & 1 & 3 \\ 0 & 0 & 2 \end{bmatrix}$$

be similar matrices? Why or why not?

22. a. If $S = Q^{-1}PQ$, show that $S^2 + 2S + I = Q^{-1}(P^2 + 2P + I)Q$.
    b. Let $f(t) = a_nt^n + a_{n-1}t^{n-1} + \cdots + a_0$ be an arbitrary polynomial. For any square matrix A, let $f(A)$ be $a_nA^n + a_{n-1}A^{n-1} + \cdots + a_0I$. Generalize part a by showing that if $S = Q^{-1}PQ$, then $f(S) = Q^{-1}f(P)Q$.

23. If $A = CDC^{-1}$, show that $A^k = CD^kC^{-1}$. If D is diagonal, what can be said about D^k?

24. Show that A and A^t have the same characteristic polynomial and hence the same eigenvalues. (*Hint:* Consider $\det(A - \lambda I)^t$.)

25. For those matrices in Problem 1 which have all real eigenvalues, find similar matrices in the form

$$\begin{bmatrix} \lambda_1 & 0 \\ 0 & \lambda_2 \end{bmatrix} \quad \text{or} \quad \begin{bmatrix} \lambda & 1 \\ 0 & \lambda \end{bmatrix}$$

as in Example 10.

26. Suppose that a certain student is taking two subjects, mathematics and sociology, in a six-week term. Each day, he studies only one of these subjects; he concludes that if he studies sociology one day, he is equally likely to study either subject the next, but if he studies mathematics on a given day, he is twice as likely to study mathematics on the next day as he is to study sociology. Over the course of the term, approximately what percentage of the days will he have studied mathematics? (*Hint:* Show that transition matrix for this problem is

$$A = \begin{bmatrix} \frac{1}{2} & \frac{1}{3} \\ \frac{1}{2} & \frac{2}{3} \end{bmatrix}$$

and proceed as in Example 11.)

27. a. Show that, for a diagonalizable matrix A, the trace of A is the sum of the eigenvalues. (This is true, in fact, for matrices in general.)
    b. Show that, for any matrix A, $\det A$ is the product of the eigenvalues. (*Hint:* Consider the constant term in the characteristic polynomial.)
    c. Show that, for a diagonalizable matrix A, $\operatorname{tr} A$ equals the negative of the coefficient of the next-to-highest degree term in the characteristic polynomial.
28. Prove the 2×2 case of Theorem 6.6: For a 2×2 symmetric matrix A, the eigenvalues must be real. (*Hint:* Let

$$A = \begin{bmatrix} a & b \\ b & c \end{bmatrix}$$

Then $\det(A - \lambda I) = (a - \lambda)(c - \lambda) - b^2$. Write this in the form $\alpha\lambda^2 + \beta\lambda + \gamma$ and show that the "discriminant" $\beta^2 - 4\alpha\gamma \geq 0$, which guarantees by the quadratic formula that the roots are real.)
29. Complete all details of Theorem 6.9, including showing that

$$\begin{bmatrix} \lambda & 1 \\ 0 & \lambda \end{bmatrix}$$

cannot be similar to a diagonal matrix.

6.3 Diagonalizing Symmetric Matrices with Orthogonal Matrices

In this section, we examine more closely the results of Section 6.2 in the case of symmetric matrices and present an application to conic sections.

According to Theorem 6.8, a given $n \times n$ matrix A can be *diagonalized* provided that there is a basis for $\mathbf{R}^n$ consisting of eigenvectors of A. Moreover, this diagonalization is realized by computing $C^{-1}AC$, where the columns of the matrix C are the eigenvectors in that basis.

As we have observed, such diagonalization is always possible in case A is symmetric; moreover, this case is especially interesting since one can construct an eigenvector basis which is *orthonormal*.

In the following example, we review the procedure for diagonalizing a symmetric matrix by means of an orthonormal basis of eigenvectors.

Example 1

Let

$$A = \begin{bmatrix} 4 & 0 & 0 \\ 0 & 1 & 3 \\ 0 & 3 & 1 \end{bmatrix}$$

The characteristic polynomial for A is

$$-(\lambda - 4)^2(\lambda + 2)$$

so we have an eigenvalue 4 of multiplicity 2 and an eigenvalue -2 of multiplicity 1. For the eigenvalue $\lambda = 4$, we obtain the eigenspace

$$\{(r, s, s)\} = \{r(1, 0, 0) + s(0, 1, 1)\}$$

and hence the orthonormal eigenspace basis

$$\left\{(1, 0, 0), \frac{1}{\sqrt{2}}(0, 1, 1)\right\}$$

For the eigenvalue $\lambda = -2$, the eigenspace is

$$\{(0, -t, t)\}$$

with orthonormal basis

$$\left\{\frac{1}{\sqrt{2}}(0, -1, 1)\right\}$$

Therefore, an orthonormal basis for $\mathbf{R}^3$ consisting of eigenvectors of A is

$$\left\{(1, 0, 0), \frac{1}{\sqrt{2}}(0, 1, 1), \frac{1}{\sqrt{2}}(0, -1, 1)\right\}$$

and we use these basis vectors for the columns of a matrix C, obtaining

$$C = \begin{bmatrix} 1 & 0 & 0 \\ 0 & 1/\sqrt{2} & -1/\sqrt{2} \\ 0 & 1/\sqrt{2} & 1/\sqrt{2} \end{bmatrix}$$

According to Theorem 6.8,

$$C^{-1}AC = \begin{bmatrix} 4 & 0 & 0 \\ 0 & 4 & 0 \\ 0 & 0 & -2 \end{bmatrix}$$

and we see the eigenvalues, in appropriate order, displayed on the diagonal of this diagonal matrix. □

Checking the result of Example 1 by actually computing C^{-1} and then the product $C^{-1}AC$ would normally involve a certain amount of effort. However, this work is less than might be expected because the columns of C constitute an orthonormal set. Such matrices have the peculiar property that C^{-1} is simply C^t, the transpose of C.

Definition An *orthogonal* matrix is a square matrix whose columns constitute an orthonormal set of vectors. (A better term for such a matrix would be "orthonormal," but "orthogonal" is well-established.)

Theorem 6.10 A matrix C is orthogonal if and only if C is nonsingular and $C^{-1} = C^t$.

Proof. If C is orthogonal, we need to show that $CC^t = I$, or, what is simpler but equivalent, that $C^tC = I$. Consider the various entries of C^tC: $(C^tC)_{ij}$ is just the usual dot product of the i-th row of C^t and the j-th column of C, according to matrix multiplication. But this is the dot product of the i-th *column* of C itself and the j-th column of C. Because the columns form an orthonormal set, we get

$$(C^tC)_{ij} = \begin{cases} 1, & \text{if } i = j \\ 0, & \text{if } i \neq j \end{cases}$$

Thus $C^tC = I$. Reversing the argument above will give us the converse.

Example 2
The matrix

$$C = \begin{bmatrix} 1/\sqrt{3} & 1/\sqrt{6} & 1/\sqrt{2} \\ 1/\sqrt{3} & -2/\sqrt{6} & 0 \\ 1/\sqrt{3} & 1/\sqrt{6} & -1/\sqrt{2} \end{bmatrix}$$

is seen to be orthogonal. (Verify.) Thus its inverse is just its transpose:

$$C^{-1} = C^t = \begin{bmatrix} 1/\sqrt{3} & 1/\sqrt{3} & 1/\sqrt{3} \\ 1/\sqrt{6} & -2/\sqrt{6} & 1/\sqrt{6} \\ 1/\sqrt{2} & 0 & -1/\sqrt{2} \end{bmatrix}$$ □

Apart from their role in diagonalizing symmetric matrices, orthogonal matrices possess some interesting properties of their own, which we shall enumerate shortly.

In light of Theorem 6.10, the situation for symmetric matrices can be summarized as follows.

Theorem 6.11 If A is an $n \times n$ symmetric matrix, then there is an orthogonal matrix C such that C^tAC is a diagonal matrix. The diagonal entries of C^tAC are just the eigenvalues of A, each appearing according to its multiplicity as a root of the characteristic polynomial of A. The columns of C constitute an orthonormal basis for $\mathbf{R}^n$ and they are eigenvectors of A in the same order as the corresponding eigenvalues appear on the diagonal of C^tAC. Moreover, each eigenspace has dimension equal to the multiplicity of its eigenvalue.

We do not offer a proof for this theorem; much of it is simply a recasting of our results in Section 6.2, with C^t taking the role of C^{-1}. Remaining considerations, such as the connection between the multiplicities of eigenvalues and the dimensions of their eigenspaces and the fact that all eigenvalues of a symmetric matrix must be real, will be addressed in our concluding sections and appendices.

Before considering an interesting application which makes a difficult problem in analytic geometry fairly simple, we record some facts about orthogonal matrices.

Theorem 6.12 Let A be orthogonal.

a. The *rows* of A constitute an orthonormal set.

b. If B is orthogonal and the same size as A, then AB is orthogonal.

c. A^{-1} is orthogonal.

d. $\det A = \pm 1$.

e. If A is $n \times n$, then $(A\mathbf{x}) \cdot (A\mathbf{y}) = \mathbf{x} \cdot \mathbf{y}$, for any $\mathbf{x}$, $\mathbf{y}$ in $\mathbf{R}^n$ (written in column form); consequently, $|A\mathbf{x}| = |\mathbf{x}|$, for any $\mathbf{x}$ in $\mathbf{R}^n$.

f. If A is 2×2 and $\det A = 1$, then A represents a linear map $f\colon \mathbf{R}^2 \to \mathbf{R}^2$, which has the effect on $\mathbf{R}^2$ of rotating all points through a fixed angle about the origin.

Proof. Parts a through d are in varying degrees routine and we leave them to the exercises, with some hints. For part e, we calculate

$$(A\mathbf{x}) \cdot (A\mathbf{y}) = (A\mathbf{x})^t(A\mathbf{y}) = \mathbf{x}^tA^tA\mathbf{y} = \mathbf{x}^tI\mathbf{y} = \mathbf{x}^t\mathbf{y} = \mathbf{x} \cdot \mathbf{y}$$

In particular, $|A\mathbf{x}| = \sqrt{(A\mathbf{x}) \cdot (A\mathbf{x})} = \sqrt{\mathbf{x} \cdot \mathbf{x}} = |\mathbf{x}|$.

For part f, let f be the linear map given by $f(\mathbf{x}) = A\mathbf{x}$, as usual. Let

$$\mathbf{u}_1 = f\left(\begin{bmatrix} 1 \\ 0 \end{bmatrix}\right) \quad \text{and} \quad \mathbf{u}_2 = f\left(\begin{bmatrix} 0 \\ 1 \end{bmatrix}\right)$$

Thus $\mathbf{u}_1$ is the first column of A and $\mathbf{u}_2$ is the second column. Now by part e, $\mathbf{u}_1$ and $\mathbf{u}_2$ must be unit vectors, for they must have the same length as the unit vectors (1, 0) and (0, 1). Moreover, by the orthogonality of A, $\mathbf{u}_1 \cdot \mathbf{u}_2 = 0$; that is, $\mathbf{u}_1$ and $\mathbf{u}_2$ are perpendicular. Since $\mathbf{u}_1$ has unit length, we must have

$$\mathbf{u}_1 = \begin{bmatrix} \cos\alpha \\ \sin\alpha \end{bmatrix}$$

for some angle α. We represent this in Figure 6.2.

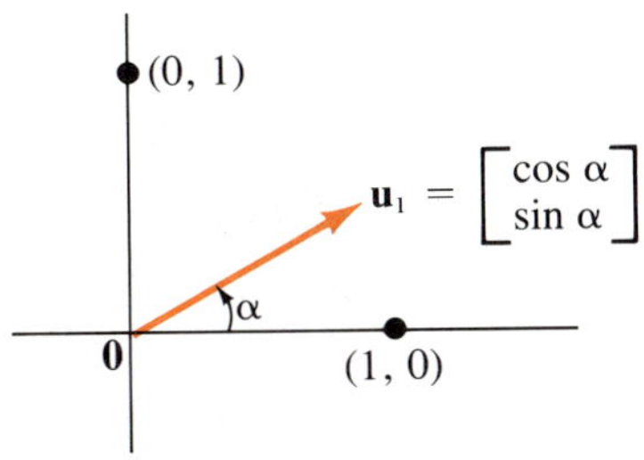

Figure 6.2

Now, there are only two possible locations for the unit vector $\mathbf{u}_2$, perpendicular to $\mathbf{u}_1$ (Figure 6.3).

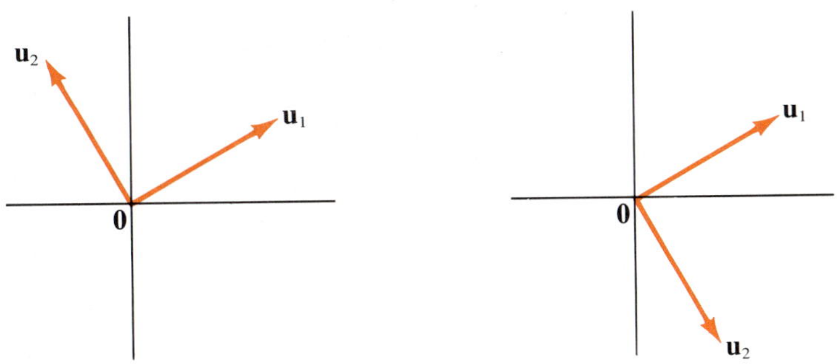

Case I: $\mathbf{u}_2 = \begin{bmatrix} -\sin\alpha \\ \cos\alpha \end{bmatrix}$ (a)

Case II: $\mathbf{u}_2 = \begin{bmatrix} \sin\alpha \\ -\cos\alpha \end{bmatrix}$ (b)

Figure 6.3

As a consequence, we have

$$A = \begin{bmatrix} \cos\alpha & -\sin\alpha \\ \sin\alpha & \cos\alpha \end{bmatrix} \quad \text{or} \quad A = \begin{bmatrix} \cos\alpha & \sin\alpha \\ \sin\alpha & -\cos\alpha \end{bmatrix}$$

But the condition $\det A = 1$ rules out the second case, so we get

$$A = \begin{bmatrix} \cos\alpha & -\sin\alpha \\ \sin\alpha & \cos\alpha \end{bmatrix}$$

Now, a map g that rotates all points in $\mathbf{R}^2$ through an angle α about the origin is indeed a linear map, for, as Figure 6.4 shows, LM1 and LM2 are satisfied by g.

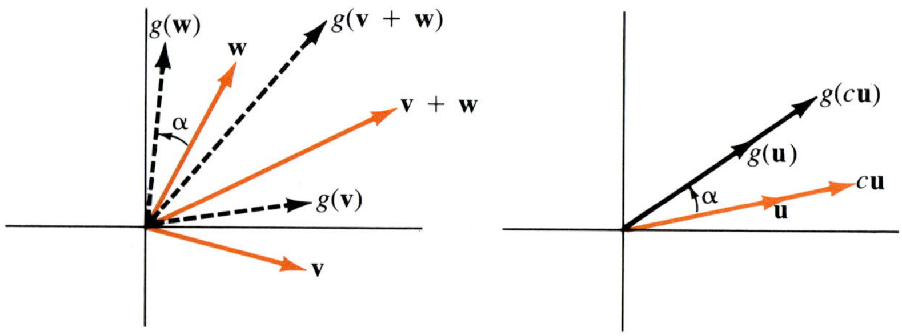

LM1: $g(\mathbf{v} + \mathbf{w}) = g(\mathbf{v}) + g(\mathbf{w})$
(a)

LM2: $g(c\mathbf{u}) = cg(\mathbf{u})$
(b)

Figure 6.4

Our matrix A is the matrix in the natural basis for such a rotation, so we are finished with the proof.

(An orthogonal 2×2 matrix having determinant equal to -1 represents a rotation followed by a reflection across a line, as might be suggested by our case II of Figure 6.3, but we shall not argue that here. See the exercises.)

For an interesting application of the methods we have developed so far, we turn our attention to conic sections in $\mathbf{R}^2$ and consider the problem of identifying them as to type if they are not in standard position.

Consider the quadratic equation

$$ax^2 + bxy + cy^2 = d \tag{1}$$

We will show that the graph of this equation is generally either an ellipse or a hyperbola.

We set

$$A = \frac{1}{2}\begin{bmatrix} 2a & b \\ b & 2c \end{bmatrix}$$

Note that A is symmetric.*

Let

$$\mathbf{u} = \begin{bmatrix} x \\ y \end{bmatrix}$$

Then (1) can be written

$$\mathbf{u}^t A\mathbf{u} = d \tag{2}$$

If λ_1 and λ_2 are eigenvalues of A, we can diagonalize A:

$$C^tAC = D = \begin{bmatrix} \lambda_1 & 0 \\ 0 & \lambda_2 \end{bmatrix}$$

where C is a suitable orthogonal matrix. Then $A = CDC^t$ and (2) becomes

$$\mathbf{u}^tCDC^t\mathbf{u} = d$$

or

$$(C^t\mathbf{u})^tD(C^t\mathbf{u}) = d \tag{3}$$

Let

$$\begin{bmatrix} x' \\ y' \end{bmatrix} = C^t\mathbf{u}$$

From (3), we now get

$$[x' \quad y']\begin{bmatrix} \lambda_1 & 0 \\ 0 & \lambda_2 \end{bmatrix}\begin{bmatrix} x' \\ y' \end{bmatrix} = d$$

which gives us

$$\lambda_1(x')^2 + \lambda_2(y')^2 = d \tag{4}$$

---

*In multivariable calculus, the matrix

$$\begin{bmatrix} 2a & b \\ b & 2c \end{bmatrix}$$

is called the *Hessian matrix* for the function given by $F(x, y) = ax^2 + bxy + cy^2$.

This is recognizable as the standard form for an equation of an ellipse or a hyperbola. Now we recall that $C^t = C^{-1}$ changes coordinates in the natural basis to coordinates in the basis consisting of the columns of C. Thus (x', y') are coordinates in this alternate basis. Moreover, the orthogonal matrix C can be chosen so that it has $\det C = 1$, and hence corresponds to a rotation. We have thus derived a result from analytic geometry: *By a rotation of axes, the general quadratic equation (1) can be changed to standard form (4), relative to the new axes.* We compute an example.

Example 3

To graph the equation $x^2 - xy + y^2 = 6$, we first obtain

$$A = \frac{1}{2}\begin{bmatrix} 2 & -1 \\ -1 & 2 \end{bmatrix}$$

The eigenvalues of

$$2A = \begin{bmatrix} 2 & -1 \\ -1 & 2 \end{bmatrix}$$

are just twice the eigenvalues of A (because $(2A)\mathbf{x} = \lambda\mathbf{x}$ is equivalent to $A\mathbf{x} = (\lambda/2)\mathbf{x}$ and we choose to find these to avoid, for a while, dealing with fractions).

$$\det(2A - \lambda I) = \det\begin{bmatrix} 2-\lambda & -1 \\ -1 & 2-\lambda \end{bmatrix} = (2-\lambda)^2 - 1 = 0$$

yields the roots 1 and 3. Thus the eigenvalues for A are half these: $\lambda_1 = \frac{1}{2}$, $\lambda_2 = \frac{3}{2}$. Consequently, upon rotating the graph in question through a certain angle, its equation will be

$$\left(\frac{1}{2}\right)(x')^2 + \left(\frac{3}{2}\right)(y')^2 = 6$$

which can be written

$$\frac{(x')^2}{12} + \frac{(y')^2}{4} = 1$$

This is an equation for an ellipse having major axis of length $2\sqrt{12} = 4\sqrt{3}$ and minor axis of length $2\sqrt{4} = 4$. The major axis lies along the x' axis. Now to locate the graph we must determine the angle of rotation required to bring about this standard form. We must find the rotation matrix C, by finding, as before, an orthonormal basis of eigenvectors.

For $\lambda_1 = \frac{1}{2}$, we obtain the eigenspace as the nullspace of

$$A - \left(\frac{1}{2}\right)I = \begin{bmatrix} \frac{1}{2} & -\frac{1}{2} \\ -\frac{1}{2} & \frac{1}{2} \end{bmatrix}$$

The echelon form is

$$\begin{bmatrix} 1 & -1 \\ 0 & 0 \end{bmatrix}$$

The nullspace is $\{(t, t)\} = \{t(1, 1)\}$; an orthonormal basis for this eigenspace is $\{1/\sqrt{2}, \quad 1/\sqrt{2})\}$.

For $\lambda_2 = \frac{3}{2}$, the same sort of work gives us the eigenspace $\{(-t, t)\} = \{t(-1, 1)\}$, because the nullspace in question is that of

$$\begin{bmatrix} 1 & 1 \\ 0 & 0 \end{bmatrix}$$

An orthonormal basis is $\{(-1/\sqrt{2}, 1/\sqrt{2})\}$.

Using these basis vectors as columns, we obtain

$$C = \begin{bmatrix} 1/\sqrt{2} & -1/\sqrt{2} \\ 1/\sqrt{2} & 1/\sqrt{2} \end{bmatrix}$$

Now, C must have the form

$$\begin{bmatrix} \cos\alpha & -\sin\alpha \\ \sin\alpha & \cos\alpha \end{bmatrix}$$

Thus by inspection we conclude that the required rotation is through an angle of $\alpha = 45°$ (Figure 6.5). □

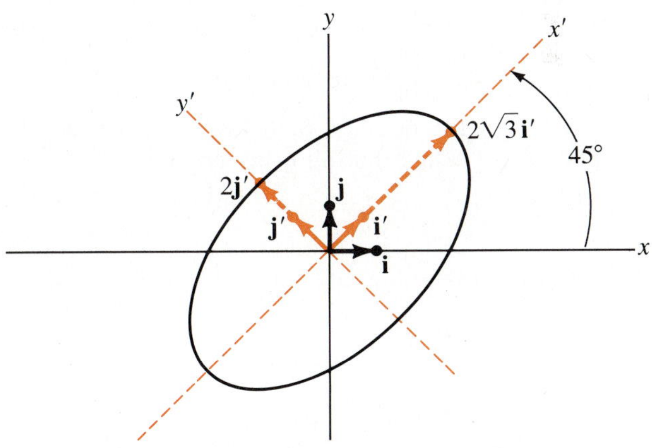

$\mathbf{i} = (1, 0) \qquad \mathbf{i}' = (1/\sqrt{2}, 1/\sqrt{2})$

$\mathbf{j} = (0, 1) \qquad \mathbf{j}' = (-1/\sqrt{2}, 1/\sqrt{2})$

Upon rotation, $x^2 - xy + y^2 = 6$ becomes $\dfrac{(x')^2}{12} + \dfrac{(y')^2}{4} = 1$.

Figure 6.5

Note: In determining the orthogonal rotation matrix C in Example 3, bear in mind that we need to have $\det C = 1$ (not $\det C = -1$) in order that C represent a rotation. This was why we wrote the second eigenspace as $\{t(-1, 1)\}$ and not as $\{t(1, -1)\}$. However, we could have used the latter and reversed the order of the eigenvalues. This would have given the equally valid results

$$\frac{(x')^2}{4} + \frac{(y')^2}{12} = 1$$

$$C = \begin{bmatrix} 1/\sqrt{2} & 1/\sqrt{2} \\ -1/\sqrt{2} & 1/\sqrt{2} \end{bmatrix}$$

$$\alpha = 315° \text{ (or } -45°\text{)}$$

There are other legitimate alternatives as well.

We return to our discussion of the equation $ax^2 + bxy + cy^2 = d$, where we assume that $d > 0$. We can recognize from (4) that the graph is

a. an ellipse if $\lambda_1 > 0$ and $\lambda_2 > 0$ (a circle if $\lambda_1 = \lambda_2$).

b. a hyperbola if λ_1 and λ_2 have opposite signs.

c. an empty graph if $\lambda_1 < 0$ and $\lambda_2 < 0$.

Since $\det D = \lambda_1\lambda_2$, the graph, therefore, is

a. an ellipse or empty if $\det D > 0$.

b. a hyperbola if $\det D < 0$.

But $\det A = \det(CDC^t) = \det(CDC^{-1}) = \det D$. Thus we have established the following criteria.

Theorem 6.13 Let

$$A = \frac{1}{2}\begin{bmatrix} 2a & b \\ b & 2c \end{bmatrix}$$

Then $ax^2 + bxy + cy^2 = d$, $d > 0$, gives

a. an ellipse or the empty set if $\det A > 0$.

b. a hyperbola if $\det A < 0$.

(The case for $d < 0$ can be handled by multiplying the original equation by -1.)

Example 4

We may use Theorem 6.13 to determine independently the nature of the graph of the equation of Example 3. In that example,

$$A = \frac{1}{2}\begin{bmatrix} 2 & -1 \\ -1 & 2 \end{bmatrix}$$

so

$$\begin{aligned} \det A &= \frac{1}{4}\det\begin{bmatrix} 2 & -1 \\ -1 & 2 \end{bmatrix} \qquad \text{(Why?)} \\ &= \frac{3}{4} > 0 \end{aligned}$$

Thus the graph is either empty or an ellipse. Since it is not empty (it contains the point $(0, \sqrt{6})$, for example), it must be an ellipse. □

Recall from analytic geometry that, in the rotation of axes, the expression $b^2 - 4ac$ is called the *discriminant* of the equation $ax^2 + bxy + cy^2 = d$, and that its sign indicates the type of conic section the equation represents.

Corollary 6.13 For $ax^2 + bxy + cy^2 = d$, $d > 0$,

a. if $b^2 - 4ac < 0$, the graph is an ellipse or empty.

b. if $b^2 - 4ac > 0$, the graph is a hyperbola.

c. if $b^2 - 4ac = 0$, the graph is a pair of parallel lines or is empty.

Proof. For parts a and b, observe that $b^2 - 4ac = -4 \det A$, so the result follows from Theorem 6.13. For part c, if $b^2 = 4ac$, we have first the possibility that either a or c is 0, but not the possibility that both are 0. If $a = 0$, then b must be 0, making our equation have the form $cy^2 = d$ which gives two horizontal lines. Similarly, if $c = 0$ we obtain two vertical lines from $ax^2 = d$. Now we consider the case in which neither a nor c is 0, and from the condition $b^2 = 4ac$ we determine that a and c must have the same sign. If $a, c > 0$, we have $b = \pm 2\sqrt{a}\sqrt{c}$, where the sign is determined by the sign of b. Thus our equation becomes either $(\sqrt{a}\,x + \sqrt{c}\,y)^2 = d$ or $(\sqrt{a}\,x - \sqrt{c}\,y)^2 = d$. In either case, taking square roots of both sides results in equations of two parallel lines. Finally, if we have $a, c < 0$, we obtain in a similar way either $-(\sqrt{|a|}\,x + \sqrt{|c|}\,y)^2 = d$ or $-(\sqrt{|a|}\,x - \sqrt{|c|}\,y)^2 = d$. Since $d > 0$, the result in either case is an empty graph.

One final remark: we have omitted and will not discuss the more general equation $ax^2 + bxy + cy^2 + dx + ey + f = 0$. From this equation it is possible to get a parabola, as well as the graphs we've already discussed.

We summarize the procedure for graphing the equation $ax^2 + bxy + cy^2 = d$ as follows.

1. Find eigenvalues λ_1, λ_2 for the matrix

$$A = \begin{bmatrix} a & b/2 \\ b/2 & c \end{bmatrix}$$

   The desired graph will then have equation $\lambda_1(x')^2 + \lambda_2(y')^2 = d$ relative to the x', y' plane, which is the result of rotating the xy plane through an angle α.

2. To find α, first find an orthonormal basis of eigenvectors corresponding to λ_1, λ_2. Let C be the matrix whose columns are those basis vectors, in that order. Adjust the bases so that $\det C = 1$. Then C will be of the form

$$C = \begin{bmatrix} \cos\alpha & -\sin\alpha \\ \sin\alpha & \cos\alpha \end{bmatrix}$$

   from which α may be determined.

If only the *type* of conic section is required, and not its graph, we need compute only the discriminant $b^2 - 4ac$ and apply Corollary 6.13.

We conclude this section with two illustrative examples.

Example 5

For $x^2 + 4xy - 2y^2 = 6$, the appropriate matrix is

$$A = \frac{1}{2}\begin{bmatrix} 2 & 4 \\ 4 & -4 \end{bmatrix} = \begin{bmatrix} 1 & 2 \\ 2 & -2 \end{bmatrix}$$

Note that $\det A = -6$, so, by Theorem 6.13, the graph should be a hyperbola. Moreover, the discriminant is greater than 0 ($b^2 - 4ac = 24$) which also indicates that the graph is a hyperbola. The characteristic polynomial is found to be $(\lambda + 3)(\lambda - 2)$, so we have eigenvalues $\lambda_1 = -3$ and $\lambda_2 = 2$. The corresponding eigenspaces are $\{t(1, -2)\}$ and $\{t(2, 1)\}$, respectively. For our orthogonal matrix C we therefore take

$$C = \begin{bmatrix} 1/\sqrt{5} & 2/\sqrt{5} \\ -2/\sqrt{5} & 1/\sqrt{5} \end{bmatrix}$$

Relative to the rotated axes, our graph has the equation

$$(-3)(x')^2 + (2)(y')^2 = 6$$

or

$$\frac{(y')^2}{3} - \frac{(x')^2}{2} = 1$$

which we indeed recognize as representing a hyperbola, with vertices on the y' axis a distance of $\sqrt{3}$ units from the origin. The original natural basis $\mathbf{i} = (1, 0)$, $\mathbf{j} = (0, 1)$ is rotated to become $\mathbf{i}' = (1/\sqrt{5}, -2/\sqrt{5})$, $\mathbf{j}' = (2/\sqrt{5}, 1/\sqrt{5})$, because these vectors are the columns of C. The angle α of rotation is seen to be $\alpha = \arctan(-2) \approx -63.4°$ (Figure 6.6). □

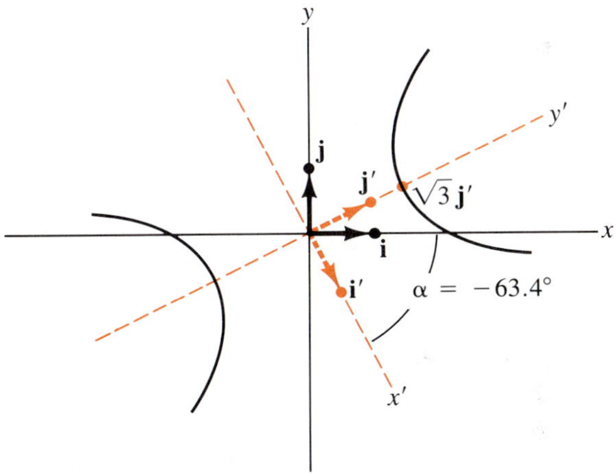

$\mathbf{i} = (1, 0)$ $\quad \mathbf{i}' = (1/\sqrt{5}, -2/\sqrt{5})$

$\mathbf{j} = (0, 1)$ $\quad \mathbf{j}' = (2/\sqrt{5}, 1/\sqrt{5})$

Upon rotation, $x^2 + 4xy - 2y^2 = 6$ becomes $\dfrac{(y')^2}{3} - \dfrac{(x')^2}{2} = 1.$

Figure 6.6

Example 6

$9x^2 + 12xy + 4y^2 = 36$ has discriminant $b^2 - 4ac = 144 - 144 = 0$. Hence, by Corollary 6.13, we have two lines or an empty graph. Actually, our equation converts to $(3x + 2y)^2 = 36$ which gives us $3x + 2y = \pm 6$, or the two parallel lines

$$y = -\frac{3}{2}x + 3 \qquad y = -\frac{3}{2}x - 3$$

It is interesting, however, to confirm this result by using the matrix method. We have

$$A = \frac{1}{2}\begin{bmatrix} 18 & 12 \\ 12 & 8 \end{bmatrix} = \begin{bmatrix} 9 & 6 \\ 6 & 4 \end{bmatrix}$$

The characteristic polynomial is $\lambda^2 - 13\lambda$, which gives eigenvalues $\lambda_1 = 0$, $\lambda_2 = 13$. (It could have been predicted that 0 is an eigenvalue. Why?) We obtain

$$C = \begin{bmatrix} 2/\sqrt{13} & 3/\sqrt{13} \\ -3/\sqrt{13} & 2/\sqrt{13} \end{bmatrix}$$

so the natural basis vectors are rotated to

$$\mathbf{i}' = \frac{1}{\sqrt{13}}(2, -3) \quad \text{and} \quad \mathbf{j}' = \frac{1}{\sqrt{13}}(3, 2)$$

With reference to the rotated axes we have the equation

$$(0)(x')^2 + (13)(y')^2 = 36$$

or

$$y' = 6/\sqrt{13} \qquad y' = -6/\sqrt{13}$$

The angle of rotation is $\arctan(-3/2) \approx -56.3°$ (Figure 6.7). □

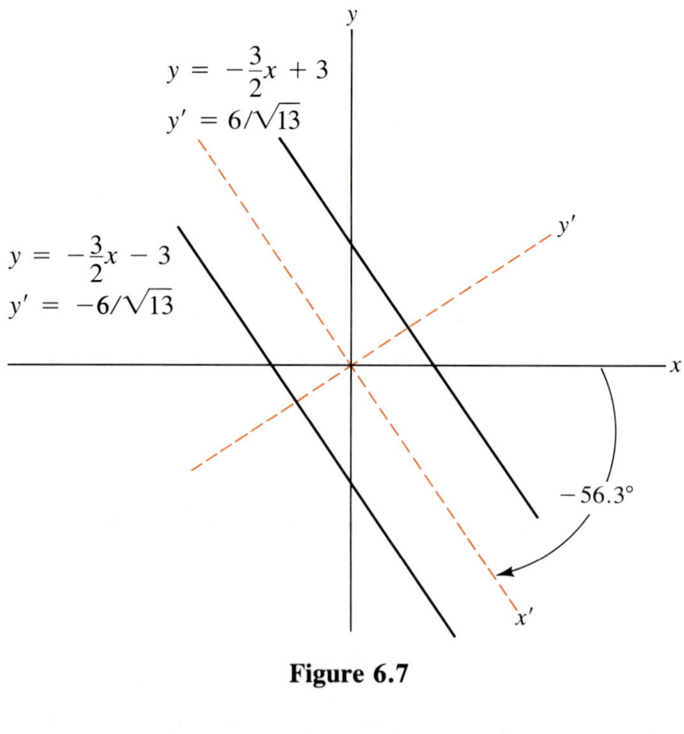

Figure 6.7

6.3 Problem Set

1. For each of the following symmetric matrices, follow the method of Section 6.2: find an orthonormal basis of eigenvectors, the corresponding orthogonal change-of-basis matrix, and the diagonal matrix similar to the given matrix.

a. $\begin{bmatrix} 1 & 2 \\ 2 & 1 \end{bmatrix}$ b. $\begin{bmatrix} 2 & -1 \\ -1 & 2 \end{bmatrix}$

c. $\begin{bmatrix} 0 & 0 & 2 \\ 0 & 3 & 0 \\ 2 & 0 & 0 \end{bmatrix}$ d. $\begin{bmatrix} -1 & -3 & 0 \\ -3 & -1 & 0 \\ 0 & 0 & 2 \end{bmatrix}$

e. $\begin{bmatrix} 1 & -1 & -5 & -3 \\ -1 & 1 & -3 & -5 \\ -5 & -3 & 1 & -1 \\ -3 & -5 & -1 & 1 \end{bmatrix}$

2. Follow the method of Examples 5 and 6 to analyze the graphs of these equations.

a. $x^2 - 4xy + y^2 = 6$

b. $9x^2 - 24xy + 16y^2 = 144$

c. $2x^2 - 12xy - 3y^2 = 84$

d. $x^2 + 4xy - 2y^2 = 12$

e. $9x^2 - 4xy + 5y^2 = 144$

3. Prove the remaining parts of Theorem 6.12. (*Hint:* Consider AA^t; use facts about $(AB)^t$ and $(AB)^{-1}$; and consider $\det(AA^t)$.)

4. Justify the two cases cited in the proof of Theorem 6.12, part f (see Figure 6.3).

5. If f: $\mathbf{R}^2 \to \mathbf{R}^2$ gives a rotation about the origin through an angle α and g: $\mathbf{R}^2 \to \mathbf{R}^2$ gives a rotation through an angle β, then $g \circ f$ ought to give a rotation through the angle $\alpha + \beta$. Justify this by computing the matrix product of the matrix for g and the matrix for f to obtain as the matrix for $g \circ f$ the following

$$\begin{bmatrix} \cos(\alpha + \beta) & -\sin(\alpha + \beta) \\ \sin(\alpha + \beta) & \cos(\alpha + \beta) \end{bmatrix}$$

6. If the discriminant $b^2 - 4ac$ is 0 for $ax^2 + bxy + cy^2 = 0$, show that $\lambda = 0$ must be an eigenvalue for

$$A = \frac{1}{2}\begin{bmatrix} 2a & b \\ b & 2c \end{bmatrix}$$

7. In the proof of Corollary 6.13, why must $b^2 - 4ac = -4 \det A$?

8. Show that an equally valid answer for Example 5 would have

$$\frac{(x')^2}{3} - \frac{(y')^2}{2} = 1 \qquad C = \begin{bmatrix} 2/\sqrt{5} & -1/\sqrt{5} \\ 1/\sqrt{5} & 2/\sqrt{5} \end{bmatrix} \qquad \alpha \approx 26.6^\circ$$

Obtain this alternate solution mathematically, not from Figure 6.6.

9. Let A be a 2×2 orthogonal matrix having determinant equal to -1. Argue that A represents a rotation about the origin followed by a reflection as follows. Let

$$C = \begin{bmatrix} 0 & 1 \\ 1 & 0 \end{bmatrix}$$

a. Show that C corresponds to a reflection about the line $y = x$ in $\mathbf{R}^2$ by showing that it reverses the roles of $\mathbf{i} = (1, 0)$ and $\mathbf{j} = (0, 1)$.

b. Let $B = CA$. Show that $\det B = 1$ (and hence B corresponds to a rotation).

c. Show that $A = CB$, and explain why the conclusion follows.

10. In Example 5, the choice

$$C = \begin{bmatrix} -1/\sqrt{5} & 2/\sqrt{5} \\ 2/\sqrt{5} & 1/\sqrt{5} \end{bmatrix}$$

will also yield the diagonalization

$$C^tAC = \begin{bmatrix} -3 & 0 \\ 0 & 2 \end{bmatrix}$$

and hence will give the equation

$$\frac{(y')^2}{3} - \frac{(x')^2}{2} = 1$$

in the new coordinate system caused by C. Since $\det C = -1$, show that C corresponds to the same rotation of axes as in Problem 8 above, followed by a reflection about $y = x$. Refer to Figure 6.6 for confirmation.

7

Complex Vector Spaces and Differential Equations

7.1 Complex Numbers

The real number system was developed as an extension of the rationals, to guarantee that every set of numbers which is bounded above has a *least* upper bound, a property that the rational numbers lack. The reals, in turn, are flawed in that not every real polynomial has a root. The simplest such is $x^2 + 1$; that is, the equation

$$x^2 + 1 = 0 \tag{1}$$

has no solution among the real numbers.

This flaw is remedied with the introduction of the "imaginary" number i, which has the property that $i^2 = -1$, so that $i = \sqrt{-1}$. Then i, as well as its negative, $-i$, is a solution to equation (1).

Definition A *complex number* is one of the form $z = x + yi$, where x and y are real and $i^2 = -1$. We denote by **C** the set of all complex numbers. The number x is called the *real part* of z, and y is the *imaginary part* of z. If $y = 0$, then z is real; if not, then z is *imaginary*. Addition and multiplication are defined in a natural way:

$$\begin{aligned}(x + yi) + (u + vi) &= (x + u) + (y + v)i \\ (x + yi)(u + vi) &= xu + yui + xvi + yvi^2 \\ &= (xu - yv) + (yu + xv)i\end{aligned}$$

Associated with each complex number z is its *conjugate*, $\bar{z} = x - yi$. The *modulus* or *absolute value* of z is the real number $|z| = \sqrt{z\bar{z}} = \sqrt{x^2 + y^2}$.

Example 1

a. $(2 - 5i) + (-6 + 8i) = -4 + 3i$

b. $(2 - 5i)(-6 + 8i) = 28 + 46i$

c. $\overline{2 - 5i} = 2 + 5i$

d. $|2 - 5i| = \sqrt{29}$ □

Complex numbers are an invaluable tool in describing and solving a wide variety of mathematical problems. (For example, in the next section we will use them to prove that symmetric matrices have only real eigenvalues.) In this section, we will describe ways to interpret complex numbers and their interactions geometrically. First, however, we establish some elementary properties of the complex number system.

We begin by observing that rules N1–9 of Section 1.1, obeyed by the real numbers, also apply to the complex numbers. Property N8—the existence of multiplicative inverses—bears special attention. The reciprocal (that is, multiplicative inverse) of the nonzero complex number z is

$$\frac{1}{z} = \frac{1}{z}\frac{1}{\bar{z}}\bar{z} = \frac{1}{z\bar{z}}\bar{z} = \frac{1}{|z|^2}\bar{z}$$

Properties of the conjugate are collected in the next theorem, whose proof is routine.

Theorem 7.1

a. $\overline{z_1 \pm z_2} = \bar{z}_1 \pm \bar{z}_2$

b. $\overline{z_1 z_2} = \bar{z}_1 \bar{z}_2$

c. $\overline{\left(\dfrac{z_1}{z_2}\right)} = \dfrac{\bar{z}_1}{\bar{z}_2}$

It can be confirmed that with addition and multiplication as defined above, **C** is a vector space (with real scalars). It is isomorphic to $\mathbf{R}^2$, under the isomorphism $f(x + yi) = (x, y)$. Moreover, "modulus" in **C** corresponds to "length" in $\mathbf{R}^2$; indeed, we use the same notation, $|\ |$. For example, $|3 - 4i| = |(3, -4)| = 5$. Thus the Triangle Inequality for $\mathbf{R}^2$, proved in Section 3.2, may be recast as a complex triangle inequality and requires no further proof. We list it and two companion equations in the following theorem.

Theorem 7.2

a. $|z_1 z_2| = |z_1||z_2|$

b. $\left|\dfrac{z_1}{z_2}\right| = \dfrac{|z_1|}{|z_2|}$

c. $|z_1 + z_2| \leq |z_1| + |z_2|$ (the Triangle Inequality)

Example 2

We find the reciprocal of $z = 2 - 3i$. Now, $\bar{z} = 2 + 3i$, and

$$\begin{aligned} \frac{1}{z} &= \frac{1}{2 - 3i} \\ &= \frac{1}{(2 - 3i)(2 + 3i)}(2 + 3i) \\ &= \frac{1}{13}(2 + 3i) \\ &= \frac{2}{13} + \frac{3}{13}i \end{aligned}$$

By computing

$$(2 - 3i)\left(\frac{2}{13} + \frac{3}{13}i\right)$$

it can be verified that we have indeed found the multiplicative inverse of z. □

A *complex polynomial* $p(z) = a_0 + a_1z + a_2z^2 + \cdots + a_nz^n$ is one whose coefficients $a_0, a_1, \ldots, a_n$ are all complex numbers; a *complex root* is a number z_0 such that $p(z_0) = 0$. For example, one root of the complex polynomial $p(z) = 1 + 3i - iz + z^2$ is $1 - i$, because

$$\begin{aligned} p(1 - i) &= 1 + 3i - i(1 - i) + (1 - i)^2 \\ &= 1 + 3i - i + i^2 + 1 - 2i + i^2 \\ &= 2 + 2i^2 \\ &= 0 \end{aligned}$$

Returning to the subject raised in this section's opening paragraph, we now state without proof the Fundamental Theorem of Algebra.

Theorem 7.3 *Fundamental Theorem of Algebra.* Every nonconstant complex polynomial (and hence every real polynomial) has at least one complex root. Moreover, imaginary roots of a real polynomial occur in pairs of the form $a \pm bi$, where a and b are real.

Now z_0 is a root of $p(z)$ if and only if $z - z_0$ is a factor of $p(z)$; that is, if and only if $p(z) = (z - z_0)q(z)$ for some polynomial $q(z)$ of degree $n - 1$. It follows by induction that every nth-degree complex polynomial is expressible as a product of n (not necessarily distinct) first-degree factors.

Example 3
We factor the polynomial $z^4 - 1$:

$$\begin{aligned} z^4 - 1 &= (z^2 + 1)(z^2 - 1) \\ &= (z - i)(z + i)(z - 1)(z + 1) \end{aligned}$$ □

Note that the theoretical existence of roots of a real polynomial does not carry with it a mechanism for obtaining those roots. In the case that $n = 2$ we do have such a device, the *quadratic formula*: the real polynomial $p(z) = ax^2 + bz + c$ has roots given by

$$z = \frac{-b \pm \sqrt{b^2 - 4ac}}{2a}$$

Similar, although much more complicated, formulas give the roots of real polynomials of degree three and four. However, it can be proved that it is *impossible* to obtain an algebraic formula for the roots of a general polynomial of degree five or more.

Complex numbers may be viewed geometrically by identifying the number $z = x + yi$ with the vector (x, y) in the Cartesian plane. For example, the number $3 - 2i$ is located at the point $(3, -2)$.

Addition of complex numbers, being essentially the same as addition of vectors in $\mathbf{R}^2$, obeys the parallelogram law (Figure 7.1).

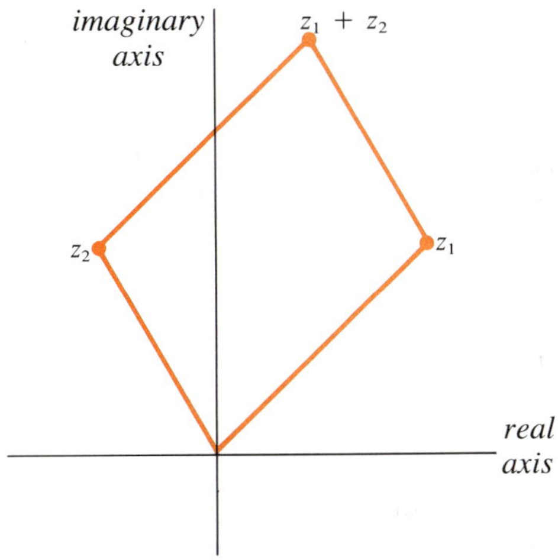

Figure 7.1

To interpret multiplication of complex numbers geometrically, we first put the numbers to be multiplied in *polar form*; that is, we find, for each $z = x + yi$, real numbers r and θ such that $x = r \cos \theta$ and $y = r \sin \theta$. Then we have

$$z = r(\cos \theta + i \sin \theta)$$

If $r \geq 0$, then r must be the modulus $|z|$ of z. In that case, θ, called an *argument* of z, will measure in radians an angle with initial side on the positive x axis and with terminal side on the ray from 0 to z. If $r < 0$, then θ must be an argument of z as described above, plus π. Although each number has infinitely many arguments, the *principal* argument of a nonzero number satisfies $0 \leq \theta < 2\pi$.

Example 4
We may write $1 + i$ in polar form variously as

$$\sqrt{2}\left(\cos\frac{\pi}{4} + i\sin\frac{\pi}{4}\right)$$

$$\sqrt{2}\left(\cos\frac{17\pi}{4} + i\sin\frac{17\pi}{4}\right)$$

$$-\sqrt{2}\left(\cos\frac{5\pi}{4} + i\sin\frac{5\pi}{4}\right)$$

and so on (Figure 7.2). □

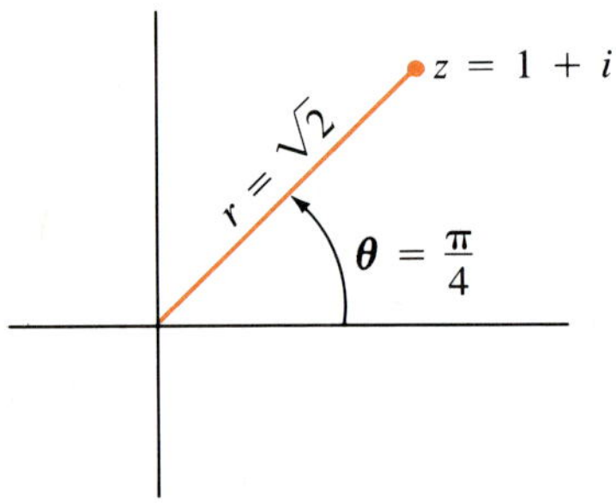

Figure 7.2

The proof of the following trigonometric identity, using the familiar formulas for $\cos(\alpha + \beta)$ and $\sin(\alpha + \beta)$, is left to the exercises.

Theorem 7.4 $(\cos\alpha + i\sin\alpha)(\cos\beta + i\sin\beta) = \cos(\alpha + \beta) + i\sin(\alpha + \beta)$

Let $z_1 = r_1(\cos\theta_1 + i\sin\theta_1)$ and $z_2 = r_2(\cos\theta_2 + i\sin\theta_2)$. Then $z_1z_2 = r_1r_2(\cos(\theta_1 + \theta_2) + i\sin(\theta_1 + \theta_2))$. This last formula gives us an interpretation of multiplication: to multiply two complex numbers, *multiply* their moduli then *add* their arguments.

Example 5
Let $z_1 = 1 + i$ and $z_2 = -1 + \sqrt{3}\,i$. To locate z_1z_2, we first put the numbers in polar form:

$$z_1 = \sqrt{2}\left(\cos\frac{\pi}{4} + i\sin\frac{\pi}{4}\right) \quad \text{and} \quad z_2 = 2\left(\cos\frac{2\pi}{3} + i\sin\frac{2\pi}{3}\right)$$

Now

$$z_1z_2 = 2\sqrt{2}\left(\cos\frac{11\pi}{12} + i\sin\frac{11\pi}{12}\right)$$

(Figure 7.3). □

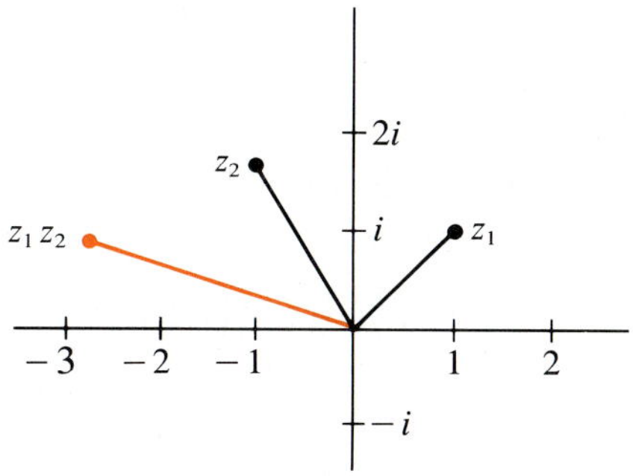

Figure 7.3

The next theorem is an easy consequence of Theorem 7.4.

Theorem 7.5 *DeMoivre's Theorem.* If n is an integer and $z = r(\cos\theta + i\sin\theta)$, then

$$z^n = r^n(\cos n\theta + i\sin n\theta)$$

Example 6
To illustrate DeMoivre's Theorem, let

$$z = -1 + \sqrt{3}\,i = 2\left(\cos\frac{2\pi}{3} + i\sin\frac{2\pi}{3}\right)$$

Then

$$\begin{aligned} z^3 &= 2^3\left(\cos 3\cdot\frac{2\pi}{3} + i\sin 3\cdot\frac{2\pi}{3}\right) \\ &= 8(\cos 2\pi + i\sin 2\pi) \\ &= 8 \end{aligned}$$

□

We can see from Example 6 that $-1 + \sqrt{3}\,i$ is a *cube root* of 8. It can be verified that two other cube roots are $-1 - \sqrt{3}\,i$ and 2. In general, the nth roots of $z = r(\cos\theta + i\sin\theta)$ (where $r > 0$) are all numbers of the form

$$w = r^{1/n}\left(\cos\frac{\theta + 2k\pi}{n} + i\sin\frac{\theta + 2k\pi}{n}\right)$$

where k is an integer. There are exactly n distinct such numbers, corresponding to $k = 0, 1, 2, \ldots, n - 1$.

Example 7

We find all fourth roots of -16. Let $z = -16 = 16(\cos\pi + i\sin\pi)$. Its fourth roots are, for $k = 0, 1, 2, 3$, the numbers

$$\begin{aligned} w &= 16^{1/4}\left(\cos\frac{\pi + 2k\pi}{4} + i\sin\frac{\pi + 2k\pi}{4}\right) \\ &= 2\left(\cos\frac{\pi}{4} + i\sin\frac{\pi}{4}\right), \quad 2\left(\cos\frac{3\pi}{4} + i\sin\frac{3\pi}{4}\right), \\ &\quad\ 2\left(\cos\frac{5\pi}{4} + i\sin\frac{5\pi}{4}\right), \quad 2\left(\cos\frac{7\pi}{4} + i\sin\frac{7\pi}{4}\right) \\ &= \pm\sqrt{2} \pm \sqrt{2}\, i \end{aligned}$$

□

Example 8

The square roots of

$$1 + i = \sqrt{2}\left(\cos\frac{\pi}{4} + i\sin\frac{\pi}{4}\right)$$

have modulus $\sqrt[4]{2}$ and arguments $\pi/8$ and $5\pi/8$. □

Note the relationship between a number z and its nth roots: the roots are uniformly distributed on the circle of center 0 and radius $|z|^{1/n}$; the principal argument of one of the roots is $1/n$ times the principal argument of z.

7.1 Problem Set

1. Compute $z_1 + z_2$, $z_1 z_2$, and z_1/z_2. Express in the form $x + yi$.

   a. $z_1 = 2 + 5i$, $z_2 = -1 + i$ b. $z_1 = 4 - i$, $z_2 = 3i$

   c. $z_1 = 2 + i$, $z_2 = 2 - i$ d. $z_1 = i$, $z_2 = 1 - 2i$

2. Write each number in polar form.
   a. $2 - 2i$ b. $\sqrt{3} + i$
   c. $-i$ d. $-\frac{1}{2} + \frac{1}{2}\sqrt{3}\, i$
3. Write each number in the form $x + yi$.
   a. $2\left(\cos \frac{5\pi}{6} + i \sin \frac{5\pi}{6}\right)$
   b. $-1\left(\cos\left(-\frac{\pi}{3}\right) + i \sin\left(-\frac{\pi}{3}\right)\right)$
   c. $3(\cos 5\pi + i \sin 5\pi)$
   d. $4\left(\cos \frac{7\pi}{4} + i \sin \frac{7\pi}{4}\right)$
4. Factor these polynomials completely.
   a. $z^2 + 2z + 2$
   b. $z^4 + 16$ (*Hint:* See Example 7.)
   c. $z^3 - z^2 + z$
   d. $z^4 + 2z^2 + 1$
   e. $z^3 + z - 2$ (*Hint:* A little ingenuity and long division are required.)
5. Find these powers using polar form.
   a. $(1 + i)^2$ b. $(\sqrt{3} - i)^4$
   c. $\left(\frac{1}{2}\sqrt{3} + \frac{1}{2}i\right)^7$ d. $(1 + \sqrt{3}\, i)^6$
   e. $(1 - i)^{-3}$
6. Find
   a. the fifth roots of i.
   b. the cube roots of $-1 + i$.
   c. the square roots of $1 + \sqrt{3}\, i$.
   d. the ninth roots of $-i$.
7. Prove Theorem 7.1.
8. Prove Theorem 7.2, parts a and b.
9. Prove Theorem 7.4.
10. Prove the following variant of the triangle inequality, and interpret it geometrically:
$$|z_1 - z_2| \geq \big||z_1| - |z_2|\big|$$
11. Find a basis for **C** when it is considered as a vector space with real scalars.

7.2

Complex Vector Spaces

We have occasionally remarked in passing that it is entirely possible to permit something other than real numbers to be the scalars in vector-space theory. To have virtually all the theory we have developed remain standing, we need only know that the scalars—whatever they are—form a mathematical system satisfying number-properties N1–9. According to Section 7.1, the complex numbers form such a system.

In this section, we shall outline some of the consequences of allowing scalars to be complex. One primary goal is finally to establish that a real symmetric matrix has all of its eigenvalues real.

Let us begin our examination of the possibility of complex scalars by considering the complex analogues of the most familiar real vector spaces.

We may let $\mathbf{C}^n$ denote the set of all ordered n-tuples of complex numbers and define multiplication by scalars and addition in the same way as is the case for $\mathbf{R}^n$. It can easily be verified that we have the required closure properties and also properties VS1–8; in short, we have a vector space.

Example 1

In $\mathbf{C}^3$, for the scalar $2 - 3i$ and the vector $(1 + i, 3, 5i)$, we have the scalar multiple

$$(2 - 3i)(1 + i, 3, 5i) = (5 - i\,, 6 - 9i, 15 + 10i)$$

which is obtained by multiplying each coordinate of the vector by the scalar. □

Matrices, too, may have complex entries. With operations defined as for real entries and scalars, we can see that our basic rules M1–11 are obeyed, since these hinge only on N1–9. Thus it follows readily that the set of all $m \times n$ matrices provides another example of a vector space with complex scalars.

Example 2

In the complex vector space of all 2×2 complex matrices, we compute a sample matrix product in the usual way and obtain

$$\begin{bmatrix} 1-i & 2 \\ 3i & 2-i \end{bmatrix}\begin{bmatrix} 2-2i & 2i & 1+i \\ 3 & -2i & 1-2i \end{bmatrix}$$
$$= \begin{bmatrix} 6-4i & 2-2i & 4-4i \\ 12+3i & -8-4i & -3-2i \end{bmatrix}$$

□

Without unnecessarily belaboring the point, let us also remark that we can solve systems of equations having complex coefficients in the customary way by applying elementary row operations to the augmented complex matrix. Of course, the solution(s), if any, may involve complex numbers. Furthermore, determinants and all their uses are still available: we may employ Cramer's Rule in complex situations; we may test by determinants for nonsingularity of matrices; or we may use the adjoint method to find a matrix inverse, and so on.

Definition

Let A be a matrix having complex entries. The *conjugate* $\overline{A}$ *of* A is that matrix obtained by replacing each entry of A with its complex conjugate. Thus

$$(\overline{A})_{ij} = \overline{A_{ij}}$$

Example 3

If

$$A = \begin{bmatrix} 1-3i & 2+i \\ 3i & 5 \end{bmatrix}$$

then

$$\overline{A} = \begin{bmatrix} 1+3i & 2-i \\ -3i & 5 \end{bmatrix}$$

□

Since ordered n-tuples can be presented either as $n \times 1$ matrices or as $1 \times n$ matrices, we automatically have the notion of *conjugate of a vector* in $\mathbf{C}^n$, obtained by conjugating each coordinate.

Our next theorem shows how conjugating a matrix interacts with other familiar operations on matrices. Its proof is an immediate consequence of the pertinent definitions and is left to the exercises.

Theorem 7.6 For complex matrices A and B of appropriate sizes and any complex scalar c,

a. $\overline{(A + B)} = \overline{A} + \overline{B}$

b. $\overline{(AB)} = \overline{A}\,\overline{B}$

c. $\overline{(cA)} = \overline{c}\overline{A}$

d. $\overline{(A^t)} = (\overline{A})^t$

e. $\overline{(A^{-1})} = (\overline{A})^{-1}$, if A is nonsingular

As we have seen, the set of all $m \times n$ complex matrices, with the usual operations, but with complex scalars, provides us with a convenient example of a system meeting all the requirements of a vector space. So does $\mathbf{C}^n$, for any n. We now record the definition formally.

Definition A system satisfying the definition of vector space in Section 4.1, but with the compex numbers playing the role of scalars, is called a *complex vector space*.

For complex vector spaces, we may, without adjustment, employ our earlier definitions of *linear independence, spanning, basis, subspace,* and *dimension*. The theorems which proceed from these concepts, therefore, carry over unaltered to the complex setting.

There are concepts from the real situation, however, which require modification. Consider the possibility of defining, in $\mathbf{C}^n$, a "scalar product" by the rule

$$(x_1, x_2, \ldots, x_n) \cdot (y_1, y_2, \ldots, y_n) = x_1y_1 + x_2y_2 + \cdots + x_ny_n$$

(as we did in $\mathbf{R}^n$). Then property SP4 is violated. Recall that SP4 asserts that $\mathbf{x} \cdot \mathbf{x}$ is nonnegative (and hence real) for all $\mathbf{x}$. But

$$\begin{aligned}(1, 3i) \cdot (1, 3i) &= 1 - 9 \\ &= -8 \qquad \text{(and not a } \textit{positive} \text{ real number)}\end{aligned}$$

and

$$(1, 1 + i) \cdot (1, 1 + i) = 1 + 2i \qquad \text{(not even } \textit{real}\text{)}$$

This unfortunate state of affairs can be corrected, however, by a minor modification in the formula defining our scalar product.

Definition

In $\mathbf{C}^n$, let the *dot product* $\mathbf{x} \cdot \mathbf{y}$ of $\mathbf{x} = (x_1, x_2, \ldots, x_n)$ and $\mathbf{y} = (y_1, y_2, \ldots, y_n)$ be defined by

$$\mathbf{x} \cdot \mathbf{y} = x_1\bar{y}_1 + x_2\bar{y}_2 + \cdots + x_n\bar{y}_n$$

Even under this definition, SP1 must be slightly adjusted, as our next theorem declares. Its proof is routine and is left as an exercise.

Theorem 7.7

Dot product in $\mathbf{C}^n$ satisfies the following:

a. $\mathbf{x} \cdot \mathbf{y} = \overline{\mathbf{y} \cdot \mathbf{x}}$

b. $(\mathbf{x} + \mathbf{y}) \cdot \mathbf{z} = \mathbf{x} \cdot \mathbf{z} + \mathbf{y} \cdot \mathbf{z}$

c. $(c\mathbf{x}) \cdot \mathbf{y} = c(\mathbf{x} \cdot \mathbf{y})$, c a scalar

d. $\mathbf{x} \cdot \mathbf{x} > 0$ if $\mathbf{x} \neq \mathbf{0}$

Equally routine are the proofs of the following corollaries.

Corollary 7.7a

In $\mathbf{C}^n$, for vectors $\mathbf{x}$, $\mathbf{y}$ and scalars a, b, c, d,

a. $\mathbf{x} \cdot (c\mathbf{y}) = \bar{c}(\mathbf{x} \cdot \mathbf{y})$

b. $(a\mathbf{x} + b\mathbf{y}) \cdot (c\mathbf{u} + d\mathbf{v})$
$$= a\bar{c}(\mathbf{x} \cdot \mathbf{u}) + b\bar{c}(\mathbf{y} \cdot \mathbf{u}) + a\bar{d}(\mathbf{x} \cdot \mathbf{v}) + b\bar{d}(\mathbf{y} \cdot \mathbf{v})$$

Note that when a scalar from the *second* factor is repositioned as a coefficient of a term, it must be conjugated.

Corollary 7.7b

In $\mathbf{C}^n$,

a. $\mathbf{x} \cdot \mathbf{y} = \mathbf{x}\bar{\mathbf{y}}^t$, if $\mathbf{x}$ and $\mathbf{y}$ are in $1 \times n$ format

b. $\mathbf{x} \cdot \mathbf{y} = \mathbf{x}^t\bar{\mathbf{y}}$, if $\mathbf{x}$ and $\mathbf{y}$ are in $n \times 1$ format

Example 4

To illustrate Corollary 7.7b (which will be useful later in establishing Theorem 7.13) we consider the vectors

$$\mathbf{x} = [1 + 2i \quad 3 - i] \quad \text{and} \quad \mathbf{y} = [-2i \quad 4 + 2i]$$

Now,

$$\bar{\mathbf{y}}^t = \begin{bmatrix} 2i \\ 4 - 2i \end{bmatrix}$$

so we find

$$\begin{aligned} \mathbf{x} \cdot \mathbf{y} = \mathbf{x}\bar{\mathbf{y}}^t &= [1 + 2i \quad 3 - i]\begin{bmatrix} 2i \\ 4 - 2i \end{bmatrix} \\ &= (-4 + 2i) + (10 - 10i) \\ &= 6 - 8i \end{aligned}$$

On the other hand, if we write

$$\mathbf{x} = \begin{bmatrix} 1 + 2i \\ 3 - i \end{bmatrix} \quad \text{and} \quad \mathbf{y} = \begin{bmatrix} -2i \\ 4 + 2i \end{bmatrix}$$

then

$$\mathbf{x}^t = [1 + 2i \quad 3 - i] \quad \text{and} \quad \bar{\mathbf{y}} = \begin{bmatrix} 2i \\ 4 - 2i \end{bmatrix}$$

and the formula $\mathbf{x} \cdot \mathbf{y} = \mathbf{x}^t\bar{\mathbf{y}}$ will clearly produce the same result as above. It is easy to confirm that our definition of dot product in this section yields the same answer as either of the matrix products we employed here. □

In light of the modification made for dot product, we redefine the more general notion of inner product.

Definition

In a complex vector space V, an *inner product* is a way of multiplying any two vectors $\mathbf{x}$ and $\mathbf{y}$ to produce a complex scalar, denoted $\langle \mathbf{x}, \mathbf{y} \rangle$, so that the following rules are obeyed:

CSP1 $\langle \mathbf{x}, \mathbf{y} \rangle = \overline{\langle \mathbf{y}, \mathbf{x} \rangle}$

CSP2 $\langle \mathbf{x} + \mathbf{y}, \mathbf{z} \rangle = \langle \mathbf{x}, \mathbf{z} \rangle + \langle \mathbf{y}, \mathbf{z} \rangle$

CSP3 $\langle c\mathbf{x}, \mathbf{y} \rangle = c\langle \mathbf{x}, \mathbf{y} \rangle$, c a scalar

CSP4 $\langle \mathbf{x}, \mathbf{x} \rangle > 0$ if $\mathbf{x} \neq \mathbf{0}$

We take note of the fact that the rules given in Corollary 7.7a carry over to inner products in general.

Example 5

As in the real case, integrals can be used to provide an example of an inner product. To consider this possibility, we must first examine complex-valued functions and how they might be integrated. A complex-function f with real domain can be written in the form $f(x) = u(x) + v(x)i$, where each of u and v is a real-valued function with real domain. Then an integral can be defined by

$$\int_a^b f(x)\,dx = \int_a^b u(x)\,dx + \left(\int_a^b v(x)\,dx\right)i$$

where the integrals on the right are the usual first-year calculus integrals, and will exist if the real part u and the imaginary part v are continuous on $[a, b]$, for instance. It is not difficult to see that the customary integral properties hold—at least the ones needed to verify that the rule

$$\langle f, g\rangle = \int_a^b f(x)\,\overline{g(x)}\,dx$$

defines an inner product on the complex vector space of all complex-valued functions with real and imaginary parts continuous on $[a, b]$.

□

A complex vector space equipped with an inner product will be called, as in the real case, an *inner product space*. We observe here (omitting the proof, which is routine) that the concepts of *orthogonal, orthonormal, projection,* and the *Gram–Schmidt method* can be lifted intact from the real vector space setting and inserted here. We single out *length*, which will carry the same definition and properties as before, and the *Schwarz Inequality*, which relates inner product and length, for special attention.

Definition

In a complex inner product space V, the *length* of a vector $\mathbf{x}$ is given by

$$|\mathbf{x}| = \sqrt{\langle \mathbf{x}, \mathbf{x}\rangle}$$

The following three propositions dealing with length can be proved by making appropriate adjustments in the arguments for their counterparts—adjustments necessitated by the complex version of SP1. We ask for the proofs in the exercises.

Lemma 7.8 In a complex inner product space V, if $\mathbf{x} \in V$ and c is any scalar, then

$$|c\mathbf{x}| = |c||\mathbf{x}|$$

where $|c\mathbf{x}|$ and $|\mathbf{x}|$ are vector lengths and $|c|$ is the absolute value or modulus of c.

Theorem 7.8 *The Schwarz Inequality.* In a complex inner product space V,

$$|\langle \mathbf{x}, \mathbf{y} \rangle| \leq |\mathbf{x}||\mathbf{y}|$$

Corollary 7.8 *The Triangle Inequality.* In a complex inner product space V,

$$|\mathbf{x} + \mathbf{y}| \leq |\mathbf{x}| + |\mathbf{y}|$$

Before turning our attention to the question of eigenvalues and topics explored in Sections 6.2 and 6.3, we illustrate the above theorems and corollaries with an example.

Example 6

In $\mathbf{C}^3$ we use the standard dot product and calculate certain vector lengths, dot products, projections, and illustrate the Schwarz and Triangle Inequalities. Let $\mathbf{x} = (1 + 3i, i, 2 - i)$ and $\mathbf{y} = (5, 2i, 2 + 3i)$. We have

$$\begin{aligned} \mathbf{x} \cdot \mathbf{y} &= (1 + 3i)(5) + (i)(-2i) + (2 - i)(2 - 3i) \\ &= (5 + 15i) + (2) + (1 - 8i) \\ &= 8 + 7i \end{aligned}$$

The length of $\mathbf{x}$ is given by

$$\begin{aligned} |\mathbf{x}| &= \sqrt{\mathbf{x} \cdot \mathbf{x}} \\ &= \sqrt{(1 + 3i)(1 - 3i) + (i)(-i) + (2 - i)(2 + i)} \\ &= \sqrt{10 + 1 + 5} \\ &= \sqrt{16} \\ &= 4 \end{aligned}$$

Similarly, the length of $\mathbf{y}$ is

$$|\mathbf{y}| = \sqrt{\mathbf{y} \cdot \mathbf{y}} = \sqrt{42}$$

We find

$$\begin{aligned}\text{proj}_{\mathbf{y}}\mathbf{x} &= \frac{\mathbf{x}\cdot\mathbf{y}}{\mathbf{y}\cdot\mathbf{y}}\mathbf{y}\\ &= \frac{8+7i}{42}(5, 2i, 2+3i)\\ &= \frac{1}{42}(40+35i, -14+16i, -5+38i)\end{aligned}$$

Recall that **y** and $\mathbf{x} - \text{proj}_{\mathbf{y}}\mathbf{x}$ are guaranteed to be orthogonal. We compute

$$\mathbf{x} - \text{proj}_{\mathbf{y}}\mathbf{x} = \frac{1}{42}(2+91i, 14+26i, 89-80i)$$

Then

$$\begin{aligned}\mathbf{y}\cdot(\mathbf{x} - \text{proj}_{\mathbf{y}}\mathbf{x}) &= \frac{1}{42}(10-455i+52+28i-62+427i)\\ &= 0\end{aligned}$$

as expected. For Schwarz's Inequality, we find

$$|\mathbf{x}\cdot\mathbf{y}| = \sqrt{64+49} = \sqrt{113} \le \sqrt{672} = \sqrt{16}\sqrt{42} = |\mathbf{x}||\mathbf{y}|$$

For the Triangle Inequality, we calculate

$$\begin{aligned}|\mathbf{x}+\mathbf{y}| &= |(6+3i, 3i, 4+2i)|\\ &= \sqrt{45+9+20}\\ &= \sqrt{74}\\ &\approx 8.602\end{aligned}$$

and

$$|\mathbf{x}| + |\mathbf{y}| = 4 + \sqrt{42} \approx 10.48$$

We thus have

$$|\mathbf{x}+\mathbf{y}| \le |\mathbf{x}| + |\mathbf{y}|, \text{ as required.}$$ □

For the notions of eigenvalues and eigenvectors, no modifications of the real definitions are necessary.

Definition

For a complex $n \times n$ matrix A, a complex number λ is called an *eigenvalue* of A, and a vector **x** in $\mathbf{C}^n$ (in $n \times 1$ format) is called a corresponding *eigenvector* provided that **x** is nonzero and $A\mathbf{x} = \lambda\mathbf{x}$.

Likewise, the *method for finding* eigenvalues and eigenvectors is unchanged for complex spaces.

Theorem 7.9 For a complex $n \times n$ matrix A, the eigenvalues of A are precisely the solutions of the characteristic equation $\det(A - \lambda I) = 0$; the eigenvectors corresponding to λ are the nontrivial solutions of the matrix equation $(A - \lambda I)\mathbf{x} = \mathbf{0}$.

Example 7

Let

$$A = \begin{bmatrix} 1 & -1 \\ 1 & 1 \end{bmatrix}$$

Then

$$A - \lambda I = \begin{bmatrix} 1 - \lambda & -1 \\ 1 & 1 - \lambda \end{bmatrix}$$

The characteristic equation is $(1 - \lambda)^2 + 1 = 0$ or $\lambda^2 - 2\lambda + 2 = 0$. By the quadratic formula, the eigenvalues are $1 + i$ and $1 - i$. Let us find the eigenspace for each eigenvalue.

For $\lambda = 1 + i$, we must find the nullspace of

$$A - \lambda I = \begin{bmatrix} -i & -1 \\ 1 & -i \end{bmatrix}$$

for which the echelon form is

$$\begin{bmatrix} 1 & -i \\ 0 & 0 \end{bmatrix}$$

We obtain the eigenspace $\{t(i, 1) | t \text{ arbitrary complex}\}$. Similarly, for $\lambda = 1 - i$, the echelon form for $A - \lambda I$ is

$$\begin{bmatrix} 1 & i \\ 0 & 0 \end{bmatrix}$$

The eigenspace is $\{t(-i, 1) | t \text{ arbitrary complex}\}$. □

Note that in our example, the set $\{(i, 1), (-i, 1)\}$ of eigenvectors is linearly independent, even if complex scalars are allowed. This is because the selected vectors correspond to different eigenvalues, the same as in the real case. We record the pertinent reason here.

Theorem 7.10 If $\mathbf{u}_1, \mathbf{u}_2, \ldots, \mathbf{u}_m$ are eigenvectors of a complex matrix A corresponding to *distinct* eigenvalues $\lambda_1, \lambda_2, \ldots, \lambda_m$, respectively, then $\{\mathbf{u}_1, \mathbf{u}_2, \ldots, \mathbf{u}_m\}$ is linearly independent.

Proof. This proof follows that of Theorem 6.4.

As we have seen in Section 7.1, a polynomial of degree n having complex coefficients has a full complement of n complex roots, albeit not necessarily distinct. Thus the complex version of Corollary 6.4 follows.

Corollary 7.10 If the eigenvalues $\lambda_1, \lambda_2, \ldots, \lambda_n$ of an $n \times n$ complex matrix A are distinct, and $\mathbf{u}_1, \mathbf{u}_2, \ldots, \mathbf{u}_n$ are corresponding eigenvectors, respectively, then $\{\mathbf{u}_1, \mathbf{u}_2, \ldots, \mathbf{u}_n\}$ is a basis for $\mathbf{C}^n$.

In Example 7, the set $\{(i, 1), (-i, 1)\}$ is thus a basis for $\mathbf{C}^2$.

Because of the change we made in dot product for the complex case, symmetric matrices must also be modified if they are to perform their roles with regard to eigenvalues and diagonalizability in the complex setting. In fact, both transposes and symmetric matrices will give way to the following alternate forms.

Definition If A is a complex matrix, then the *conjugate transpose* of A is given by $A^* = \overline{A}^t$. A square matrix A is called *Hermitian* provided that $A = A^*$.

Properties for conjugate transpose are entirely analogous to those for transpose, and we record some of these without proof.

Theorem 7.11 For complex matrices A and B of appropriate dimensions,

a. $(A^*)^* = A$

b. $(A + B)^* = A^* + B^*$

c. $(kA)^* = \bar{k}A^*$, *for any complex scalar* k

d. $(AB)^* = B^*A^*$

e. $(A^*)^{-1} = (A^{-1})^*$, if A is nonsingular

Example 8

If

$$A = \begin{bmatrix} 2-3i & 5i & 3+i \\ 1+2i & 3 & -i \end{bmatrix}$$

then

$$A^* = \begin{bmatrix} 2+3i & 1-2i \\ -5i & 3 \\ 3-i & i \end{bmatrix}$$

□

Example 9

By inspection, the matrix

$$A = \begin{bmatrix} 5 & 1+2i \\ 1-2i & 3 \end{bmatrix}$$

is Hermitian, since the conjugate transpose of A is clearly A itself. □

In Section 6.2, we asserted that all the eigenvalues of a (real) symmetric matrix must be real. We are finally in a position to confirm that assertion.

Lemma 7.12 For an $n \times n$ Hermitian matrix A and any vector $\mathbf{x}$ in $\mathbf{C}^n$ (written as an $n \times 1$ matrix), $\mathbf{x}^*A\mathbf{x}$ is real.

Proof. We calculate $(\mathbf{x}^*A\mathbf{x})^* = \mathbf{x}^*A^*(\mathbf{x}^*)^* = \mathbf{x}^*A\mathbf{x}$, by Theorem 7.11. But we note that $\mathbf{x}^*A\mathbf{x}$ is 1×1; that is, it is a complex number. Therefore our calculation actually shows that $\overline{(\mathbf{x}^*A\mathbf{x})} = \mathbf{x}^*A\mathbf{x}$, so $\mathbf{x}^*A\mathbf{x}$ is real.

Theorem 7.12 All eigenvalues of a Hermitian matrix are real.

Proof. Suppose we have $A\mathbf{x} = \lambda\mathbf{x}$, where $\mathbf{x}$ is a nonzero vector. Then

$$\mathbf{x}^*A\mathbf{x} = \mathbf{x}^*(\lambda\mathbf{x}) = \lambda(\mathbf{x}^*\mathbf{x})$$

If A is Hermitian, Lemma 7.12 tells us that $\mathbf{x}^*A\mathbf{x}$ is real and also that $\mathbf{x}^*\mathbf{x}$ is real (which follows from putting $A = I$). Consequently, since $\mathbf{x}^*\mathbf{x}$ is not zero (Why?), λ is the quotient of two real numbers and hence is real.

Example 10

We consider the eigenvalues of the matrix

$$A = \begin{bmatrix} 5 & 1 + 2i \\ 1 - 2i & 3 \end{bmatrix}$$

which was observed to be Hermitian in Example 9. We find the characteristic polynomial to be

$$\begin{aligned} \det(A - \lambda I) &= \det\begin{bmatrix} 5 - \lambda & 1 + 2i \\ 1 - 2i & 3 - \lambda \end{bmatrix} \\ &= \lambda^2 - 8\lambda + 10 \end{aligned}$$

The quadratic formula specifies that the roots of the polynomial $a\lambda^2 + b\lambda + c$ are given by

$$\lambda = \frac{-b \pm \sqrt{b^2 - 4ac}}{2a}$$

We have real roots provided that $b^2 - 4ac \geq 0$, and in our case this is so since $b^2 - 4ac = 24$. □

An *arbitrary* 2×2 Hermitian matrix can easily be shown to have real eigenvalues by using the quadratic formula as we did above. (See the exercises.)

Corollary 7.12 The eigenvalues of a *real* symmetric matrix are real.

Proof. For a real matrix A, symmetry gives us

$$A = A^t = \overline{A}^t = A^*$$

Thus A is Hermitian and hence has only real eigenvalues, by Theorem 7.12.

For the remainder of this section, we will use only the standard dot product in $\mathbf{C}^n$.

Our next theorem is the complex version of Theorem 6.5.

Theorem 7.13 If A is Hermitian and $\mathbf{u}_1$ and $\mathbf{u}_2$ are eigenvectors of A corresponding to distinct eigenvalues λ_1 and λ_2, respectively, then $\mathbf{u}_1$ and $\mathbf{u}_2$ are orthogonal.

Proof. Note that since A is Hermitian, we have $A = A^* = \overline{A}^t$, so $A^t = \overline{A}$. We can follow the pattern of our proof of Theorem 6.5, writing vectors in column format; we have $A\mathbf{u}_1 = \lambda_1\mathbf{u}_1$ and $A\mathbf{u}_2 = \lambda_2\mathbf{u}_2$. Now,

$$\begin{aligned}\lambda_1(\mathbf{u}_1 \cdot \mathbf{u}_2) &= (\lambda_1\mathbf{u}_1) \cdot \mathbf{u}_2 = (\lambda_1\mathbf{u}_1)^t\overline{\mathbf{u}}_2 \\ &= (A\mathbf{u}_1)^t\overline{\mathbf{u}}_2 = (\mathbf{u}_1^tA^t)\overline{\mathbf{u}}_2 = \mathbf{u}_1^t(A^t\overline{\mathbf{u}}_2) \\ &= \mathbf{u}_1^t(\overline{A}\overline{\mathbf{u}}_2) = \mathbf{u}_1^t\overline{(A\mathbf{u}_2)} = \mathbf{u}_1^t\overline{(\lambda_2\mathbf{u}_2)} \\ &= \mathbf{u}_1 \cdot (\lambda_2\mathbf{u}_2) = \overline{\lambda}_2(\mathbf{u}_1 \cdot \mathbf{u}_2) = \lambda_2(\mathbf{u}_1 \cdot \mathbf{u}_2)\end{aligned}$$

(We can make the last step because an eigenvalue of a Hermitian matrix is real.) We therefore obtain $(\lambda_1 - \lambda_2)(\mathbf{u}_1 \cdot \mathbf{u}_2) = 0$; but $\lambda_1 - \lambda_2 \neq 0$, so orthogonality of $\mathbf{u}_1$ and $\mathbf{u}_2$ follows.

We conclude this brief treatment of complex vector spaces by showing that Hermitian matrices have the same "diagonalizability" as real symmetric matrices. The role of orthogonal matrices will be played by *unitary* matrices.

Definition A complex matrix C is *unitary* provided that it is nonsingular and $C^{-1} = C^*$. Note that in the real case, unitary and orthogonal matrices are the same.

Theorem 7.14 A matrix C is unitary if and only if its columns (or rows) constitute an orthonormal set of vectors.

Proof. Since this is similar to the real case, we leave the details as an exercise.

We can now recast our diagonalizability results for real symmetric matrices in terms of complex Hermitian matrices. We need only rephrase Theorem 6.11 as follows.

Theorem 7.15 If A is an $n \times n$ Hermitian matrix, then there is a unitary matrix C such that C^*AC is a diagonal matrix. The diagonal entries of C^*AC are just the eigenvalues of A, each appearing according to its multiplicity as a root of the characteristic polynomial of A. The columns of C constitute an orthonormal basis for $\mathbf{C}^n$ and they are eigenvectors of A in the same order as the corresponding eigenvalues appear on the diagonal of C^*AC. Moreover, each eigenspace has dimension equal to the multiplicity of its eigenvalue.

The diagonalizing unitary matrix C is formed exactly as before: we first find a basis for each eigenspace of A. Next, we apply the Gram–Schmidt method to each basis, and then we combine these to form an orthonormal basis for $\mathbf{C}^n$, the vectors of which are used as the columns of the unitary matrix C.

Carrying out this diagonalization process can prove to be lengthy, so we present only a 2×2 example.

Example 11

Let

$$A = \begin{bmatrix} 2 & 1+i \\ 1-i & 3 \end{bmatrix}$$

Note that A is Hermitian. We find a unitary matrix C such that C^*AC is diagonal.

The characteristic polynomial is $\det(A - \lambda I) = \lambda^2 - 5\lambda + 4$, which yields the real eigenvalues $\lambda_1 = 1$ and $\lambda_2 = 4$. We find eigenspaces, as in the real case, by solving $(A - \lambda I)\mathbf{x} = \mathbf{0}$: for $\lambda_1 = 1$, we solve

$$\begin{bmatrix} 1 & 1+i \\ 1-i & 2 \end{bmatrix}\begin{bmatrix} x_1 \\ x_2 \end{bmatrix} = \begin{bmatrix} 0 \\ 0 \end{bmatrix}$$

and obtain the eigenspace $\{r(-1 - i, 1)\}$. A basis for this is $\{(-1 - i, 1)\}$. For $\lambda_2 = 4$, we solve

$$\begin{bmatrix} -2 & 1+i \\ 1-i & -1 \end{bmatrix}\begin{bmatrix} x_1 \\ x_2 \end{bmatrix} = \begin{bmatrix} 0 \\ 0 \end{bmatrix}$$

and obtain the eigenspace $\{s(1/2 + i/2, 1)\}$. A basis is $\{(1/2 + i/2, 1)\}$. Since each eigenspace is only one-dimensional, the Gram-Schmidt process reduces simply to converting the single eigenvector to a unit vector. The resulting orthonormal basis for $\mathbf{C}^2$ is

$$\left\{\frac{1}{\sqrt{3}}\begin{bmatrix} -1 - i \\ 1 \end{bmatrix}, \frac{1}{\sqrt{6}}\begin{bmatrix} 1 + i \\ 2 \end{bmatrix}\right\}$$

and these vectors constitute the columns of the desired unitary matrix C. It can be verified that C is unitary and that C^*AC is a diagonal matrix with diagonal entries 1 and 4. □

We conclude this section by observing that we have been able to recast vector-space theory in the more general complex setting with very little effort. Yet in this more general setting, our main result, Theorem 7.12, which has a remarkably straight-forward proof, provides final confirmation of our long-standing assertion that all eigenvalues of a real symmetric matrix must be real. It is often the case in mathematics that the resolution of a more general problem is easier than the solving of a special case.

7.2 Problem Set

1. Calculate $\mathbf{x} \cdot \mathbf{y}$, the standard dot product; verify Schwarz's Inequality.

   a. $\mathbf{x} = (1 - i, 3 + i)$; $\mathbf{y} = (2 + 2i, i)$

   b. $\mathbf{x} = (i, -3i, 2 + i)$; $\mathbf{y} = (1, 5, 6i)$

   c. $\mathbf{x} = (2 - i, 4 - 3i)$; $\mathbf{y} = (1 + 2i, 3 + 4i)$

2. Which of these sets of vectors are orthogonal? Which are orthonormal?

   a. $\left\{\begin{bmatrix} i \\ 0 \end{bmatrix}, \begin{bmatrix} 0 \\ 1 + i \end{bmatrix}\right\}$

   b. $\left\{\begin{bmatrix} \frac{1 + i}{\sqrt{11}} \\ \frac{3}{\sqrt{11}} \end{bmatrix}, \begin{bmatrix} \frac{1 - i}{\sqrt{3}} \\ \frac{-1}{\sqrt{3}} \end{bmatrix}\right\}$

   c. $\left\{\begin{bmatrix} \frac{2 - i}{\sqrt{6}} \\ \frac{1}{\sqrt{6}} \\ 0 \end{bmatrix}, \begin{bmatrix} \frac{2 - i}{\sqrt{30}} \\ \frac{-5}{\sqrt{30}} \\ 0 \end{bmatrix}, \begin{bmatrix} 0 \\ 0 \\ i \end{bmatrix}\right\}$

3. Show that each of these matrices is a unitary matrix.

a. $\begin{bmatrix} 0 & 0 & 1 \\ i & 0 & 0 \\ 0 & -i & 0 \end{bmatrix}$ b. $\frac{1}{2}\begin{bmatrix} 1+i & 1-i \\ 1+i & -1+i \end{bmatrix}$

c. $\frac{1}{5\sqrt{2}}\begin{bmatrix} 3+4i & 4-3i \\ 3-4i & 4+3i \end{bmatrix}$ d. $\frac{1}{|x|\sqrt{2}}\begin{bmatrix} x & \bar{x} \\ -ix & i\bar{x} \end{bmatrix}$, $x \neq 0$

e. $\begin{bmatrix} \frac{i}{\sqrt{3}} & \frac{1}{\sqrt{3}} & \frac{-i}{\sqrt{3}} \\ \frac{i}{2} & \frac{-i}{2} & \frac{1+i}{2} \\ \frac{i}{2\sqrt{10}} & \frac{-4+3i}{2\sqrt{10}} & \frac{-3+i}{2\sqrt{10}} \end{bmatrix}$

4. In each of the following, find a unitary matrix C such that C^*AC is diagonal and write the diagonal matrix C^*AC.

a. $A = \begin{bmatrix} 2 & i \\ -i & 2 \end{bmatrix}$ b. $A = \begin{bmatrix} 2 & 2+i \\ 2-i & 0 \end{bmatrix}$

c. $A = \begin{bmatrix} 3 & 3-i \\ 3+i & 2 \end{bmatrix}$ d. $A = \begin{bmatrix} 2 & 0 & 0 \\ 0 & 2 & 1+i \\ 0 & 1-i & 3 \end{bmatrix}$

e. $A = \begin{bmatrix} 5 & 0 & 0 \\ 0 & 3 & 3-i \\ 0 & 3+i & 2 \end{bmatrix}$

5. Show that all the main-diagonal entries of a Hermitian matrix are real.

6. If A is a complex matrix, show that AA^* has on its main diagonal only nonnegative real numbers.

7. Show that the eigenvalues of

$$\begin{bmatrix} a & i \\ -i & a \end{bmatrix}$$

where a is real, are $a + 1$ and $a - 1$.

8. Prove Theorem 7.6.

9. Prove Theorem 7.7 and its two corollaries.

10. Prove Theorem 7.8 (the Schwarz Inequality) and its corollary (the Triangle Inequality). (*Hint:* To adjust the proofs given in Section 3.2, keep in mind the dot-product rules in Corollary 7.7a and the

fact that, for a complex number c, $|c| = \sqrt{c\bar{c}}$. Thus the goal in proving Schwarz's Inequality should be to obtain $\langle \mathbf{x}, \mathbf{x} \rangle \langle \mathbf{y}, \mathbf{y} \rangle \geq \langle \mathbf{x}, \mathbf{y} \rangle \overline{\langle \mathbf{x}, \mathbf{y} \rangle}$.)

11. Prove Theorem 7.11.
12. Prove Theorem 7.14.
13. Let

$$A = \begin{bmatrix} r & u + iv \\ u - iv & s \end{bmatrix}$$

where r, s, u, and v are all real. Show that this arbitrary 2×2 Hermitian matrix has real eigenvalues by showing that the discriminant $b^2 - 4ac$, in the quadratic formula

$$\lambda = \frac{-b \pm \sqrt{b^2 - 4ac}}{2a}$$

for the roots of $a\lambda^2 + b\lambda + c$, must be nonnegative. (*Hint:* Show that the discriminant for the characteristic polynomial of A must be $(r + s)^2 - 4(rs - u^2 - v^2)$ and determine that this expression must be nonnegative.)

7.3 Differential Equations

Differential equations—equations involving derivatives—express our most fundamental natural laws and relationships. Their analysis is indispensible in the study of virtually all the scientific disciplines. Our purpose in this section is to see how the linear algebra developed thus far may be employed to describe, catalog, and solve certain of these equations.

In Section 4.1 we introduced $\mathcal{F}$, the vector space of all functions $\mathbf{R} \to \mathbf{R}$. Let $\mathcal{C}$ be the subspace of $\mathcal{F}$ consisting of those functions which have continuous derivatives. Let $D: \mathcal{C} \to \mathcal{F}$ be the derivative operator. (The term *operator* refers to a function whose domain and range are sets of functions.) For example, $D(x^3) = 3x^2$.

It is shown in introductory calculus books that

$$D(u + v) = Du + Dv \quad \text{and} \quad D(cu) = cDu$$

Thus D is a *linear* map.

The simplest differential equation we consider is $y' = 0$, which we may write as

$$Dy = 0 \tag{1}$$

We recall the important calculus theorem that a function with derivative zero must be constant. Thus the set of all solutions to (1)—the *kernel* of D, in other words—is the set of all constant functions. We record this fact as the next theorem.

Theorem 7.16 $Dy = 0$ if and only if $y = C$ (constant).

If we let $\mathbf{1}$ represent the constant function $\mathbf{1}(x) = 1$, we may write the set of solutions to equation (1) as $\text{Sp}\{\mathbf{1}\}$, a one-dimensional subspace of $\mathscr{C}$.

A *first-order homogeneous equation with constant coefficient* is one of the form $y' + ay = 0$ (a is constant). Such an equation may be written

$$(D + aI)y = 0 \tag{2}$$

where I is the identity operator, given by $I(f) = f$. Note that since D and I are each linear, so is the linear combination $D + aI$, by Theorems 5.8 and 5.9. By direct substitution we find that $y = e^{-ax}$ solves (2). (It should be noted that there is no single systematic approach toward solving differential equations in general. In fact, in many important cases, although the *existence* of solutions can be established, these solutions cannot be *expressed* in terms of familiar, elementary functions, and must be described indirectly.) To find the *general* solution—a single expression incorporating all the solutions—suppose that $y = ue^{-ax}$ is a solution, where u is some unknown differentiable function. Then

$$\begin{aligned} 0 &= (D + aI)ue^{-ax} \\ &= D(ue^{-ax}) + (aI)(ue^{-ax}) \\ &= u'e^{-ax} - aue^{-ax} + aue^{-ax} \\ &= u'e^{-ax} \end{aligned}$$

Since e^{-ax} is never zero, this equation is equivalent to $u' = 0$, or $u = C$ (constant). Therefore we have the following theorem.

Theorem 7.17 $(D + aI)y = 0$ if and only if $y = Ce^{-ax}$ for some constant C.

Put another way, Theorem 7.17 says that the kernel of $D + aI$ is the one-dimensional subspace $\mathrm{Sp}\{e^{-ax}\}$ of $\mathscr{C}$.

Example 1

The equation $y' + 8y = 0$ is of type (2). Its general solution is therefore $y = Ce^{-8x}$. □

We next treat the *nonhomogeneous* first-order equation $y' + ay = f(x)$, or

$$(D + aI)y = f(x) \tag{3}$$

Multiplying both sides by e^{ax} gives the equivalent equation

$$e^{ax}y' + e^{ax}ay = e^{ax}f(x)$$

which the Product Rule for derivatives converts to

$$D(ye^{ax}) = e^{ax}f(x)$$

This means that ye^{ax} is the indefinite integral $\int e^{ax}f(x)\,dx$, so we have established the next theorem.

Theorem 7.18 The general solution to $(D + aI)y = f(x)$ is $y = e^{-ax}\int e^{ax}f(x)\,dx$.

Example 2

The equation $y' - y = 1$ is of type (3). Its general solution is $y = e^x \int e^{-x} \cdot 1\,dx = e^x(-e^{-x} + C) = Ce^x - 1$. □

Example 3

The equation $y' + 3y = x$ is also of type (3). Its general solution is

$$\begin{aligned} y &= e^{-3x}\int e^{3x}x\,dx \\ &= e^{-3x}\left(x\frac{1}{3}e^{3x} - \frac{1}{9}e^{3x} + C\right) \\ &= \frac{1}{3}x - \frac{1}{9} + Ce^{-3x} \end{aligned}$$

(Here we used integration by parts.) □

The fourth type of equation to be considered has the form

$$(D + aI)(D + bI)y = 0 \tag{4}$$

To solve (4), we first see that by Theorem 7.17 it is equivalent to

$$(D + bI)y = Ce^{-ax}$$

Now this equation is of type (3), so it in turn is equivalent to

$$\begin{aligned} y &= e^{-bx} \int e^{bx} C e^{-ax}\,dx \\ &= Ce^{-bx} \int e^{(b-a)x}\,dx \\ &= \begin{cases} Ce^{-bx}\left(\dfrac{1}{b-a} e^{(b-a)x} + C_0\right) & \text{if } b \neq a \\ Ce^{-bx}(x + C_0) & \text{if } b = a \end{cases} \end{aligned}$$

Collecting together and renaming the constants appearing above leads to this theorem.

Theorem 7.19 The general solution to $(D + aI)(D + bI)y = 0$ is

$$y = C_1 e^{-ax} + C_2 e^{-bx} \quad \text{if } b \neq a$$
$$y = C_1 e^{-bx} + C_2 x e^{-bx} \quad \text{if } b = a$$

In the first case of Theorem 7.19, the solution set is the two-dimensional space $\mathrm{Sp}\{e^{-ax}, e^{-bx}\}$; in the second it is $\mathrm{Sp}\{e^{-bx}, xe^{-bx}\}$.

A composition of linear combinations of operators, linear by Theorem 5.11, may be expanded like a product of polynomials. Following the rules for composition, addition, and scalar multiplication of functions, we could write, for example,

$$\begin{aligned} (D + 4I)(D - 5I) &= (D + 4I)D + (D + 4I)(-5I) \\ &= DD + (4I)D + D(-5I) + (4I)(-5I) \\ &= D^2 + 4D - 5D - 20I^2 \\ &= D^2 - D - 20I \end{aligned}$$

Thus the differential equation $(D + 4I)(D - 5I)y = 0$ could be written $(D^2 - D - 20I)y = 0$ or $y'' - y' - 20y = 0$.

Conversely, a linear operator in polynomial form $D^n + a_{n-1}D^{n-1} + \cdots + a_1 D + a_0 I$ may be factored into a composition of polynomials of lower order. Note that these factors commute. For example,

$$\begin{aligned} D^3 - 2D^2 - D + 2I &= (D + I)(D - I)(D - 2I) \\ &= (D - 2I)(D + I)(D - I) \end{aligned}$$

and so on.

Example 4
The equation $y'' - 5y' + 6y = 0$ may be written $(D^2 - 5D + 6I)y = (D - 3I)(D - 2I)y = 0$. Its solution is $y = C_1e^{3x} + C_2e^{2x}$. □

These four types of equations culminate in our most general homogeneous case, the nth-order linear equation with constant coefficients, or, simply, a *polynomial* equation.

Theorem 7.20 The general solution to

$$(D + a_1I)^{n_1}(D + a_2I)^{n_2} \cdots (D + a_kI)^{n_k}y = 0 \tag{5}$$

(where $n_1 + n_2 + \cdots + n_k = n$) is the n-dimensional space

$$\begin{aligned} \text{Sp}\{&e^{-a_1x}, xe^{-a_1x}, x^2e^{-a_1x}, \ldots, x^{n_1-1}e^{-a_1x}, \\ &e^{-a_2x}, xe^{-a_2x}, \ldots, x^{n_2-1}e^{-a_2x}, e^{-a_3x}, \ldots, x^{n_k-1}e^{-a_kx}\} \end{aligned}$$

The proof of the theorem, which involves mathematical induction, is omitted.

Example 5
The equation $y'' + 4y' + 4y = 0$ is written $(D + 2I)^2y = 0$. Its solution is $y = C_1e^{-2x} + C_2xe^{-2x}$. □

Example 6
The general solution to $(D + 4I)^3y = 0$ is $y = C_1e^{-4x} + C_2xe^{-4x} + C_3x^2e^{-4x}$. □

Example 7
The general solution to $(D + 5I)^2(D - 2I)^3y = 0$ is $y = C_1e^{-5x} + C_2xe^{-5x} + C_3e^{2x} + C_4xe^{2x} + C_5x^2e^{2x}$. □

Example 8

To solve $(D^4 + 4D^3 + 4D^2)y = 0$, we factor to get $D^2(D + 2I)^2y = 0$. The solution is then $y = C_1 + C_2x + C_3e^{-2x} + C_4xe^{-2x}$. □

Let L denote the (linear) polynomial operator

$$(D + a_1I)^{n_1}(D + a_2I)^{n_2} \cdots (D + a_kI)^{n_k}$$

Our objective now is to solve the nonhomogeneous polynomial equation

$$Ly = f(x) \tag{6}$$

The solution to (6) may be expressed as a composition of n indefinite integrals by generalizing the technique used to solve (4). If, however, a single solution $y = y_0$ can be found (by whatever means), one can immediately obtain the general solution. For if u is any solution to (5), $Lu = 0$; but then $L(u + y_0) = Lu + Ly_0 = 0 + f(x) = f(x)$, so $u + y_0$ solves (6). Conversely, if $Ly = f(x)$, then $L(y - y_0) = Ly - Ly_0 = f(x) - f(x) = 0$, so $u = y - y_0$ solves (5). That is, we have the following theorem.

Theorem 7.21 The general solution to equation (6) is $y = u + y_0$, where u is the *general* solution to equation (5) and y_0 is a *particular* solution to equation (6).

Example 9

To use the method above to solve $y'' - 5y' + 6y = \sin x$, we first need a particular solution. Experimenting with linear combinations of $\sin x$ and $\cos x$, we discover that $y = 0.1 \sin x + 0.1 \cos x$ is a solution. Combining this fact with the result of Example 4, we have the general solution

$$y = C_1e^{3x} + C_2e^{2x} + 0.1 \sin x + 0.1 \cos x.$$ □

Up to this point we have assumed that all numbers referred to are real; however, imaginary numbers may arise as roots of polynomials. For example, $(D^2 + 4D + 13I)y = 0$ factors as

$$(D + (2 - 3i)I)(D + (2 + 3i)I)y = 0$$

We solve this complex equation in exactly the same way as in the real case, bearing in mind that, by definition, $e^{x+iy} = e^x(\cos y + i \sin y)$.

Thus the solution to $y'' + 4y' + 13y = 0$ is

$$\begin{aligned} y &= C_1e^{-(2-3i)x} + C_2e^{-(2+3i)x} \\ &= C_1e^{-2x}(\cos 3x + i \sin 3x) + C_2e^{-2x}(\cos(-3x) + i \sin(-3x)) \\ &= (C_1 + C_2)e^{-2x}\cos 3x + i(C_1 - C_2)e^{-2x}\sin 3x \\ &= K_1e^{-2x}\cos 3x + K_2e^{-2x}\sin 3x \end{aligned}$$

This function will solve our equation if K_1 and K_2 are any *complex* numbers; in particular, it will also be a solution when K_1 and K_2 are *real*.

Generalizing the discussion above shows that the second-order equation

$$(D + (a - bi)I)(D + (a + bi)I)y = 0$$

has general solution

$$y = C_1e^{-ax}\cos bx + C_2e^{-ax}\sin bx$$

The Fundamental Theorem of Algebra tells us that

(i) Every complex polynomial factors completely as a product of first-order factors, and, if all the coefficients are real,

(ii) imaginary roots occur in pairs of the form $a \pm bi$ (a, b real).

It follows that equations (5) and (6) may be extended to include the case where some or all of the roots $-a_1, -a_2, \ldots, -a_k$, are imaginary; that is, to *all* polynomial differential equations.

Example 10

To solve $(D^4 + 2D^2 - 8D + 5I)y = 0$. we must first factor:

$$(D - I)^2(D + (1 + 2i)I)(D + (1 - 2i)I)y = 0$$

Now the general solution is

$$y = C_1e^x + C_2xe^x + C_3e^{-x}\cos 2x + C_4e^{-x}\sin 2x$$ □

Our final objective will be to solve a *system* of linear differential equations, such as

$$\begin{aligned} y_1' &= ay_1 + by_2 \\ y_2' &= cy_1 + dy_2 \end{aligned}$$

We see that if $b = c = 0$—that is, if the coefficient matrix

$$A = \begin{bmatrix} a & b \\ c & d \end{bmatrix}$$

is diagonal—then the system reduces to

$$\begin{aligned} y_1' &= ay_1 \\ y_2' &= dy_2 \end{aligned}$$

and the solution is an immediate consequence of Theorem 7.17:

$$(y_1, y_2) = (k_1e^{ax}, k_2e^{dx}) \qquad (k_1, k_2 \text{ constant})$$

In general, we consider the system

$$\begin{aligned} y_1' &= A_{11}y_1 + A_{12}y_2 + \cdots + A_{1n}y_n \\ y_2' &= A_{21}y_1 + A_{22}y_2 + \cdots + A_{2n}y_n \\ &\cdots \\ y_n' &= A_{n1}y_1 + A_{n2}y_2 + \cdots + A_{nn}y_n \end{aligned}$$

which we write in matrix form as

$$\mathbf{y}' = A\mathbf{y} \tag{7}$$

where

$$\mathbf{y} = \begin{bmatrix} y_1 \\ y_2 \\ \vdots \\ y_n \end{bmatrix} \quad \text{and} \quad \mathbf{y}' = \begin{bmatrix} y_1' \\ y_2' \\ \vdots \\ y_n' \end{bmatrix}$$

and A is the $n \times n$ coefficient matrix. If A is diagonal, we have the solution

$$(y_1, y_2, \ldots, y_n) = (k_1e^{A_{11}x}, k_2e^{A_{22}x}, \ldots, k_ne^{A_{nn}x})$$

If A is *not* diagonal, we are still able to solve the system, provided that A is *diagonalizable*; that is, if A is similar to a diagonal matrix. The solution is then found in the next theorem.

Theorem 7.22 Suppose A is diagonalizable. Then the general solution to the linear system

$$\mathbf{y}' = A\mathbf{y}$$

is the space $\text{Sp}\{e^{\lambda_1x}\mathbf{u}_1, e^{\lambda_2x}\mathbf{u}_2, \ldots, e^{\lambda_nx}\mathbf{u}_n\}$, where $\lambda_1, \lambda_2, \ldots, \lambda_n$ are the real (but not necessarily distinct) eigenvalues of A, and $\{\mathbf{u}_1, \mathbf{u}_2, \ldots, \mathbf{u}_n\}$ is a basis of $\mathbf{R}^n$ consisting of corresponding eigenvectors of A.

Proof. Let $A = CDC^{-1}$, where D is a diagonal matrix whose diagonal entries are eigenvalues of A, and where the columns of C are corresponding eigenvectors $\mathbf{u}_1, \mathbf{u}_2, \ldots, \mathbf{u}_r$ (see Theorem 6.8). Now equation (1) becomes

$$\mathbf{y}' = CDC^{-1}\mathbf{y}$$

or

$$C^{-1}\mathbf{y}' = DC^{-1}\mathbf{y}$$

Setting $\mathbf{w} = C^{-1}\mathbf{y}$, we have the equivalent equation

$$\mathbf{w}' = D\mathbf{w}$$

which may be written as the system

$$\begin{aligned} w_1' &= \lambda_1 w_1 \\ w_2' &= \lambda_2 w_2 \\ &\cdots \\ w_n' &= \lambda_n w_n \end{aligned}$$

The general solution to this system is

$$\mathbf{w} = (w_1, w_2, \ldots, w_n) = (k_1 e^{\lambda_1 x}, k_2 e^{\lambda_2 x}, \ldots, k_n e^{\lambda_n x})$$

where $k_1, k_2, \ldots, k_n$ are arbitrary constants. Now $\mathbf{y} = C\mathbf{w}$, so the desired general solution is

$$\mathbf{y} = C\mathbf{w} = \sum_{i=1}^{n} k_i e^{\lambda_i x}\mathbf{u}_i \qquad (k_1, k_2, \ldots, k_n \text{ constants})$$

The set of all such n-tuples of functions may alternately be denoted $\mathrm{Sp}\{e^{\lambda_1 x}\mathbf{u}_1, e^{\lambda_2 x}\mathbf{u}_2, \ldots, e^{\lambda_n x}\mathbf{u}_n\}$, and the proof is now complete.

We conclude with an illustration of the theorem.

Example 11

We will solve the system

$$\begin{aligned} y_1' &= y_1 - y_2 - y_3 \\ y_2' &= \quad\;\; 2y_2 + y_3 \\ y_3' &= \qquad\qquad y_3 \end{aligned}$$

We found earlier (Example 5 of Section 6.2) that the coefficient matrix

$$A = \begin{bmatrix} 1 & -1 & -1 \\ 0 & 2 & 1 \\ 0 & 0 & 1 \end{bmatrix}$$

has eigenvalues $\lambda_1 = 1$, $\lambda_2 = 1$, $\lambda_3 = 2$ with corresponding eigenvectors $\mathbf{u}_1 = (1, 0, 0)$, $\mathbf{u}_2 = (0, 1, -1)$, $\mathbf{u}_3 = (1, -1, 0)$. Thus the space of solutions to the system is $\text{Sp}\{e^x(1, 0, 0), e^x(0, 1, -1), e^{2x}(1, -1, 0)\}$, which may also be described as all triples of functions (y_1, y_2, y_3) of the form

$$\begin{aligned} y_1 &= k_1e^x + k_3e^{2x} \\ y_2 &= k_2e^x - k_3e^{2x} \\ y_3 &= -k_2e^x \end{aligned} \qquad (8)$$

To confirm that we indeed have a solution, we compute

$$y_1' = k_1e^x + 2k_3e^{2x}$$

and see that it is the same as

$$y_1 - y_2 - y_3 = k_1e^x + k_3e^{2x} - k_2e^x + k_3e^{2x} + k_2e^x$$

Similarly we may find that, for functions satisfying conditions (8),

$$y_2' = 2y_2 + y_3 \quad \text{and} \quad y_3' = y_3$$ □

7.3 Problem Set

In Problems 1–4, find the general solution of the given equation.

1. a. $y' - 2y = 0$
   b. $y' + 6y = 0$
   c. $y' + \frac{3}{4}y = 0$
   d. $y' + y \ln 5 = 0$
   e. $10y' + y = 0$
2. a. $y' - 2y = 3$
   b. $y' + 5y = \sin x$
   c. $y' + 6y = e^x$
   d. $y' - y = x$
   e. $y' + y = e^{2x}$
3. a. $y'' - y' - 6y = 0$
   b. $y'' - 3y' - 4y = 0$
   c. $y'' - 6y' + 9y = 0$
   d. $y'' = 0$
   e. $2y'' + 3y' + y = 0$
   f. $3y'' - y' = 0$
   g. $y'' + 2y' + y = 0$
4. a. $(D^3 + D^2 - D - I)y = 0$
   b. $(D + I)^2(D - 2I)^3y = 0$
   c. $(D^4 + 7D^3 + 12D^2)y = 0$
   d. $(D^3 + 2D^2 + 2D)y = 0$
   e. $y'' + y = 0$
   f. $y'' + y = 2e^x$

5. a. Find the (unique) solution to $y' + 2y = 0$ which satisfies $y(0) = 2$.

   b. Find the solution to $y'' - 3y' - 4y = 0$ which satisfies $y(0) = 3$ and $y'(0) = 2$.

6. a. Use Theorem 7.18 *twice* to solve

$$(D + aI)(D + bI)y = f(x)$$

   b. Use the result from (a) to solve $y'' + 7y' + 10y = 1$. Note that the solution, $y = 0.1 + C_1e^{-2x} + C_2e^{-5x}$, could also be found by Theorem 7.21.

7. Use the method of *separation of variables* to solve equation (2) as follows. Begin with the equation

$$\frac{dy}{dx} + ay = 0$$

Write it as

$$dy + ay\,dx = 0 \quad \text{or} \quad \frac{1}{y}dy = -a\,dx$$

(The "variables" are now "separated.") Integrate each side, and solve for y.

8. Let $L: V \to W$ be a linear map and let $\mathbf{w} \in$ range L. The *pre-image of* $\mathbf{w}$ *under* L is the set $f^{-1}(\mathbf{w}) = \{\mathbf{v} \in V \mid f(\mathbf{v}) = \mathbf{w}\}$. Show that $f^{-1}(\mathbf{w})$ is an affine subspace of V. Explain how this fact generalizes Theorem 7.21.

9. Solve the following systems of differential equations.

   a. $$\begin{aligned} y_1' &= y_1 + 2y_2 \\ y_2' &= 2y_1 + y_2 \end{aligned}$$

   b. $$\begin{aligned} y_1' &= y_1 + 2y_2 \\ y_2' &= -y_1 + y_3 \\ y_3' &= -y_1 + y_3 \end{aligned}$$

APPENDICES

A

MATHEMATICAL INDUCTION

It is probably no exaggeration to say that any proof in mathematics which contains the phrase, "continue this process," would, if recast formally, involve the method of *mathematical induction*. There have been several instances of this in our text, where we either used such phrases in a proof or omitted the proof entirely. In this appendix we shall examine the method of proof by mathematical induction and apply it to some of the theorems we've already presented and to some new ones as well.

In the terminology of logic, two modes of reasoning are *induction* and *deduction*. The former involves reasoning from the specific to the general and is logically invalid. (*Example:* Two of the 25 students in my mathematics class are from Georgia; therefore all 25 students must be from Georgia.) The latter, on the other hand, is the reverse—reasoning from the general to the specific—and is valid logically. (*Example:* If all 25 students in my class are known to be from Georgia, then any two of them are from Georgia.) It is interesting to note that "conclusions" drawn from statistical sampling are examples of inductive reasoning; that is why there is a "margin of error."

Mathematical induction, despite its name, is not a case of invalid inductive reasoning. It is based on *deductions* drawn from a fundamental axiom of the system of positive integers.

> *The Axiom of Induction.* Let S be a set of positive integers with two properties:
>
> a. 1 belongs to S.
>
> b. If k belongs to S, then $k + 1$ necessarily belongs to S as well.
>
> Then, S must contain *all* positive integers.

We reiterate that this is an axiom; to a mathematician, it (or an equivalent axiom) forms a part of the very definition of the system of integers. Let us illustrate the use of this axiom in establishing a simple formula.

Example 1

For any positive integer n,

$$1 + 2 + \cdots + n = \frac{n}{2}(n + 1) \tag{1}$$

Now we could check this formula for as many specific cases as we wish:

$$1 = \frac{1}{2}(1 + 1)$$

$$1 + 2 = \frac{2}{2}(2 + 1)$$

$$1 + 2 + 3 = \frac{3}{2}(3 + 1)$$

$$\vdots$$

It seems to work. Should we therefore conclude that the proposition is valid? No! That would be an instance of inductive reasoning. Instead, consider this: Let S be the set of positive integers n which satisfy equation (1). We want to deduce that S contains all positive integers, of course. The axiom will permit this conclusion, provided that conditions a and b hold. Condition a is simple; in fact, we have seen above that 1 qualifies for membership in S by the calculation $1 = \frac{1}{2}(1 + 1)$. Note that condition b, however, is of a fundamentally different sort: "If . . . , then. . . . " It has a hypothesis and a conclusion. We *assume* that k belongs to S and *deduce from that assumption* that $k + 1$ belongs there, too. So, *assume* that

$$1 + 2 + \cdots + k = \frac{k}{2}(k + 1) \qquad \text{for some } k$$

What *follows* about the qualification for the membership of $k + 1$? We see that

$$\begin{aligned} 1 + 2 + \cdots + (k + 1) & \\ &= 1 + 2 + \cdots + k + (k + 1) \\ &= \frac{k}{2}(k + 1) + (k + 1) && \text{(by the assumption)} \\ &= (k + 1)\left(\frac{k}{2} + 1\right) && \text{(by factoring)} \\ &= (k + 1)\left(\frac{k + 2}{2}\right) \\ &= \frac{k + 1}{2}((k + 1) + 1) \end{aligned}$$

Thus $k + 1$ meets the criterion for membership in S, *if* k does. We conclude, by virtue of the axiom, that S contains *all* positive integers; that is, the formula is valid for any positive integer n. □

In actual practice, we will usually not be quite so formal about the format of a mathematical induction proof. Instead, we will organize things more or less along these lines:

To prove a certain proposition involving the positive integer n is valid for all n, we show

I. The proposition is true for $n = 1$.

II. If the proposition is true for $n = k$, then it must be true for $n = k + 1$.

There are other summation formulas like Example 1 which we could use as models, but we put those in the exercises and give here examples of a different sort.

Example 2

Given a line segment of unit length, a straight-edge (unmarked), and a compass, we show that, for any positive integer n, a line segment of length $\sqrt{n}$ can be constructed in a finite number of steps. (We are allowed to use only straight-edge and compass constructions, as in plane geometry.) Step I is automatic, since we are given a line segment of length 1. Now, we *assume* we can construct, in a finite number of constructions, a line segment of length $\sqrt{k}$. Recall from plane geometry that, by straight-edge and compass, we can draw a line perpendicular to another line at any point on the second line. First, we draw our given unit-length line segment (Figure A.1). At one end, construct a second line perpendicular to that line segment. With compass, mark off a segment on the second line of length $\sqrt{k}$. (Remember the inductive hypothesis says that we can construct a line segment of length $\sqrt{k}$ in finitely many steps; with compass, we can duplicate it.) Now, use the straight-edge to draw the hypotenuse of the right triangle of which we have already two legs. By Pythagoras's theorem, this hypotenuse will have length $\sqrt{k+1}$. We assumed only finitely many steps yielded $\sqrt{k}$; from that point, we performed only finitely many more and got $\sqrt{k+1}$. So we are finished. □

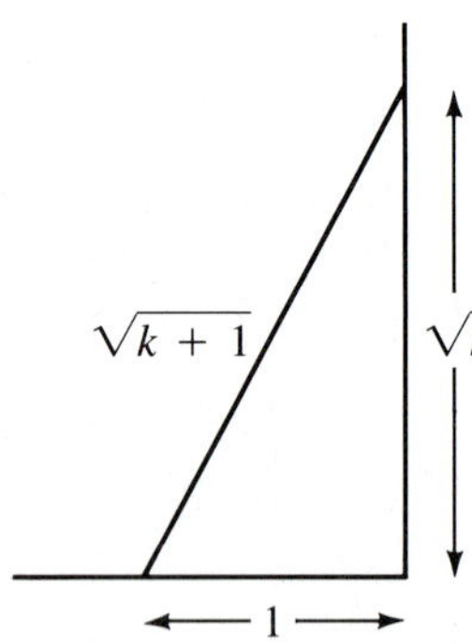

Figure A.1

Let us apply mathematical induction to some propositions from this text. What we don't do here, for the most part, we put in the exercises or in Appendix B.

Example 3
(Corollary 1.8) A product $A_1A_2 \cdots A_n$ of nonsingular matrices is nonsingular and its inverse is given by $(A_1A_2 \cdots A_n)^{-1} = A_n^{-1} \cdots A_2^{-1}A_1^{-1}$. The case for $n = 1$ is a tautology; in fact, we have nothing of content unless we begin with $n = 2$, and that case is Theorem 1.8 itself. So, we assume the proposition to be true for some k, $k \geq 2$. Thus we have the formula $(A_1A_2 \cdots A_k)^{-1} = A_k^{-1} \cdots A_2^{-1}A_1^{-1}$. Now

$$(A_1A_2 \cdots A_{k+1}) = (A_1A_2 \cdots A_k)A_{k+1}$$

So, by Theorem 1.8 (which is the case $n = 2$)

$$\begin{aligned}(A_1A_2 \cdots A_{k+1})^{-1} &= A_{k+1}^{-1}(A_1A_2 \cdots A_k)^{-1} \\ &= A_{k+1}^{-1}A_k^{-1} \cdots A_2^{-1}A_1^{-1} \qquad \text{(by the assumption)}\end{aligned}$$

Hence the formula is valid for $k + 1$ if it is for k, and we are done. (Note that what we have done implies the existence of $(A_1A_2 \cdots A_{k+1})^{-1}$ and hence the fact that $A_1A_2 \cdots A_{k+1}$ is nonsingular follows.) □

Example 4
(Theorem 2.5) Recall that a lower-triangular matrix is one having all zeros above the main diagonal; that is, $A_{ij} = 0$ if $i < j$. We show that the determinant of such a matrix is the product of the entries on the main diagonal.

The proposition is clearly correct for a 1×1 matrix, since the determinant is just the single entry, which is the sole main-diagonal entry, as well. Assume that the proposition is correct for all $k \times k$ lower-triangular matrices. Let A be a $(k + 1) \times (k + 1)$ lower-triangular matrix. We calculate det A by the definition, which is expansion about the first row. Note, however, that all entries on the first row, except possibly the first, are zero. We get $\det A = A_{11} \det(A^{11}) + 0 + \cdots + 0$, because of this. Now, the minor matrix A^{11} is itself lower-triangular, but of size $k \times k$. Applying the inductive hypothesis to $\det(A^{11})$ gives us $\det(A^{11}) = A_{22}A_{33} \cdots A_{k+1,k+1}$. Therefore

$$\det A = A_{11}A_{22}A_{33} \cdots A_{k+1,k+1}$$

as desired. □

Example 5
As another example of a mathematical induction proof, we establish, in a way different from the discussion in Section 6.2, that a symmetric matrix is diagonalizable by an orthogonal matrix (see Sections 6.2 and 6.3 to review the terminology). Recall that, by virtue of Section 7.2, all eigenvalues of a symmetric matrix are real. Our proposition is this:

If A is a symmetric matrix, then there is an orthogonal matrix C such that C^tAC is diagonal.

If A is 1×1, then it is already diagonal and we can write $A = I^tAI$, if we wish, where I is the 1×1 identity matrix.

Assume now that the proposition is true for $k \times k$ matrices satisfying the stated conditions. Let A be $(k + 1) \times (k + 1)$ and symmetric; let λ_1 be any eigenvalue of A; let $\mathbf{x}_1$ be an eigenvector corresponding to λ_1. We have $A\mathbf{x}_1 = \lambda_1\mathbf{x}_1$. We may assume that $|\mathbf{x}_1| = 1$. By Theorem 4.16, $\mathbf{x}_1$ is contained in some basis for $\mathbf{R}^{k+1}$ and we can also, by Section 6.1, employ the Gram–Schmidt process to make that an *orthonormal* basis for $\mathbf{R}^{k+1}$. So, let this orthonormal basis be $\{\mathbf{x}_1, \mathbf{u}_2, \ldots, \mathbf{u}_{k+1}\}$ and let P be the $(k + 1) \times (k + 1)$ orthogonal matrix with these vectors (in order) as columns. Now, the first column of AP is $A\mathbf{x}_1$, which is $\lambda_1\mathbf{x}_1$. The first column of P^tAP consists of the dot products of $\lambda_1\mathbf{x}_1$ with $\mathbf{x}_1, \mathbf{u}_2, \ldots, \mathbf{u}_{k+1}$; therefore P^tAP has the form

$$P^tAP = \left[\begin{array}{c|c} \lambda_1 & \cdots \\ \hline 0 & \\ \vdots & A_1 \\ 0 & \end{array}\right]$$

where A_1 is $k \times k$ and the entries to the right of λ_1 are (for the moment) not known. However, P^tAP must be symmetric (Why?), so these entries are, in fact, all zero. Moreover, A_1 must be symmetric.

Thus we have

$$P^tAP = \left[\begin{array}{c|ccc} \lambda_1 & 0 & \cdots & 0 \\ \hline 0 & & & \\ \vdots & & A_1 & \\ 0 & & & \end{array}\right]$$

with A_1 symmetric and $k \times k$. Our inductive hypothesis, therefore, applies to A_1. We get a $k \times k$ orthogonal matrix Q_1 such that $Q_1^tA_1Q_1$ is diagonal. Let

$$Q = \left[\begin{array}{c|ccc} 1 & 0 & \cdots & 0 \\ \hline 0 & & & \\ \vdots & & Q_1 & \\ 0 & & & \end{array}\right]$$

Then Q is $(k + 1) \times (k + 1)$ and orthogonal. Finally, let $C = PQ$. Since the product of two orthogonal matrices is orthogonal (Section 6.3), C is orthogonal. Now, it follows from the way matrices multiply that

$$C^tAC = (PQ)^tA(PQ) = Q^tP^tAPQ$$

$$= Q^t \left[\begin{array}{c|ccc} \lambda_1 & 0 & \cdots & 0 \\ \hline 0 & & & \\ \vdots & & A_1 & \\ 0 & & & \end{array}\right] Q$$

$$= \left[\begin{array}{c|ccc} \lambda_1 & 0 & \cdots & 0 \\ \hline 0 & & & \\ \vdots & & D_1 & \\ 0 & & & \end{array}\right]$$

where D_1 is the diagonal $k \times k$ matrix $Q_1^tA_1Q_1$. Thus C^tAC is diagonal. (It follows in the usual way that the diagonal entries of C^tAC are the eigenvalues of A and that the columns of C constitute an orthonormal basis of eigenvectors, although we could have done that in the induction as well.) □

Another proposition not proved in the main text is Theorem 1.5, which asserts that any matrix is row-equivalent to an echelon matrix. We make that our next example of mathematical induction.

Example 6

If A is an $m \times n$ matrix, then there is an $m \times n$ echelon matrix C such that A is row-equivalent to C.

We use induction on n, the number of columns. If A is $m \times 1$, and all entries are zero, we are done; A is already an echelon matrix. If A is $m \times 1$ and not all entries are zero, a row-interchange will bring a nonzero entry to the top, if one isn't already there. Multiplying the first row by the reciprocal of the entry there will put a 1 in the top position. Adding appropriate multiples of this 1 to the other rows will make all other entries zero, as required.

Now, assume that any matrix of k columns is row-equivalent to an echelon matrix and let A be $m \times (k + 1)$. As in the one-column case, certain row operations will change the first column of A to either

$$\begin{bmatrix} 1 \\ 0 \\ \vdots \\ 0 \end{bmatrix} \quad \text{or} \quad \begin{bmatrix} 0 \\ 0 \\ \vdots \\ 0 \end{bmatrix}$$

Thus A is row-equivalent to

$$\left[\begin{array}{c|c} 0 \text{ or } 1 & \cdots \\ \hline 0 & \\ \vdots & A_1 \\ 0 & \end{array}\right]$$

where A_1 is a certain $(m - 1) \times k$ matrix. Because of the 0's in the first column, a row operation on A_1 amounts to a row operation on the entire matrix, and vice versa. By the inductive hypothesis, A_1 can be brought to echelon form by such row operations. Thus A is row-equivalent to

$$\left[\begin{array}{c|c} 0 \text{ or } 1 & \cdots \\ \hline 0 & \\ \vdots & C_1 \\ 0 & \end{array}\right]$$

where C_1 is an $(m - 1) \times k$ echelon matrix. We would be finished except for having to deal with the first row, beyond the first entry, where there might be some nonzero entries above some leading ones in C_1. But, appropriate row operations of Type III will change those to zeros (without altering the first column), creating, overall, an echelon matrix row-equivalent to A. □

The question of *uniqueness* of the echelon form is somewhat thornier. (Such uniqueness was also an assertion of Theorem 1.5.) We outline a method of attacking that issue, but omit some details.

Example 7

If A is an $m \times n$ matrix and each of C and D is an echelon matrix row-equivalent to A, then $C = D$. Note that we need to show this: If each of C and D is an echelon matrix and C and D are row-equivalent to each other, then $C = D$. We use induction on n, the number of columns. If C and D are $m \times 1$, then C and D must *both* be

$$\begin{bmatrix} 1 \\ 0 \\ \vdots \\ 0 \end{bmatrix}$$

or *both* be

$$\begin{bmatrix} 0 \\ 0 \\ \vdots \\ 0 \end{bmatrix}$$

(Why?) Now, assume the proposition is true for cases involving k columns; let C and D be $m \times (k + 1)$. As above, the first columns of C and D must *both* be

$$\begin{bmatrix} 1 \\ 0 \\ \vdots \\ 0 \end{bmatrix} \quad \text{or} \quad \begin{bmatrix} 0 \\ 0 \\ \vdots \\ 0 \end{bmatrix}$$

In the second situation, the remaining columns of C and D form $m \times k$ *echelon* matrices (Why?) which are row-equivalent (Why?) and therefore are equal. Even in the first situation, we can see that we have things as depicted in the following sketch.

$$C = \left[\begin{array}{c|c} 1 & \cdots \\ \hline 0 & \\ \vdots & C_1 \\ 0 & \end{array}\right] \qquad D = \left[\begin{array}{c|c} 1 & \cdots \\ \hline 0 & \\ \vdots & D_1 \\ 0 & \end{array}\right]$$

where C_1 and D_1 are $(m - 1) \times k$ echelon matrices which are row-equivalent. Thus C_1 and D_1 are equal. There remains the question of why the first-row entries (after the leading one) must be equal in C and D. We leave that to the reader. □

A

Problem Set

1. Establish, by mathematical induction, the validity of these formulas.

   a. $1^2 + 2^2 + \cdots + n^2 = \frac{1}{6}n(n + 1)(2n + 1)$

   b. $1^3 + 2^3 + \cdots + n^3 = \frac{1}{4}n^2(n + 1)^2$

2. Show, by mathematical induction, that a set having n elements has 2^n subsets. (The empty set and the entire set count as subsets of a set.)

3. Explain why the definition of determinant in Section 2.1 is called an "inductive" definition.

4. Theorem 1.9, part d, shows that $(AB)^t = B^tA^t$, assuming matrices A and B are of appropriate dimensions. Generalize this to obtain the equation $(A_1A_2 \cdots A_n)^t = A_n^t \cdots A_2^tA_1^t$.

5. Generalize Theorem 2.9 to obtain

$$\det(A_1A_2 \cdots A_r) = (\det A_1)(\det A_2) \cdots (\det A_r)$$

6. Show, by mathematical induction, these generalizations of vector-space properties VS5 and VS6:

   a. For a scalar a and vectors $\mathbf{u}_1, \mathbf{u}_2, \ldots, \mathbf{u}_n$,

$$a(\mathbf{u}_1 + \mathbf{u}_2 + \cdots + \mathbf{u}_n) = a\mathbf{u}_1 + a\mathbf{u}_2 + \cdots + a\mathbf{u}_n$$

   b. For scalars $a_1, a_2, \ldots, a_n$ and a vector $\mathbf{u}$,

$$(a_1 + a_2 + \cdots + a_n)\mathbf{u} = a_1\mathbf{u} + a_2\mathbf{u} \cdots + a_n\mathbf{u}$$

7. Prove Theorem 5.1: For a linear map f: $U \rightarrow V$, vectors $\mathbf{u}_1, \mathbf{u}_2, \ldots, \mathbf{u}_n$ and scalars $a_1, a_2, \ldots, a_n$,

$$f(a_1\mathbf{u}_1 + a_2\mathbf{u}_2 + \cdots + a_n\mathbf{u}_n)$$
$$= a_1f(\mathbf{u}_1) + a_2f(\mathbf{u}_2) + \cdots + a_nf(\mathbf{u}_n)$$

8. Show that, if a map $f: U \rightarrow V$ has property LM1, then, for any positive integer n and any vector $\mathbf{u} \in U$,

$$f(n\mathbf{u}) = nf(\mathbf{u})$$

9. Show that the determinant of an upper triangular matrix is the product of the entries on the main diagonal.
10. Show, by mathematical induction, that, for an $n \times n$ identity matrix I, $\det I = 1$.
11. Complete all details of Example 7.
12. Prove, by mathematical induction, DeMoivre's Theorem for complex numbers:

$$[r(\cos\theta + i\sin\theta)]^n = r^n(\cos n\theta + i\sin n\theta)$$

(*Hint:* You will need the trigonometric identities $\sin(\alpha + \beta) = \sin\alpha\cos\beta + \cos\alpha\sin\beta$ and $\cos(\alpha + \beta) = \cos\alpha\cos\beta - \sin\alpha\sin\beta$.)

B

More about Determinants

To justify some assertions about determinants, notably Theorem 2.2, whose proof was omitted, we find it appropriate to formulate an alternate description of determinant, a description which many take to be the definition. To do this, we must digress at first into a consideration of *permutations*.

Definition

A *permutation* of the integers $1, 2, \ldots, n$ is an ordered n-tuple $(i_1, i_2, \ldots, i_n)$ where the value of each i_k is one of the integers $1, 2, \ldots, n$ and there is no repetition of these values; that is, $i_r \neq i_s$ if $r \neq s$.

Example 1

(3, 4, 1, 2) is a permutation of the integers 1, 2, 3, 4; here, $i_1 = 3$, $i_2 = 4$, $i_3 = 1$, and $i_4 = 2$. □

Definition

In a permutation $(i_1, i_2, \ldots, i_n)$ an *inversion* is an instance of a larger integer preceding (but not necessarily *immediately* preceding) a smaller integer; that is, an instance of $i_r > i_s$ when $r < s$. Thus an inversion is an instance of the integers' being out of their natural order.

Example 2

In the permutation (3, 4, 1, 2) in Example 1, there are several inversions: 3 precedes 1, 3 precedes 2, 4 precedes 1, and 4 precedes 2—all instances of the integers' being out of natural order. In all, there are four inversions, and, in general, what is important is the *number of inversions* in a permutation. □

Now let us consider an $n \times n$ matrix

$$A = \begin{bmatrix} A_{11}A_{12} & \cdots & A_{1n} \\ A_{21}A_{22} & \cdots & A_{2n} \\ & \cdots & \\ A_{n1}A_{n2} & \cdots & A_{nn} \end{bmatrix}$$

We wish to form a sum, the terms of which will be of the form

$$A_{1i_1}A_{2i_2} \cdots A_{ni_n}$$

where $(i_1, i_2, \ldots, i_n)$ is any permutation of the column subscripts 1, 2, . . . , n. Note that in such a term, we have a product of entries of the matrix A, one factor from the first row, the next from the second row, and so on; there are no two entries from the same row and no two from the same column. To each such term, we attach a factor of $+1$ or -1, according as the number of inversions in the column subscripts is even or odd.

Definition

For an $n \times n$ matrix A, let $d(A)$ be the sum

$$d(A) = \sum \sigma(i)A_{1i_1}A_{2i_2} \cdots A_{ni_n}$$

where in this sum there is a term for each possible permutation $(i_1, i_2, \ldots, i_n)$ of the column subscripts and

$$\sigma(i) = \begin{cases} +1, & \text{if the number of inversions is even} \\ -1, & \text{if the number of inversions is odd} \end{cases}$$

as described above.

Example 3

Let us spell out $d(A)$ for the 3×3 case. First of all, there are $3! = 6$ possible ways to permute the column subscripts 1, 2, 3, namely (1, 2, 3), (1, 3, 2), (2, 1, 3), (2, 3, 1), (3, 1, 2), and (3, 2, 1). Thus there will be six terms in the sum $d(A)$. Moreover, the number of inversions in each is, respectively, even (zero inversions), odd (one inversion), odd (one inversion), even (two inversions), even (two inversions), and odd (three inversions). Thus, for $d(A)$, we get

$$\begin{aligned} d(A) = {} & (+1)A_{11}A_{22}A_{33} + (-1)A_{11}A_{23}A_{32} + (-1)A_{12}A_{21}A_{33} \\ & + (+1)A_{12}A_{23}A_{31} + (+1)A_{13}A_{21}A_{32} + (-1)A_{12}A_{22}A_{31} \end{aligned}$$

Now, we may compare this $d(A)$ with the formula given in Section 2.1 for det A for the 3×3 case. Their equality will be confirmed. □

Example 4

For the 2×2 case, $d(A) = A_{11}A_{22} - A_{12}A_{21}$. (See the exercises.) □

Clearly, we are suggesting that $d(A)$ and det A, as defined in Section 2.1, might possibly be equal. Our next theorem provides more fuel for that conjecture.

Theorem B.1 For a square matrix A,

a. If two rows of A are interchanged, then $d(A)$ changes sign.

b. If a row of A is multiplied by a constant k, then the value of $d(A)$ is multiplied by that constant k.

c. If a scalar multiple of one row is added to another row, then the value of $d(A)$ is unchanged.

(Compare with Theorem 2.2.)

Proof.

a. If two rows, say the r-th row and the s-th row, are interchanged to create a new matrix B, then the terms in $d(B)$ will have the form

$$A_{1i_1} \cdots A_{si_s} \cdots A_{ri_r} \cdots A_{ni_n}$$

assuming that $r < s$. The accompanying sign will be determined by the number of inversions in the permutation $(i_1, \ldots, i_s, \ldots, i_r, \ldots, i_n)$. For each such term, there is an identical term in $d(A)$, except possibly for the attached sign. To see that the signs are, in fact, *opposite*, we need only show that interchanging i_s and i_r in the permutation will reverse the *oddness* or *evenness* of the number of inversions. First note that interchanging two *adjacent* integers in a permutation *changes the number of inversions by exactly one*, for such an adjacent interchange either introduces or eliminates one inversion. Now, the question is: How many *adjacent* interchanges will effect the interchanging of i_s and i_r? Starting with the permutation $(i_1, \ldots, i_s, \ldots, i_r, \ldots, i_n)$, we can move i_r successively to the left, interchanging it in turn with integers until it stands immediately to the left if i_s. That requires exchanging places with $s - r$ integers between it and i_s and with i_s itself, for a total of $s - r + 1$ adjacent interchanges. Now i_s must interchange with only those integers which were originally between, requiring $s - r$ adjacent interchanges. The grand total of adjacent interchanges is therefore

$$(s - r + 1) + (s - r) = 2(s - r) + 1$$

an *odd number*. Remember now that each adjacent interchange alters the number of inversions by one; hence the net effect is that an odd number of inversions changes to an even number and an even number to an odd number. As a consequence, the terms in $d(B)$ are precisely the negatives of the terms in $d(A)$; thus, $d(B) = -d(A)$ as required.

We leave the proof of parts b and c from the definition of $d(A)$ to the exercises. We'll confirm their validity in another way, however, using induction.

Lemma B.2 Let $1 \leq m \leq n$; let i be the permutation $(m, i_2, \ldots, i_n)$ on $1, 2, \ldots, n$ (that is, $i_1 = m$); let i' be the permutation $(i_2, \ldots, i_n)$ on the first n positive integers with m deleted. (For instance, i might be $(3, 1, 4, 2, 5)$ in which case i' denotes $(1, 4, 2, 5)$.) Then the number of inversions in i equals the number of inversions in i', *plus* $m - 1$.

Proof. Consider the case $m = 1$. We compare the number of inversions in $(1, i_2, \ldots, i_n)$ and in $(i_2, \ldots, i_n)$. Since 1 cannot appear more than once in a permutation, we have $1 \neq i_k$, for all k, so both permutations have the same number of inversions. Let's try $m = 2$ and compare $(2, i_2, \ldots, i_n)$ with $(i_2, \ldots, i_n)$. Since some i_k must be 1, and all other entries are greater than 2, there is one more inversion in the first permutation than in the second. The situation is similar for all other cases.

Theorem B.2 For an $n \times n$ matrix A, $d(A) = \det A$, as defined in Section 2.1.

Proof. (By induction) We are using the definition of det A involving expansion by cofactors across the first row of A. For the case $n = 1$, $d(A)$ and det A are trivially equal since each is simply the single entry of A. Moreover, we have already seen, by direct calculation, that $d(A)$ and det A are equal in the cases $n = 2$ and $n = 3$. For the inductive step, we assume that $d(B) = \det B$ when B is $r \times r$, and deduce that $d(A) = \det A$ for an $(r + 1) \times (r + 1)$ matrix A. In the definition of $d(A)$, we collect together in a sum all those terms in which A_{11} is a factor; we have

$$\begin{aligned}\sum \sigma(i)A_{11}A_{2i_2} \cdots A_{ni_n} &= A_{11}\sum \sigma(i)A_{2i_2} \cdots A_{ni_n} \\ &= A_{11}\sum \sigma(i')A_{2i_2} \cdots A_{ni_n}\end{aligned}$$

(by Lemma B.2 and in its notation).

Now no entry from the first row or from the first column of A can appear in the summation expression; but that expression does include a term for every possible permutation of the column subscripts $2, \ldots, n$, correctly signed as in a d-formula. In fact, the summation expression is $d(A^{11})$, where A^{11} is the 1, 1 minor matrix (as in Section 2.1). Since A^{11} is $r \times r$, then by our inductive hypothesis, $d(A^{11}) = \det A^{11}$. So far, we have $A_{11}\det(A^{11})$.

In like manner, consider in $d(A)$ all terms involving A_{12}:

$$\sum \sigma(i)A_{12}A_{2i_2} \cdots A_{ni_n} = A_{12}\sum \sigma(i)A_{2i_2} \cdots A_{ni_n}$$

In this case, we can replace $\sigma(i)$ with $-\sigma(i')$, because, by Lemma B.2, the number of inversions in i and in i' differ by 1. We continue:

$$\begin{aligned}&= A_{12}\sum -\sigma(i')A_{2i_2} \cdots A_{nin} \\ &= -A_{12}\sum \sigma(i')A_{2i_2} \cdots A_{nin}\end{aligned}$$

As before, the summation expression is just $d(A^{12})$ and, by induction, is therefore $\det(A^{12})$. Thus, we have $-A_{12}\det(A^{12})$.

If we continue as above with the terms involving A_{13}, A_{14}, and so on, we will have used all of $d(A)$, since each term of $d(A)$ must contain a factor from the first row of A. We will get

$$A_{11}\det(A^{11}) - A_{12}\det(A^{12}) + \cdots + (-1)^{1+n}A_{1n}\det(A^{1n})$$

which is, by definition, det A. Thus, $d(A) = \det A$.

We have remarked that many mathematicians use the formula for $d(A)$ as the definition of determinant. Theorem B.2 shows that the result is the same as with our definition. Note also that, in Theorem B.1, part a, we have established the first determinant property in Theorem 2.2. Now, we shall validate Theorem 2.1, which says that det A can be evaluated by expansion about any row (not just about the first).

Corollary B.2 For an $n \times n$ matrix A, and for $1 \leq i \leq n$,

$$\det A = \sum_{k=1}^{n} (-1)^{i+k} A_{ik} \det(A^{ik})$$

Proof. We shall achieve our goal by moving the i-th row of A upward until it becomes the first row, leaving all the other rows in the same order, relative to each other. Call the resulting matrix B; thus B is obtained from A by successively interchanging the i-th row with the row directly above it, which entails $i - 1$ adjacent interchanges. The effect, according to Theorem B.1, on $\det A = d(A)$ is that it gets multiplied by $(-1)^{i-1}$; thus $\det B = d(B) = (-1)^{i-1} d(A) = (-1)^{i-1} \det A$.

Now, by Theorem B.2 (since the first row of B contains $A_{i1}, A_{i2}, \ldots, A_{in}$)

$$\det B = \sum_{k=1}^{n} (-1)^{1+k} A_{ik} \det(B^{1k}) = \sum_{k=1}^{n} (-1)^{1+k} A_{ik} \det(A^{ik})$$

(The minor determinants of the first-row entries of B are just those of the i-th row entries of A.) So

$$(-1)^{i-1} \det A = \sum_{k=1}^{n} (-1)^{1+k} A_{ik} \det(A^{ik})$$

Upon multiplying both sides of this equation by $(-1)^{i-1}$, we obtain our result.

Finally, we return to the remaining two determinant properties, as enunciated in Theorem B.1, parts b and c.

Proof. (of Theorem B.1, parts b and c)

b. For an $n \times n$ matrix A, suppose that the i-th row is multiplied by a scalar k; call the new matrix B and calculate its determinant by expansion about the i-th row. Each term in the expansion will contain an entry from the i-th row of B, which is just k times the corresponding entry from A. Factoring out the factor k leaves just det A; thus det B $= k$ det A.

c. (By induction) We must begin with the 2×2 case, where direct computation will confirm that adding a scalar multiple of one row to another does not change the determinant value (see exercises). Suppose, as the inductive hypothesis, that this is also the case for, say, $m \times m$, $m \geq 2$. Let A be $(m + 1) \times (m + 1)$; let c be a scalar; let B be obtained from A by adding c times the r-th row of A to the s-th row of A, where $1 \leq r < s \leq m + 1$. We compute det B by expansion about any row *other than the r-th or s-th*, say the i-th row, $i \neq r, s$:

$$\det B = \sum_{k=1}^{n} (-1)^{i+k} B_{ik} \det(B^{ik})$$

How is the minor matrix B^{ik} related to the minor matrix A^{ik}? Clearly, adding c times a certain row of A^{ik} to a certain other row of A^{ik} yields B^{ik}. But these minor matrices are $m \times m$; by the inductive hypothesis, therefore, det B^{ik} = det A^{ik}. Moreover, the entries in the i-th row of B are the same as those in the i-th row of A. Hence

$$\det B = \sum_{k=1}^{n} (-1)^{i+k} A_{ik} \det(A^{ik}) = \det A$$

An interesting point about this induction proof is this: Why did we begin the induction with the 2×2 case? Why not start with 1×1, where the proposition is surely *vacuously true*, since any assertion about *two* rows of a 1×1 matrix is logically correct? Then, from the inductive assumption that the proposition is valid for $m \times m$, $m \geq 1$, we would need to proceed to establish the $(m + 1) \times (m + 1)$ case. We'll repeat this question in the exercises.

B
Problem Set

1. How many permutations are there of the integers 1, 2, 3, 4? List them and determine how many inversions each has.

2. Determine the number of inversions in each of the following.
   a. (3, 4, 2, 5, 1, 6)
   b. (9, 3, 2, 7, 5, 4, 6, 1, 8)
   c. (4, 3, 2, 1, 5)
3. Prove, by induction, that the number of permutations of 1, 2, . . . , n is $n!$.
4. Use the definition of $d(A)$ to confirm that $d(A) = A_{11}A_{22} - A_{12}A_{21}$, in the 2×2 case.
5. Use the definition of $d(A)$ to prove parts b and c of Theorem B.1. (*Hint:* Deduce as a corollary from part a that if A has two equal rows then $d(A) = 0$, and then use this in part c.)
6. Use the definition of $d(A)$ to show that if an $n \times n$ matrix A is upper-triangular, then $d(A) = A_{11}A_{22} \cdots A_{nn}$. Recall that an upper-triangular matrix satisfies the condition $A_{ij} = 0$ if $i > j$. Use this to eliminate all terms from $d(A)$ except one.
7. Use the definition of $d(A)$ to show that $d(A^t) = d(A)$, where A^t, the *transpose* of A, is given by $(A^t)_{ij} = A_{ji}$. (*Hint:* Let $B = A^t$. A typical term in $d(B)$ involves $B_{1i_1}B_{2i_2} \cdots B_{ni_n} = A_{i_11}A_{i_22} \cdots A_{i_nn}$. To match this with the corresponding term in $d(A)$, the factors need to be reordered so that the row subscripts are in their natural order. This, of course, creates a permutation of the column subscripts. The key lies in showing that the number of inversions in the row subscripts, as they originally appear, is equal to the number of inversions introduced in the column subscripts by the reordering, so that the signs will correspond, as well as the terms. Consider this: in $(i_1, i_2, \ldots, i_n)$, to return 1 to its leading position will require exactly as many adjacent interchanges as there are inversions caused by a larger integer preceding 1. Each such interchange introduces a new inversion in the column subscripts as it eliminates an inversion in the row subscripts. Continue this line of reasoning.)
8. For the 2×2 case, show by direct computation that adding a scalar multiple of one row of a matrix to another does not change the value of the determinant.
9. To return to the point raised in the final paragraph of this appendix about beginning the induction proof with the 2×2 case, consider the following: What is incorrect about the following "proof" that interchanging two rows of a matrix leaves the determinant *unchanged*?

   I. For a 1×1 matrix, the proposition is vacuously correct, since the matrix does not have two rows to be interchanged.

II. Assume the proposition is true for $k \times k$ matrices. Let B be a matrix obtained from a $(k + 1) \times (k + 1)$ matrix A by interchanging rows r and s, and compute det B by expansion about any row other than those interchanged. Since the minor determinants involved are $k \times k$ and come from the corresponding minor matrices of A by the interchanging of two rows, we see that det B is exactly the same as det A, since these minor determinants are unchanged. End of "proof."

10. Use mathematical induction to show that $\det \overline{A} = \overline{\det A}$; that is, in the complex setting, the determinant of the conjugate of a matrix is the conjugate of the determinant of the matrix.

Selected Answers

Problem Set 1.2

1. a. $(x, y, z) = (1 + 2t, 1 + 2t, t)$, t arbitrary
c. $(x, y) = (0, -2)$ d. no solution

2. a. $(x, y, z) = (0, 0, 0)$ b. $(x, y, z) = (-\frac{1}{2}s, -\frac{1}{2}s, s)$
e. $(x, y, z, w) = (-\frac{1}{2}s + \frac{1}{2}t, s, -2t, t)$

3. $(x, y, z, w) = (2s - t, s, -3t, t)$, s, t arbitrary

8. a. $(x, y, z, w) = (2, -3t, -3, t)$
b. $(x, y, z, w) = (-15 + t, -9 + t, -4 + t, t)$
f. no solution

Problem Set 1.3

1. $AB = \begin{bmatrix} 2 & -1 & 2 \\ 8 & 7 & 1 \end{bmatrix}$; BA not possible; $A + B$ not possible;

$(2A + C)B = \begin{bmatrix} 8 & -1 & 2 \\ 31 & 26 & 5 \end{bmatrix}$; $A + C = \begin{bmatrix} 2 & -2 & 0 \\ 6 & 3 & 5 \end{bmatrix}$;

$AD = \begin{bmatrix} -2 \\ 8 \end{bmatrix}$; $D + E$ not possible; BE not possible;

$DE = \begin{bmatrix} 4 & -2 & 1 \\ 0 & 0 & 0 \\ 12 & -6 & 3 \end{bmatrix}$; $ED = [7]$

2. a. $x = 8, y = 5$ b. $x = -1, y = 3$ c. $x = -2, y = 1$

6. Let A be $m \times n$ and B be $p \times q$. Show that $m = n = p = q$.

Problem Set 1.4

1. a. $\begin{bmatrix} 2 & -3 & 2 \\ 1 & 1 & -4 \end{bmatrix}\begin{bmatrix} x \\ y \\ z \end{bmatrix} = \begin{bmatrix} -1 \\ 2 \end{bmatrix}$ b. $\begin{bmatrix} 1 & 1 & 1 \\ 1 & -1 & 0 \\ 0 & 1 & 2 \\ 1 & 0 & -1 \end{bmatrix}\begin{bmatrix} x \\ y \\ z \end{bmatrix} = \begin{bmatrix} 1 \\ 0 \\ -1 \\ 2 \end{bmatrix}$

2. a. $\left[\begin{array}{rrr|r} 2 & -3 & 2 & -1 \\ 1 & 1 & -4 & 2 \end{array}\right]$ b. $\left[\begin{array}{rrr|r} 1 & 1 & 1 & 1 \\ 1 & -1 & 0 & 0 \\ 0 & 1 & 2 & -1 \\ 1 & 0 & -1 & 2 \end{array}\right]$

3. echelon form: d, f

5. a. For Example 2a: $(x, y) = (2, -1)$
b. For Example 2a: $(x, y, z) = (-2t, t, t)$

6. a. For the second matrix: $(x, y, z, w) = (s, -\frac{1}{2} - 2t, t, 1)$
b. For the second matrix: $(x, y, z, u, v) = (s, -2t + \frac{1}{2}u, t, -u, u)$

9. Add -2 times row two to row one; add row two to row four; interchange rows one and two; etc. Echelon form is

$$\begin{bmatrix} 1 & 0 & 0 & 0 \\ 0 & 1 & 0 & 0 \\ 0 & 0 & 1 & 0 \\ 0 & 0 & 0 & 1 \\ 0 & 0 & 0 & 0 \end{bmatrix}$$

If augmented matrix, no solution; if coefficient matrix of homogeneous system, only the trivial solution: (0, 0, 0, 0).

Problem Set 1.5

1. $[A|I] = \left[\begin{array}{rr|rr} 2 & -1 & 1 & 0 \\ -3 & 2 & 0 & 1 \end{array}\right]$; $A^{-1} = \begin{bmatrix} 2 & 1 \\ 3 & 2 \end{bmatrix}$

2. a and d are singular

3. $x = 6, y = 9$

Problem Set 1.6

1. a. $\begin{bmatrix} 1 & 0 & 7 \\ 0 & 1 & 0 \\ 0 & 0 & 1 \end{bmatrix}$

2. a. $\frac{1}{12}\begin{bmatrix} 3 & 0 & -3 \\ -2 & 4 & 2 \\ 3 & 0 & 9 \end{bmatrix}$ b. $\begin{bmatrix} -4 & 3 & 1 & -1 \\ 4 & -3 & -2 & 1 \\ -8 & 6 & 4 & -1 \\ 7 & -5 & -3 & 1 \end{bmatrix}$

c. $\frac{1}{8}\begin{bmatrix} 0 & 3 & -4 & 2 \\ 12 & -3 & 8 & -2 \\ -8 & 3 & -4 & 2 \\ -4 & -1 & 0 & 2 \end{bmatrix}$ d. $\begin{bmatrix} 19 & 15 & -6 & 12 & -5 \\ 13 & 5 & -1 & 7 & -3 \\ -2 & 7 & -4 & 1 & 0 \\ -3 & -6 & 3 & -3 & 1 \\ 4 & 1 & 0 & 2 & -1 \end{bmatrix}$

4. $E_1 = \begin{bmatrix} 1 & -2 & 0 \\ 0 & 1 & 0 \\ 0 & 0 & 1 \end{bmatrix}$ $E_2 = \begin{bmatrix} 1 & 0 & 0 \\ 0 & 1 & 0 \\ 0 & -3 & 1 \end{bmatrix}$

$E_3 = \begin{bmatrix} 1 & 0 & 0 \\ 0 & 1 & 0 \\ -1 & 0 & 1 \end{bmatrix}$ $E_4 = \begin{bmatrix} 1 & 0 & 0 \\ 0 & 1 & 0 \\ 0 & 0 & \frac{1}{4} \end{bmatrix}$ $E_5 = \begin{bmatrix} 1 & 0 & 0 \\ 0 & 1 & 3 \\ 0 & 0 & 1 \end{bmatrix}$

$E_6 = \begin{bmatrix} 1 & 0 & -6 \\ 0 & 1 & 0 \\ 0 & 0 & 1 \end{bmatrix}$ $E_7 = \begin{bmatrix} \frac{1}{2} & 0 & 0 \\ 0 & 1 & 0 \\ 0 & 0 & 1 \end{bmatrix}$ $E_8 = \begin{bmatrix} 1 & 0 & 0 \\ \frac{1}{2} & 1 & 0 \\ 0 & 0 & 1 \end{bmatrix}$

$E_9 = \begin{bmatrix} 1 & 0 & 0 \\ 0 & 1 & 0 \\ -\frac{1}{2} & 0 & 1 \end{bmatrix}$ $E_{10} = \begin{bmatrix} 0 & 1 & 0 \\ 1 & 0 & 0 \\ 0 & 0 & 1 \end{bmatrix}$ $E_{11} = \begin{bmatrix} 1 & 0 & 0 \\ 0 & 0 & 1 \\ 0 & 1 & 0 \end{bmatrix}$

Let the coefficient matrix be A. Then

$$A = E_1^{-1}E_2^{-1}E_3^{-1}E_4^{-1}E_5^{-1}E_6^{-1}E_7^{-1}E_8^{-1}E_9^{-1}E_{10}^{-1}E_{11}^{-1}$$

$$= \begin{bmatrix} 1 & 2 & 0 \\ 0 & 1 & 0 \\ 0 & 0 & 1 \end{bmatrix}\begin{bmatrix} 1 & 0 & 0 \\ 0 & 1 & 0 \\ 0 & 3 & 1 \end{bmatrix}\begin{bmatrix} 1 & 0 & 0 \\ 0 & 1 & 0 \\ 1 & 0 & 1 \end{bmatrix}\begin{bmatrix} 1 & 0 & 0 \\ 0 & 1 & 0 \\ 0 & 0 & 4 \end{bmatrix}\begin{bmatrix} 1 & 0 & 0 \\ 0 & 1 & -3 \\ 0 & 0 & 1 \end{bmatrix}$$

$$\begin{bmatrix} 1 & 0 & 6 \\ 0 & 1 & 0 \\ 0 & 0 & 1 \end{bmatrix}\begin{bmatrix} 2 & 0 & 0 \\ 0 & 1 & 0 \\ 0 & 0 & 1 \end{bmatrix}\begin{bmatrix} 1 & 0 & 0 \\ -\frac{1}{2} & 1 & 0 \\ 0 & 0 & 1 \end{bmatrix}\begin{bmatrix} 1 & 0 & 0 \\ 0 & 1 & 0 \\ \frac{1}{2} & 0 & 1 \end{bmatrix}$$

$$\begin{bmatrix} 0 & 1 & 0 \\ 1 & 0 & 0 \\ 0 & 0 & 1 \end{bmatrix}\begin{bmatrix} 1 & 0 & 0 \\ 0 & 0 & 1 \\ 0 & 1 & 0 \end{bmatrix}$$

6. a. singular b. inverse: $\begin{bmatrix} -2 & 1 \\ 3 & -1 \end{bmatrix}$

c. inverse: $\frac{1}{2}\begin{bmatrix} 1 & 1 & -1 \\ -1 & 1 & 1 \\ 1 & -1 & 1 \end{bmatrix}$

Problem Set 2.1

1. a. 8 b. 2 c. 12 d. 0 e. 0 f. 1 g. -1

4. a. 6 c. 3

Problem Set 2.2

3. a. $(x, y, z) = (1, 2, -2)$ b. $(x, y, z) = (2, -2, 1)$
c. $(x, y, z, w) = (2, 0, 2, 1)$

4. a. $(x, y, z) = (1, -3, 1)$ b. $(x, y, z, w) = (1, 0, 0, 1)$
c. $(v, w, x, y, z) = (2, -1, -1, 1, 2)$

8. c. Observe that $\text{tr}(AB)$ is the sum of all possible terms of the form $A_{rs}B_{sr}$, $r, s = 1, 2, \ldots, n$. This is also $\text{tr}(BA)$. (*Why?*)
d. Use the result of part c.

Problem Set 3.2

1. a. $|\mathbf{x}| = \sqrt{3}$; $|\mathbf{x} - \mathbf{y}| = 1$; $\cos \alpha = \sqrt{3}/2$, so $\alpha = \pi/6$ or $30°$

2. Solve $x - y + 2z = 0$

6. For the vectors in 1a: $|\mathbf{x} \cdot \mathbf{y}| = 3$; $|\mathbf{x}||\mathbf{y}| = \sqrt{3}\sqrt{4} = 2\sqrt{3}$; $3 \leq 2\sqrt{3}$

8. For the vectors in 1a: $\mathbf{x} - (\text{proj}_\mathbf{y}\mathbf{x}) = \frac{1}{4}(1, -1, -3, 1)$

Problem Set 3.3

1. a. $(x, y) = (-1, 2) + t(2, 3)$; $3x - 2y = -7$
b. $(x, y) = (2, 3) + t(-1, 2)$; $2x + y = 7$
d. $(x, y) = t(1, -7)$; $7x + y = 0$

2. a. $(x, y, z) = (1, -2, 1) + t(1, 0, 3)$; $\dfrac{x-1}{1} = \dfrac{y+2}{0} = \dfrac{z-1}{3}$;
cosines: $1/\sqrt{10}$, 0, $3/\sqrt{10}$; planes: $3y = -6$, $3x - z = 2$
c. $(x, y, z) = (2, 1, 3) + t(5, -1, -3)$; $\dfrac{x-2}{5} = \dfrac{y-1}{-1} = \dfrac{z-3}{-3}$; cosines: $5/\sqrt{35}$, $-1/\sqrt{35}$, $-3/\sqrt{35}$; planes:
$3y - z = 0$, $3x + 5z = 21$

3. b. $(x, y, z) = (3, 3, -3) + t(2, -1, 2)$
d. $(x, y, z, v, w) = (-4, 2, -1, 0, 3) + t(7, -3, 1, 2, -3)$

4. a. point of intersection: $(1, 0, 2)$ b. $-3/\sqrt{14}$
c. $(x, y, z) = (1, 0, 2) + t(0, 2, -1)$

6. a. $(1/5)\sqrt{5}$ b. $(1/2)\sqrt{102}$ d. 0

9. a. parallel: $2x - y = 0$; perpendicular: $x + 2y = 5$

10. a. parallel; distance $= (2/7)\sqrt{35}$
b. skew; distance $= (1/5)\sqrt{5}$

12. a. $(1/3)\{(-8, 0, 0) + s(-5, 3, 0) + t(2, 0, 3)\}$; $3x + 5y - 2z = -8$ b. $s(-\frac{1}{2}, 1, 0) + t(\frac{3}{2}, 0, 1)$; $2x + y - 3z = 0$
d. $(1, 0, 0) + s(-\frac{5}{3}, 1, 0) + t(\frac{2}{3}, 0, 1)$; $3x + 5y - 2z = 3$
e. $(\frac{13}{2}, 0, 0) + s(-\frac{1}{2}, 1, 0) + t(\frac{3}{2}, 0, 1)$; $2x + y - 3z = 13$

13. a. 11 b. 13

14. a. $2x - y + 2z = 1$ c. $-2x - y + 4z = 0$

16. a. $-y + 2z = 5$

17. a. $4/\sqrt{14}$ c. $10/\sqrt{17}$

Problem Set 3.4

1. Let $\mathbf{u} = (2, -1, 3) - (1, 2, 1) = (1, -3, 2)$; $\mathbf{v} = (0, 4, 3) - (1, 2, 1) = (-1, 2, 2)$. Let $\mathbf{n} = \mathbf{u} \times \mathbf{v} = -10\mathbf{i} - 4\mathbf{j} - \mathbf{k} = (-10, -4, -1)$. Then $(x, y, z) \cdot (-10, -4, -1) = (1, 2, 1) \cdot (-10, -4, -1)$ gives $10x + 4y + z = 19$

2. $y - 2z = -5$

3. a. $9x - y - 5z = 0$ c. $6x - 14y - 2z = -30$

4. a. $\cos \alpha = (-7)/2\sqrt{33}$; line of intersection: $\mathbf{x} = (6, -1, 0) + t(-3, 5, 7)$

9. a. 7 b. $\sqrt{11}$

10. a. 4 b. $\sqrt{17}/2$

11. a. 4 b. 5

Problem Set 4.1

1. b. $\mathbf{x} = t(1, 0, 2)$

3. W is an affine subspace.

4. a. subspace: yes b. subspace: yes
c. subspace: no; affine: yes d. neither

10. a. subspace: yes b. subspace: yes c. neither
d. neither e. subspace: yes f. subspace: yes
g. subspace: no; affine: yes h. neither

16. yes

17. no; not closed under scalar multiplication

Problem Set 4.2

3. b. If the three vectors are $\mathbf{u}_1, \mathbf{u}_2, \mathbf{u}_3$, then $-3\mathbf{u}_1 + 1\mathbf{u}_2 + 4\mathbf{u}_3 = (0, 0, 0, 0)$.

11. a. $\{(1, 0, 2), (0, 1, 0)\}$ b. $\{(1, 0, 1, 1), (0, 1, 1, -1)\}$
c. $\{(1, 2, 3)\}$

12. a. $\left\{\begin{bmatrix}1&0&0\\0&0&0\\0&0&0\end{bmatrix}, \begin{bmatrix}0&1&0\\0&0&0\\0&0&0\end{bmatrix}, \begin{bmatrix}0&0&1\\0&0&0\\0&0&0\end{bmatrix}, \begin{bmatrix}0&0&0\\1&0&0\\0&0&0\end{bmatrix}, \begin{bmatrix}0&0&0\\0&1&0\\0&0&0\end{bmatrix},\right.$
$\left.\begin{bmatrix}0&0&0\\0&0&1\\0&0&0\end{bmatrix}, \begin{bmatrix}0&0&0\\0&0&0\\1&0&0\end{bmatrix}, \begin{bmatrix}0&0&0\\0&0&0\\0&1&0\end{bmatrix}, \begin{bmatrix}0&0&0\\0&0&0\\0&0&1\end{bmatrix}\right\}$

b. $\left\{\begin{bmatrix}1&0&0\\0&0&0\\0&0&0\end{bmatrix}, \begin{bmatrix}0&1&0\\1&0&0\\0&0&0\end{bmatrix}, \begin{bmatrix}0&0&1\\0&0&0\\1&0&0\end{bmatrix}, \begin{bmatrix}0&0&0\\0&1&0\\0&0&0\end{bmatrix},\right.$
$\left.\begin{bmatrix}0&0&0\\0&0&1\\0&1&0\end{bmatrix}, \begin{bmatrix}0&0&0\\0&0&0\\0&0&1\end{bmatrix}\right\}$

c. $\left\{\begin{bmatrix}0&1&0\\-1&0&0\\0&0&0\end{bmatrix}, \begin{bmatrix}0&0&1\\0&0&0\\-1&0&0\end{bmatrix}, \begin{bmatrix}0&0&0\\0&0&1\\0&-1&0\end{bmatrix}\right\}$

d. $\left\{\begin{bmatrix}1&0&0\\0&0&0\\0&0&-1\end{bmatrix}, \begin{bmatrix}0&1&0\\0&0&0\\0&0&0\end{bmatrix}, \begin{bmatrix}0&0&1\\0&0&0\\0&0&0\end{bmatrix}, \begin{bmatrix}0&0&0\\1&0&0\\0&0&0\end{bmatrix},\right.$
$\left.\begin{bmatrix}0&0&0\\0&1&0\\0&0&-1\end{bmatrix}, \begin{bmatrix}0&0&0\\0&0&1\\0&0&0\end{bmatrix}, \begin{bmatrix}0&0&0\\0&0&0\\1&0&0\end{bmatrix}, \begin{bmatrix}0&0&0\\0&0&0\\0&1&0\end{bmatrix}\right\}$

e. $\left\{\begin{bmatrix}1&0&0\\0&0&0\\0&0&0\end{bmatrix}, \begin{bmatrix}0&1&0\\0&0&0\\0&0&0\end{bmatrix}, \begin{bmatrix}0&0&1\\0&0&0\\0&0&0\end{bmatrix}, \begin{bmatrix}0&0&0\\0&1&0\\0&0&0\end{bmatrix},\right.$
$\left.\begin{bmatrix}0&0&0\\0&0&1\\0&0&0\end{bmatrix}, \begin{bmatrix}0&0&0\\0&0&0\\0&0&1\end{bmatrix}\right\}$

13. $\mathbf{y}$ and $\mathbf{w}$ are linear combinations of $\mathbf{u}_1$, $\mathbf{u}_2$, $\mathbf{u}_3$. The others are not. To solve this problem in one stroke, one can reduce the following matrix to echelon form. Question: Why does this do the job?

$$\begin{bmatrix}1&-1&1&9&1&0&-1\\-1&0&-2&2&-3&1&-2\\3&2&8&14&13&2&12\end{bmatrix}$$

14. a and c span; a is a basis

15. all but a span; c is a basis

16. $\mathbf{x} = (-5)(1, 0, 2) + (4)(2, 0, 3)$

Problem Set 4.3

1. b is independent

2. a and c are independent

3. a, b, d, f, and g are independent

4. a. -1 (constant); independent b. 288 (constant); independent
c. 0 (constant); dependent because $2(x^2 - 3) + (3x - 2x^2) - 3(x - 2) = 0$ for all x d. $2e^{6x}$; independent
e. 0 (constant); dependent because $3\sin^2 x - \frac{3}{2}(2) + 3\cos^2 x = 0$ for all x

Problem Set 4.4

1. Combine the given vector with any second vector not a scalar multiple of the first (for example, {(3, 2), (1, 3)}).

3. Find any third vector not a linear combination of the given two (for example, {(4, 1, 0), (1, 2, 2), (3, 2, 1)}).

4. *Suggestion:* Begin with the first vector. Combine one by one with successive vectors, testing at each stage to be sure that independence is maintained. The resulting basis will be {(3, 2, −1), (5, −2, 1), (0, 2, 0)}.

7. dimensions are: a. 2 b. 2 c. 2 d. 6 e. 6 f. 8

8. the origin; a line through the origin; a plane through the origin; all of $\mathbf{R}^3$

9. dimensions are: a. 2 b. 2 c. 1 d. 1 e. 2 f. 0 g. 0 h. 0 i. 1

10. a. no b. yes

Problem Set 5.1

1. c, h, i, k, l are linear maps; d, e, f, g, j are affine but not linear

2. d, e, f, g, j

4. c. $\begin{bmatrix} 1 & 1 & 0 \\ 1 & -1 & 0 \\ 2 & 0 & 0 \end{bmatrix}$ d. $\begin{bmatrix} 0 & 2 \\ -1 & 1 \\ 3 & 0 \end{bmatrix}\begin{bmatrix} s \\ t \end{bmatrix} + \begin{bmatrix} 1 \\ 3 \\ 2 \end{bmatrix}$

g. $\begin{bmatrix} 1 & 1 & 1 \\ 0 & 1 & 0 \\ 1 & 0 & 1 \\ 0 & 0 & 1 \end{bmatrix}\begin{bmatrix} x \\ y \\ z \end{bmatrix} + \begin{bmatrix} 0 \\ 2 \\ 3 \\ 0 \end{bmatrix}$ h. $\begin{bmatrix} 1 & -1 & 0 \\ 0 & 1 & -1 \end{bmatrix}$

5. c. $\ker f = \left\{\begin{bmatrix} 0 \\ 0 \\ t \end{bmatrix}\right\} = \left\{t\begin{bmatrix} 0 \\ 0 \\ 1 \end{bmatrix}\right\}$; basis for $\ker f$: $\left\{\begin{bmatrix} 0 \\ 0 \\ 1 \end{bmatrix}\right\}$; $N(f) = 1$;

range $f = \left\{\begin{bmatrix} a \\ b \\ a + b \end{bmatrix}\right\} = \left\{a\begin{bmatrix} 1 \\ 0 \\ 1 \end{bmatrix} + b\begin{bmatrix} 0 \\ 1 \\ 1 \end{bmatrix}\right\}$; basis for range f:

$\left\{\begin{bmatrix} 1 \\ 0 \\ 1 \end{bmatrix}, \begin{bmatrix} 0 \\ 1 \\ 1 \end{bmatrix}\right\}$; $R(f) = 2$; Theorem 5.6: $3 = 1 + 2$

h. $\ker f = \left\{\begin{bmatrix} t \\ t \\ t \end{bmatrix}\right\} = \left\{t\begin{bmatrix} 1 \\ 1 \\ 1 \end{bmatrix}\right\}$; basis for $\ker f$: $\left\{\begin{bmatrix} 1 \\ 1 \\ 1 \end{bmatrix}\right\}$; $N(f) = 1$;

range $f = \left\{\begin{bmatrix} a \\ b \end{bmatrix}\right\} = \left\{a\begin{bmatrix} 1 \\ 0 \end{bmatrix} + b\begin{bmatrix} 0 \\ 1 \end{bmatrix}\right\} = \mathbf{R}^2$; basis for range f:

$\left\{\begin{bmatrix} 1 \\ 0 \end{bmatrix}, \begin{bmatrix} 0 \\ 1 \end{bmatrix}\right\}$; $R(f) = 2$; Theorem 5.6: $3 = 1 + 2$

i. $N(f) = 1$; $R(f) = 2$ k. $N(f) = 1$; $R(f) = 2$
l. $N(f) = 1$; $R(f) = 3$

7. range $D = \mathcal{P}_2$; $R(D) = 3$; $\ker D = \mathbf{R}$; $N(D) = 1$

8. range $I = \mathbf{R}$; $R(I) = 1$; $\ker I = \left\{a + bt + ct^2 \in \mathcal{P}_2 \middle| a = -\frac{b}{2} - \frac{c}{3}\right\}$; basis for $\ker I$: $\{-\frac{1}{2} + t, -\frac{1}{3} + t^2\}$; $N(I) = 2$

9. range $A = \{a + bt + ct^2 \in \mathcal{P}_2 | a = 0\}$; basis for range A: $\{t, t^2\}$; $R(A) = 2$; $\ker A = \{\mathbf{0}\}$; $N(A) = 0$

16. c. $f(1, 0, 0) = (1, 1, 2)$, $f(0, 1, 0) = (1, -1, 0)$, $f(0, 0, 1) = (0, 0, 0)$; $\{(1, 1, 2), (1, -1, 2), (0, 0, 0)\}$ spans range f; minimal spanning set, and hence a basis, is $\{(1, 1, 2), (1, -1, 2)\}$

h. Spanning set for range f is $\left\{\begin{bmatrix} 1 \\ 0 \end{bmatrix}, \begin{bmatrix} -1 \\ 1 \end{bmatrix}, \begin{bmatrix} 0 \\ -1 \end{bmatrix}\right\}$

Any two of these vectors constitute a basis for range f

20. a. $(-4, 4)$ b. $(14, 4)$ c. $(5x + y - z, x + y + z)$

21. a. $(16, -1)$

Problem Set 5.2

2. LM2–part of Theorem 5.8:
$$\begin{aligned}(f + g)(c\mathbf{u}) &= f(c\mathbf{u}) + g(c\mathbf{u}) \\ &= cf(\mathbf{u}) + cg(\mathbf{u}) \\ &= c(f(\mathbf{u}) + g(\mathbf{u})) \\ &= c((f + g)(\mathbf{u}))\end{aligned}$$

3. and **4.** $(f + g)\left(\begin{bmatrix} x \\ y \\ z \end{bmatrix}\right) = \begin{bmatrix} 3x - 2y \\ 3y + 2z \end{bmatrix}; A = \begin{bmatrix} 2 & -1 & 0 \\ 0 & 2 & 3 \end{bmatrix};$

$$B = \begin{bmatrix} 1 & -1 & 0 \\ 0 & 1 & -1 \end{bmatrix}; C = \begin{bmatrix} 3 & -2 & 0 \\ 0 & 3 & 2 \end{bmatrix}; A + B = C$$

10. Solve $(x - y, y - z, x + z) = (r, s, t)$ to get x, y, and z in terms of r, s, and t. The result is

$$f^{-1}(r, s, t) = (x, y, z) = \frac{1}{2}(r + s + t, -r + s + t, -r - s + t)$$

Problem Set 5.3

2. b. $\begin{bmatrix} -2 & 0 & 0 \\ 2 & 3 & 1 \end{bmatrix}$

3. b. $\mathbf{x}_A = \begin{bmatrix} 0 \\ 1 \\ 1 \end{bmatrix}; f(\mathbf{x})_B = \begin{bmatrix} -2 & 0 & 0 \\ 2 & 3 & 1 \end{bmatrix}\begin{bmatrix} 0 \\ 1 \\ 1 \end{bmatrix} = \begin{bmatrix} 0 \\ 4 \end{bmatrix};$

$(0)\begin{bmatrix} 1 \\ 1 \end{bmatrix} + (4)\begin{bmatrix} 1 \\ 0 \end{bmatrix} = \begin{bmatrix} 4 \\ 0 \end{bmatrix}; f(\mathbf{x}) = \begin{bmatrix} 4 \\ 0 \end{bmatrix}$, from the formula for f

4. d. $\begin{bmatrix} 2 & 0 & 0 \\ -1 & 0 & -1 \\ 0 & 1 & 1 \end{bmatrix}$

10. $f^{-1}\left(\begin{bmatrix} r \\ s \end{bmatrix}\right) = \begin{bmatrix} x \\ y \end{bmatrix} = \frac{1}{5}\begin{bmatrix} r + s \\ 3r - 2s \end{bmatrix}$. Relative to the natural basis,

$$M_{(f^{-1})} = \frac{1}{5}\begin{bmatrix} 1 & 1 \\ 3 & -2 \end{bmatrix}; M_f = \begin{bmatrix} 2 & 1 \\ 3 & -1 \end{bmatrix};$$

$$M_f M_{(f^{-1})} = \frac{1}{5}\begin{bmatrix} 5 & 0 \\ 0 & 5 \end{bmatrix} = \begin{bmatrix} 1 & 0 \\ 0 & 1 \end{bmatrix} = I$$

Problem Set 5.4

1. b. Echelon form is $\begin{bmatrix} 1 & 0 & 2 & 0 & 1 \\ 0 & 1 & 0 & 0 & 1 \\ 0 & 0 & 0 & 1 & 1 \end{bmatrix}$

Nullspace: $\{(-2r - s, -s, r, -s, s)\} = \{r(-2, 0, 1, 0, 0) + s(-1, -1, 0, -1, 1)\}$. Basis for nullspace: $\{(-2, 0, 1, 0, 0), (-1, -1, 0, -1, -1)\}$. $N(A) = 2$. Columns 1, 2, and 4 contain the leading ones in the echelon form; a basis for the columnspace is $\{(1, 0, 2), (-1, 1, 1), (3, 0, -1)\}$. $R(A) = 3$. Theorem 5.25 reads, in this case, $5 = 2 + 3$. Columns 1, 2, and 4 will comprise a nonzero 3×3 subdeterminant.

2. See answer to Problem 1.

4. Dimension $= 3$

8. Show that every linear combination of $\mathbf{u}$, $\mathbf{v}$, $\mathbf{w}$ is a linear combination of the second set and vice-versa. For part a, for instance, solve for r, s, t in terms of a, b, c in the equation $a\mathbf{u} + b\mathbf{v} + c\mathbf{w} = r\mathbf{u} + s\mathbf{v} + t3\mathbf{w}$; reverse and solve for r, s, t in terms of a, b, c in the equation $a\mathbf{u} + b\mathbf{v} + c3\mathbf{w} = r\mathbf{u} + s\mathbf{v} + t\mathbf{w}$.

Problem Set 6.1

1. a and c are orthogonal.

2. a. $\arccos \dfrac{5}{\sqrt{84}} \approx 57°$ b. $\arccos \dfrac{9}{\sqrt{870}} \approx 72°$

c. $\arccos(-1) = 180°$ d. $\arccos \dfrac{5}{\sqrt{567}} \approx 21°$

3. a. $\arccos \dfrac{\sqrt{15}}{4} \approx 14°$

b. $\arccos \pm\dfrac{2\sqrt{2}}{\pi} \approx 26°$ or $154°$, according as $C > 0$ or $C < 0$.

c. $\arccos \dfrac{\sqrt{6n + 3}}{n + 2}$ d. $\arccos \left(\dfrac{6}{e^2 - 1}\right)^{1/2} \approx 14°$

5. a. $\{(2, 1, 3), (-2, \frac{5}{2}, \frac{1}{2})\}$

6. b. $\left\{\sqrt{3}x, \dfrac{5\sqrt{7}}{2}x^3 - \dfrac{3\sqrt{7}}{2}x, \dfrac{63\sqrt{11}}{8}x^5 - \dfrac{35\sqrt{11}}{4}x^3 + \dfrac{15\sqrt{11}}{8}x\right\}$

10. b. The coordinates of $\mathbf{x}_1$ are: $a_1 = \dfrac{1}{\sqrt{2}}(1, 0, -1) \cdot (1, 0, -2) = \dfrac{3}{\sqrt{2}}$, $a_2 = \dfrac{1}{\sqrt{3}}(1, -1, 1) \cdot (1, 0, -2) = -\dfrac{1}{\sqrt{3}}$, and $a_3 = \dfrac{1}{\sqrt{6}}(1, 2, 1) \cdot (1, 0, -2) = -\dfrac{1}{\sqrt{6}}$. (Use the result of Problem 9.)

Problem Set 6.2

1. a. no eigenvalues
b. $\lambda = 1$ has eigenspace $\{t(1, 2)\}$; $\lambda = 2$ has eigenspace $\{s(1, 1)\}$.
e. $\lambda = -2$ has eigenspace $\{t(1, -1)\}$

2. a. $\lambda = 0$ has eigenspace $\{r(1, 1, 1)\}$; $\lambda = 1$ has eigenspace $\{s(1, 0, -1)\}$; $\lambda = 3$ had eigenspace $\{t(1, -2, 1)\}$.

3. $\left\{\frac{1}{\sqrt{3}}(1, 1, 1), \frac{1}{\sqrt{2}}(1, 0, -1), \frac{1}{\sqrt{6}}(1, -2, 1)\right\}$

11. a. Characteristic polynomial: $-(\lambda + 2)^2(\lambda - 4)$; for $\lambda = -2$, a basis for the eigenspace is $\{(-1, 1, 0), (-1, 0, 1)\}$; for $\lambda = 4$, a basis for the eigenspace is $\{(1, 1, 1)\}$; the required orthonormal basis is
$$\left\{\frac{1}{\sqrt{2}}(-1, 1, 0), \frac{1}{\sqrt{6}}(-1, -1, 2), \frac{1}{\sqrt{3}}(1, 1, 1)\right\}$$

12. a. $C = \begin{bmatrix} -1 & -1 & 1 \\ 1 & -1 & 1 \\ 0 & 2 & 1 \end{bmatrix}$. Then $C^{-1}AC = \begin{bmatrix} -2 & 0 & 0 \\ 0 & -2 & 0 \\ 0 & 0 & 4 \end{bmatrix}$.

25. All but a have all real eigenvalues. b. $\begin{bmatrix} 1 & 0 \\ 0 & 2 \end{bmatrix}$ e. $\begin{bmatrix} -2 & 1 \\ 0 & -2 \end{bmatrix}$

26. 60% of the days will have been spent on mathematics.

Problem Set 6.3

1. a. eigenvalues: $-1, 3$; $C = \frac{1}{\sqrt{2}}\begin{bmatrix} 1 & 1 \\ -1 & 1 \end{bmatrix}$; $C^tAC = \begin{bmatrix} 3 & 0 \\ 0 & -1 \end{bmatrix}$

c. eigenvalues: $3, -2, 2$; $C = \begin{bmatrix} 0 & 1/\sqrt{2} & 1/\sqrt{2} \\ 1 & 0 & 0 \\ 0 & -1/\sqrt{2} & 1/\sqrt{2} \end{bmatrix}$;

$$C^tAC = \begin{bmatrix} 3 & 0 & 0 \\ 0 & -2 & 0 \\ 0 & 0 & 2 \end{bmatrix}$$

e. eigenvalues: $0, 4, 8, -8$; $C = \frac{1}{2}\begin{bmatrix} -1 & 1 & 1 & -1 \\ 1 & -1 & 1 & -1 \\ -1 & -1 & 1 & 1 \\ 1 & 1 & 1 & 1 \end{bmatrix}$;

$$C^tAC = \begin{bmatrix} 4 & 0 & 0 & 0 \\ 0 & 0 & 0 & 0 \\ 0 & 0 & 8 & 0 \\ 0 & 0 & 0 & -8 \end{bmatrix}$$

2. a. Hessian: $\frac{1}{2}\begin{bmatrix} 2 & -4 \\ -4 & 2 \end{bmatrix}$; diagonalized: $\begin{bmatrix} -1 & 0 \\ 0 & 3 \end{bmatrix}$;

$\frac{(y')^2}{2} - \frac{(x')^2}{6} = 1$; hyperbola; $\alpha = 45°$

e. Hessian: $\frac{1}{2}\begin{bmatrix} 16 & -4 \\ -4 & 10 \end{bmatrix}$; diagonalized: $\begin{bmatrix} 4 & 0 \\ 0 & 9 \end{bmatrix}$;

$\frac{(x')^2}{36} + \frac{(y')^2}{16} = 1$; ellipse; $\alpha \approx 63.43°$

Problem Set 7.1

1. a. $1 + 6i$; $-7 - 3i$; $\frac{3}{2} - \frac{7}{2}i$ c. 4; 5; $\frac{3}{5} + \frac{4}{5}i$

2. a. $2\sqrt{2}\left(\cos\frac{7\pi}{4} + i\sin\frac{7\pi}{4}\right)$ c. $\cos\frac{3\pi}{2} + i\sin\frac{3\pi}{2}$

3. a. $-\sqrt{3} + i$ c. -3

4. a. $(z + 1 - i)(z + 1 + i)$ c. $z\left(z - \frac{1}{2} - \frac{\sqrt{3}}{2}i\right)\left(z - \frac{1}{2} + \frac{\sqrt{3}}{2}i\right)$

e. $(z - 1)(z + 1 - i)(z + 1 + i)$

5. a. $2i$ c. $-\frac{1}{2}i$ e. $-\frac{1}{4} + \frac{1}{4}i$

6. a. $\cos\frac{\pi + 4k\pi}{10} + i\sin\frac{\pi + 4k\pi}{10}$

c. $\sqrt{2}\left(\cos\frac{\pi + 6k\pi}{6} + i\sin\frac{\pi + 6k\pi}{6}\right)$

11. $\{1, i\}$

Problem Set 7.2

1. a. $\mathbf{x} \cdot \mathbf{y} = 4 + 3i - 1 = 3 + 3i$

$|\mathbf{x}| = \sqrt{2 + 10} = 2\sqrt{3}$; $|\mathbf{y}| = \sqrt{8 + 1} = 3$

Schwarz's Inequality reads $3\sqrt{2} \le 6\sqrt{3}$

2. All are orthogonal; a is not orthonormal

4. b. $C = \begin{bmatrix} \frac{-1 - i}{\sqrt{6}} & \frac{1 + i}{\sqrt{3}} \\ \frac{2}{\sqrt{6}} & \frac{2}{\sqrt{3}} \end{bmatrix}$; $C^*AC = \begin{bmatrix} -2 & 0 \\ 0 & 4 \end{bmatrix}$

Problem Set 7.3

1. a. Ce^{2x} c. $Ce^{-(3/4)x}$ e. $Ce^{-(1/10)x}$

2. a. $-(3/2) + Ce^{2x}$ c. $(1/7)e^{x} + Ce^{-6x}$ e. $(1/3)e^{2x} + Ce^{-x}$

3. a. $C_1e^{3x} + C_2e^{-2x}$ c. $C_1e^{3x} + C_2xe^{3x}$ e. $C_1e^{-(1/2)x} + C_2e^{-x}$

g. $C_1e^{-x} + C_2xe^{-x}$

4. a. $C_1e^{-x} + C_2xe^{-x} + C_3e^x$ c. $C_1 + C_2x + C_3e^{-3x} + C_4e^{-4x}$
e. $C_1\cos x + C_2\sin x$

5. a. $2e^{-2x}$ b. $e^{4x} + 2e^{-x}$

6. a. $y = e^{-bx}\int e^{(b-a)x}[\int e^{ax}f(x)dx]dx$

9. a. $y_1 = k_1e^{3x} + k_2e^{-x}$
$y_2 = k_1e^{3x} - k_2e^{-x}$

Problem Set B

1. 4! = 24 inversions of 1, 2, 3, 4. Some examples are (3, 2, 4, 1) with three inversions; (1, 4, 2, 3) with two inversions; (1, 2, 3, 4) with no inversions

2. a. six inversions

INDEX

G

H

I

J

K

L

M

V

W

A 6
B 7
C 8
D 9
E 0
F 1
G 2
H 3
I 4
J 5